Pharmaceutical Discovery, Development and Manufacturing Forum 2015

Core Programming Area at the 2015 AIChE Annual Meeting

Salt Lake City, Utah, USA
8-13 November 2015

ISBN: 978-1-5108-1870-5

Printed from e-media with permission by:

Curran Associates, Inc.
57 Morehouse Lane
Red Hook, NY 12571

proceedings.com

Some format issues inherent in the e-media version may also appear in this print version.

Printed by Curran Associates, Inc. (2016)

For permission requests, please contact AIChE
at the address below.

AIChE
120 Wall Street, FL 23
New York, NY 10005-4020

Phone: (800) 242-4363
Fax: (203) 775-5177

www.aiche.org

Additional copies of this publication are available from:

Curran Associates, Inc.
57 Morehouse Lane
Red Hook, NY 12571 USA
Phone: 845-758-0400
Fax: 845-758-2634
Email: curran@proceedings.com
Web: www.proceedings.com

TABLE OF CONTENTS

437397 Biomaterials Control Cellular Behavior through Materials Engineering

Abstract:

< style="padding-left: 47pt; text-align: center;">Biomaterials Control Cellular Behavior through Materials Engineering

Ali Khademhosseini, Ph.D.

Department of Medicine, Brigham and Women's Hospital, Harvard Medical School, Cambridge, MA

Harvard-MIT Division of Health Sciences and Technology, MIT, Cambridge, MA Wyss Institute for Biologically Inspired Engineering, Harvard University, Boston, MA alik@rics.bwh

Dr. Ali Khademhosseini, 65 Landsdowne Street, PRB 252, Cambridge, Massachusetts 02139, United States

< style="padding-left: 5pt; text-align: left;">Keywords: Biomaterials, Hydrogels, Regenerative Engineering, Tissue Engineering, Nanotechnology, Polymer Chemistry, Photocrosslinkable, Bioprinting, Photolithography

< style="padding-left: 5pt; text-align: left;">Abstract: Engineered materials that integrate advances in polymer chemistry, nanotechnology, and biological sciences have the potential to create powerful medical therapies. Our group aims to engineer tissue regenerative therapies using water-containing polymer networks called hydrogels that can regulate cell behavior. Specifically, we have developed photocrosslinkable hybrid hydrogels that combine natural biomolecules with nanoparticles to regulate the chemical, biological, mechanical and electrical properties of gels. These functional scaffolds induce the differentiation of stem cells to desired cell types and direct the formation of vascularized heart or bone tissues. Since tissue function is highly dependent on architecture, we have also used microfabrication methods, such as microfluidics, photolithography, bioprinting, and molding, to regulate the architecture of these materials. We have employed these strategies to generate miniaturized tissues. To create tissue complexity, we have also developed directed assembly techniques to compile small tissue modules into larger constructs. It is anticipated that such approaches will lead to the development of next-generation regenerative therapeutics and biomedical devices.

JENIKE & JOHANSON
SCIENCE ENGINEERING DESIGN

Powder Characterization and Trouble-Shooting Methods for Successful Process-Scale up

November 9, 2015
2015 AIChE Annual Meeting
Thomas Baxter, Sr. Consultant, tbaxter@jenike.com
James Prescott, Director/Sr. Consultant, jkprescott@jenike.com

Outline

- Common powder handling challenges during scale-up
  - Obtain reliable flow from bins, hoppers and chutes
  - Achieve target transfer and production rates
  - Reduce segregation and content uniformity (CU) variation
- Powder characterization tests
  - Characterize powder flow (cohesive strength, wall friction)
  - Assess feed rate stability (permeability, compressibility)
  - Analyze segregation potential
- Trouble-shooting methods during process scale-up
  - Address powder flow problems
  - Trouble-shoot feed rate concerns
  - Reduce the risk of segregation and CU concerns

Powder characterization during development

Powder characterization and scale-up concerns during development

- *Early development (Phase I and II)*
  - Characterize API, excipient and blend properties that affect flow and segregation; E.g., cohesive strength, particle size and shape, segregation potential of candidate formulations
  - Optimize formulation: Evaluate/optimize formulas and quantities (API, lubricant, flow aids), assessment of effect of different component suppliers/grades
  - Evaluate manufacturing process alternatives (e.g., granulation vs. direct blend)
- *Late stage development, Registration batches (Phase III), Commercial scale-up*
  - Design equipment to prevent flow and segregation/CU problems
  - Evaluate equipment and/or site transfer
  - Trouble-shoot and address flow and CU problems

JENIKE & JOHANSON SCIENCE | ENGINEERING | DESIGN 3

Powder handling challenges: Obtain reliable flow

Cohesive arch

Rathole

Flow problems can result in:

- Production delays, quality issues, increased operator intervention and exposure

Flow problems can be predicted based upon:

- Flow properties: Cohesive strength, wall friction, compressibility, permeability
- First-principle predictive models
- Key equipment and process parameters: IBC/hopper geometry, wall surface finish, outlet size, storage at rest time, % fill, temperature and % relative humidity

JENIKE & JOHANSON SCIENCE | ENGINEERING | DESIGN 4

Powder handling challenges: Achieve target production rates

E.g., Increased compression variation at higher press speeds

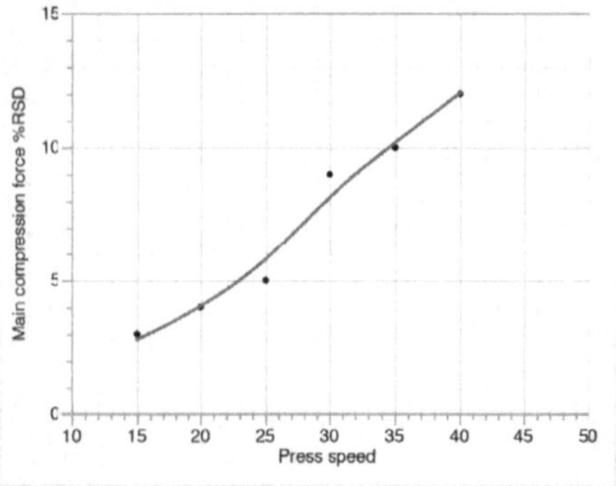

Feed rate limitations can result in:

- Below-target production rates, delays (press shut-downs), quality issues (tablet weight and hardness variation) and increased operator intervention

Flow problems can be predicted based upon:

- Flow properties: *Permeability*, PSD, compressibility, cohesive strength
- First-principle predictive models: "two-phase" (powder:air) flow modeling (*later slides*)
- Key equipment and process parameters: IBC/hopper geometry, outlet size, transfer and feed rates, venting and pressure differential in feed system

Powder handling challenges: Reduce segregation/CU variation

E.g., Segregation potential in blender to IBC transfer step

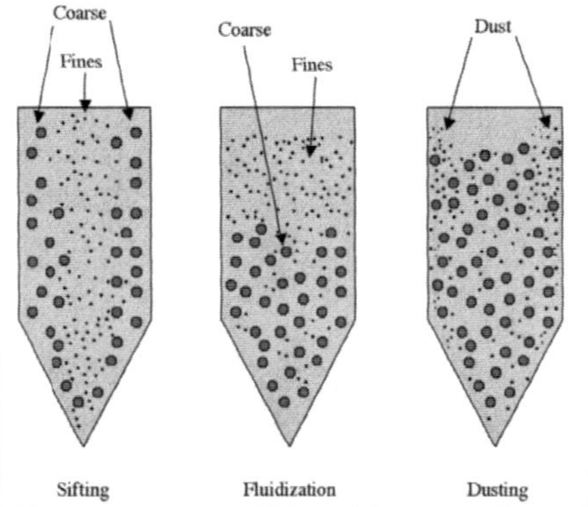

Segregation can result in:

- Quality concerns (CU variation), rejected batches, product validation and launch delays, increased operator intervention

Segregation problems can be predicted and diagnosed based upon:

- Segregation tests and flow properties (flow pattern)
- Stratified sampling data
- Key equipment and process parameters: IBC/hopper geometry (flow pattern), transfer steps, transfer rates, venting and de-acceleration methods

Batch process: Potential powder handling concerns
for direct blend process (*italics = higher risk areas*)

- Process steps upstream of blending (weigh, screen, charging)
  - Active loss during transfer after weigh-out
  - Consistent feed through screening step
  - Pneumatic or mechanical transfer of components to blender
- Blending (pre-blend, final and lubrication blend)
  - *Obtain required blend uniformity (BU)*
  - *Successful scale-up of blend processes*
  - *Reliable discharge from blender*
- Transfer of blend to intermediate bulk container (IBC) or press
  - *Potential for segregation during transfer steps (blender-to-IBC and/or IBC-to-press)*
  - *Reliable discharge from IBC*
  - Reliable transfer while minimizing dusting, build-up and active loss
  - Consistent feed into press feed frame at target compression rate
  - *Obtain required content uniformity (CU)*

JENIKE & JOHANSON SCIENCE | ENGINEERING | DESIGN 7

Continuous process: Potential powder handling concerns
for direct blend process (*italics = higher risk areas*)

- Process steps upstream of blending (weigh, screen, *store and meter*)
  - *Reliable storage and discharge from IBCs, esp. for more difficult-to-handle blend components (e.g., API, lubricant)*
  - *Reliable transfer from IBCs to gravimetric feeders (fast/controlled refills)*
  - *Obtain required accuracy from gravimetric feeder at target feed rate*
- Blending (pre-blend, final and lubrication blend)
  - Obtain required BU (at pre-blend and/or final blend stage)
  - Successful scale-up of blend processes (i.e., scale-up may be more dependent on throughput rate than physical scale-up of blender)
  - Two blenders/steps may be required (pre-blend and final/lube blend)
- Transfer of blend to press
  - Potential for segregation during transfer (often minimized vs. batch)
  - Reliable transfer of blend, *esp. potential risks of vacuum conveying* (API loss/buildup, segregation, particle/granule attrition)
  - Consistent feed into press feed frame at target compression rate (*potential concern if running at high compression rates*)
  - Obtain required CU (potential benefit of increased on-line PAT and real time release)

JENIKE & JOHANSON SCIENCE | ENGINEERING | DESIGN 8

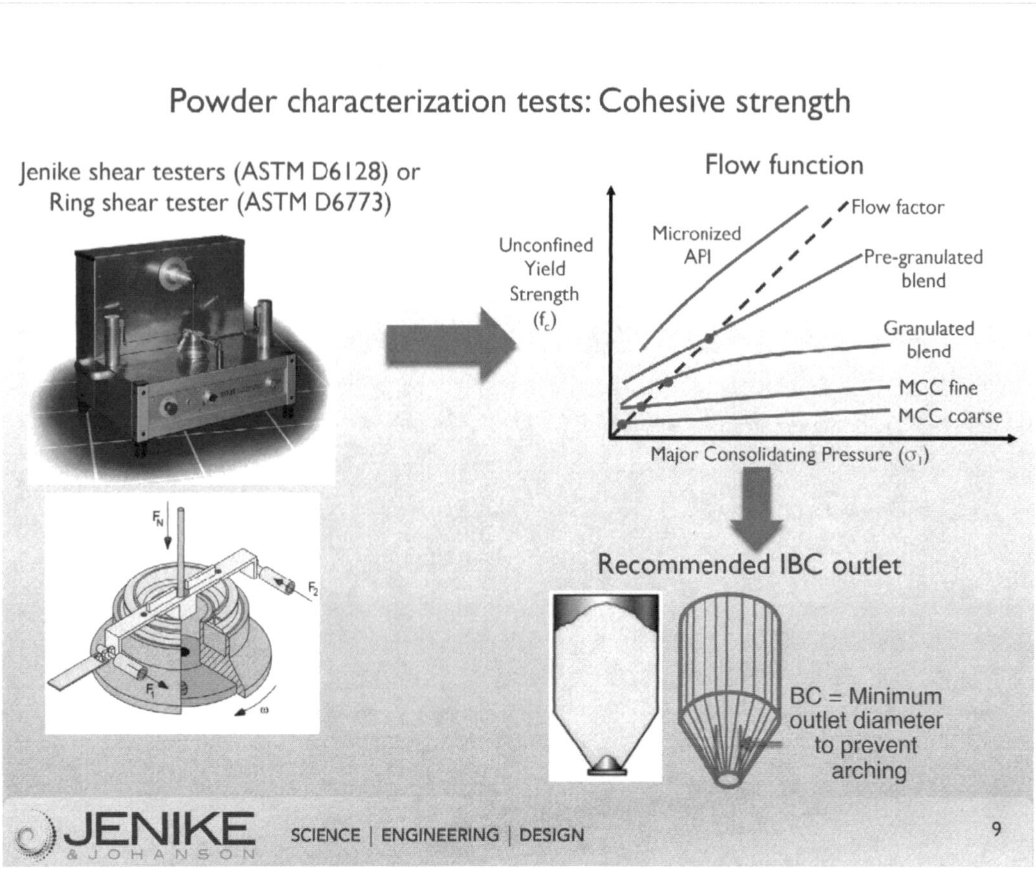
Powder characterization tests: Cohesive strength
Jenike shear testers (ASTM D6128) or
Ring shear tester (ASTM D6773)
Flow function
Unconfined
Yield
Strength
(f_c)
Micronized
API
Flow factor
Pre-granulated
blend
Granulated
blend
MCC fine
MCC coarse
Major Consolidating Pressure (σ_1)
Recommended IBC outlet
BC = Minimum
outlet diameter
to prevent
arching
JENIKE
& JOHANSON
SCIENCE | ENGINEERING | DESIGN
9

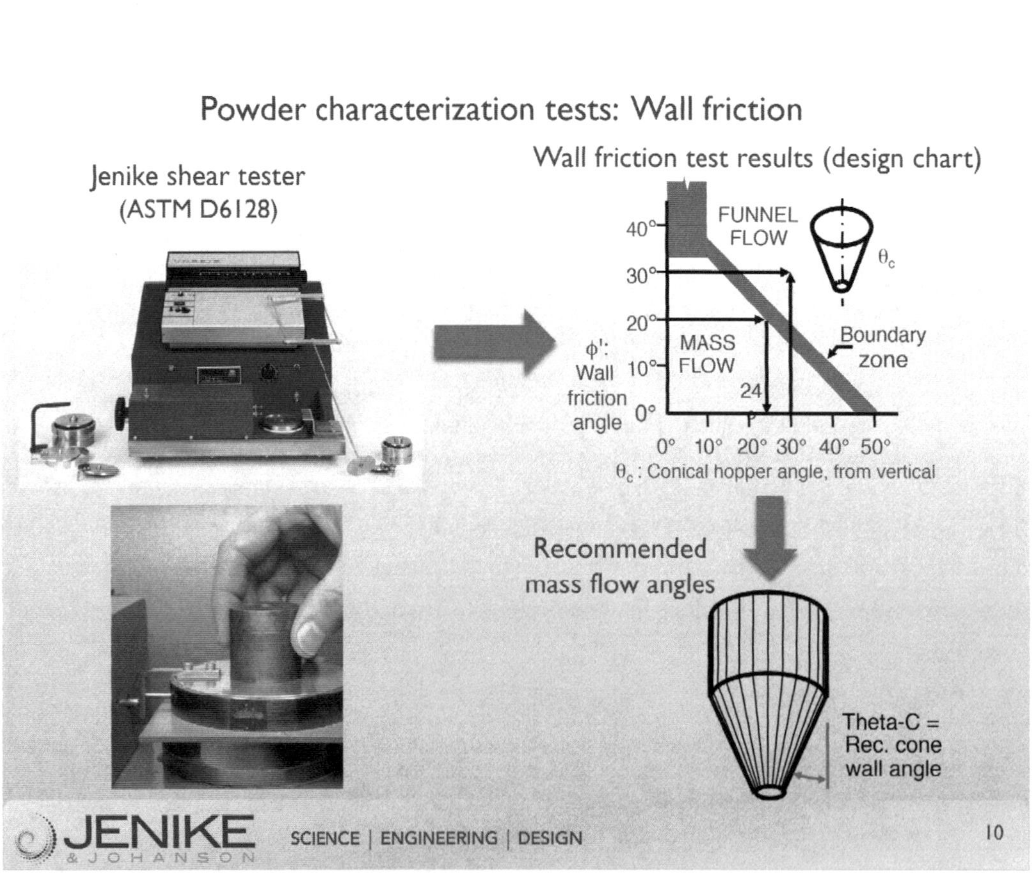
Powder characterization tests: Wall friction
Jenike shear tester
(ASTM D6128)
Wall friction test results (design chart)
40°
30°
20°
10°
0°
FUNNEL
FLOW
θ_c
MASS
FLOW
Boundary
zone
24
φ':
Wall
friction
angle
0° 10° 20° 30° 40° 50°
θ_c : Conical hopper angle, from vertical
Recommended
mass flow angles
Theta-C =
Rec. cone
wall angle
JENIKE
& JOHANSON
SCIENCE | ENGINEERING | DESIGN
10

Powder characterization tests: Permeability and compressibility

Permeability tester

Two-phase flow analysis and
Critical feed rates

Solids Pressure **Bulk Density** **Air Pressure**

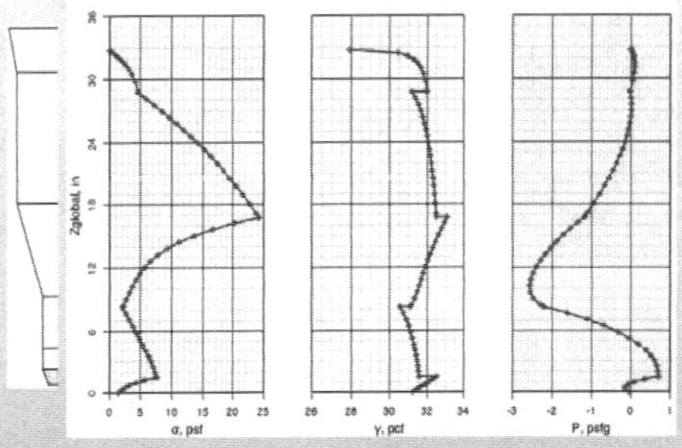

Compressibility tester(ASTM 6683)

Powder characterization tests: Fluidization segregation tests

Fluidization segregation tester

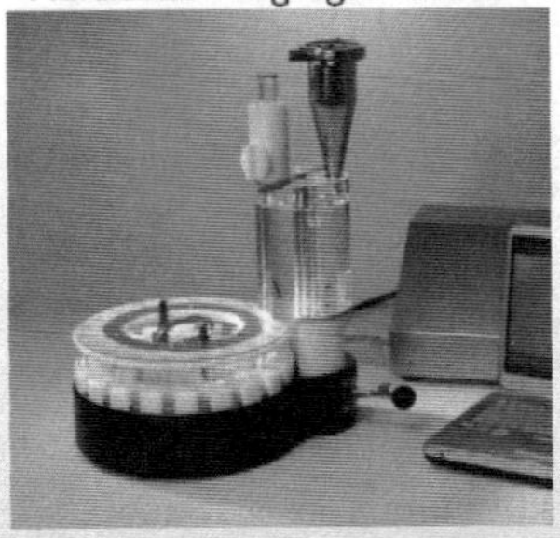

Fluidization segregation test method

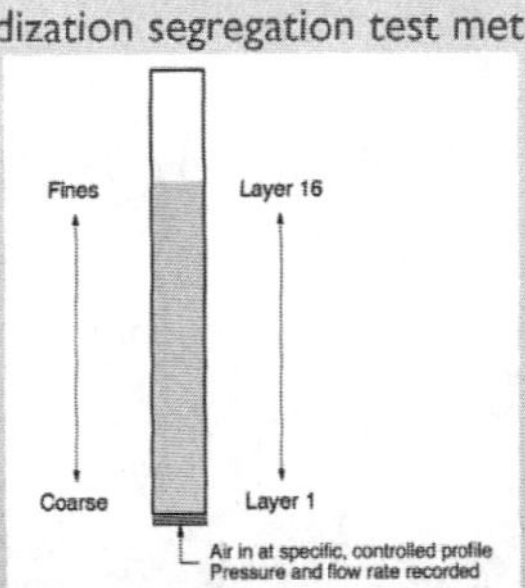

Fluidization segregation
test results

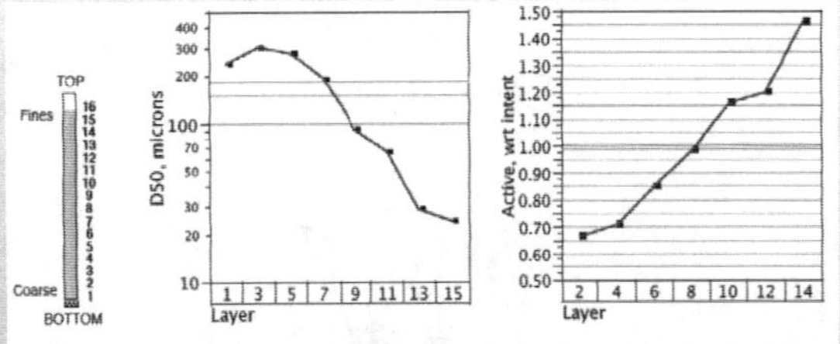

Powder characterization tests: Sifting segregation tests

Sifting segregation tester

Sifting segregation test method

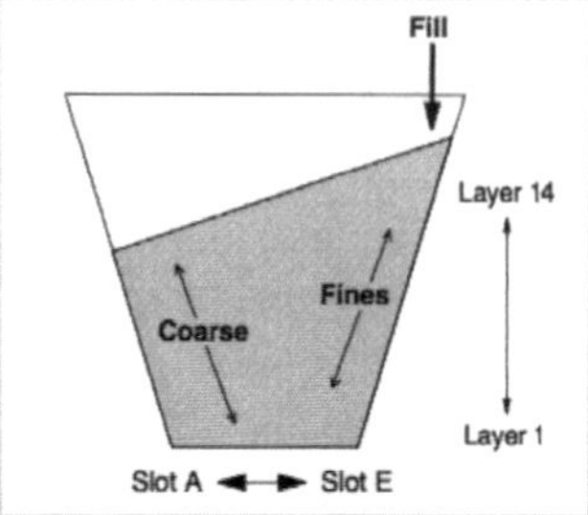

Sifting segregation test results

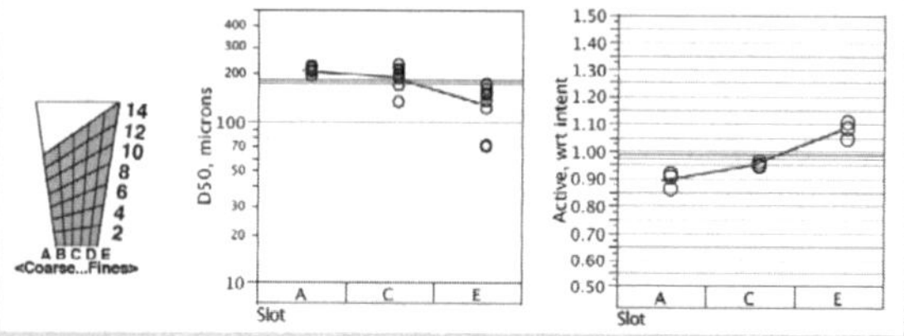

Trouble-shooting: "No Flow" and Erratic Flow Problems

Trouble-shooting methodology:

1) Conduct flow properties tests
2) Identify the flow pattern (mass flow vs. funnel flow)
3) Identify location of arch and/or rathole
4) Identify key equipment parameters
   - Bin and press parameters: capacity, wall angles, interior wall surface, outlet shape/size
   - Transfer chute: height, cross sectional area, interior surface finish, chute angles

Measure flow properties

Identify flow patterns

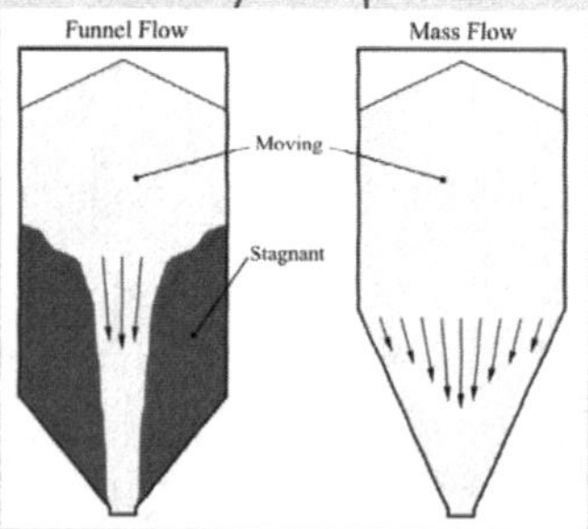

Trouble-shooting: "No Flow" and Erratic Flow Problems

- *Potential* corrective actions to prevent ratholes
  1) Design equipment for *mass flow* (eliminate stagnant material)
  2) Reduce capacity/bin level (reduce effective head)
  3) Size the outlet larger than critical rathole diameter ("Df")
  4) Use an agitator (e.g. Matcon bin, Vibroflow™)
- *Potential* corrective actions to prevent arches (mass flow)
  1) Size the outlet larger than "Bc/Bp"
  2) Use an agitator (e.g. Matcon bin, external vibrator)
  3) Change the storage conditions (Temp., RH, storage at rest)
  4) Change the material (particle size/shape, add flow aid)

Ratholing dimensions (funnel flow)

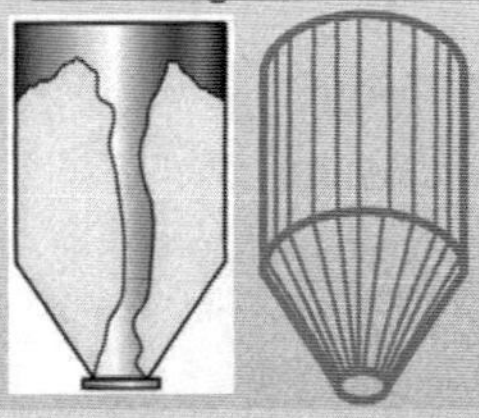

Arching or bridging (mass flow)

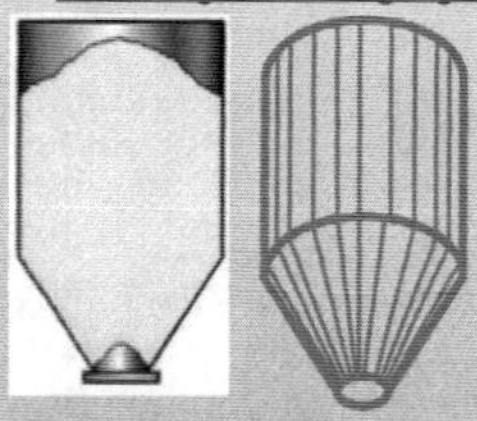

JENIKE & JOHANSON SCIENCE | ENGINEERING | DESIGN 15

Trouble-shooting: Feed Rate Problems

- Trouble-shooting methodology
  1) Identify the flow pattern
  2) Conduct flow properties tests (permeability, density)
  3) Identify key process parameters: rate, material level, transfer rate to equipment, desecration time
  4) Identify key equipment parameters
     - Bin and press parameters: capacity, wall angles, interior wall surface, outlet shape/size
     - Transfer chute: height, cross sectional area, interior surface finish, chute angles
     - *Venting/vacuums in the feed system*
  5) Two-phase flow modeling of potential corrective actions

Permeability tester

JENIKE & JOHANSON SCIENCE | ENGINEERING | DESIGN 16

Trouble-shooting: Feed Rate Problems

- *Potential* corrective actions for feed rate limitations
  1) Design equipment for mass flow
  2) Increase outlet size (if possible)
  3) Reduce capacity/bin level (reduce effective head/consolidation pressure)
  4) Account for pressure differentials and vacuums (e.g., press vacuum)
  5) Reduce production rate (tableting rate)
  6) Add air permeation (passive or active air permeation, see *analysis below*)
  7) Change the material to increase permeability (increase particle size)

Two-phase flow analysis of press feed system without venting (red) and with venting (green)

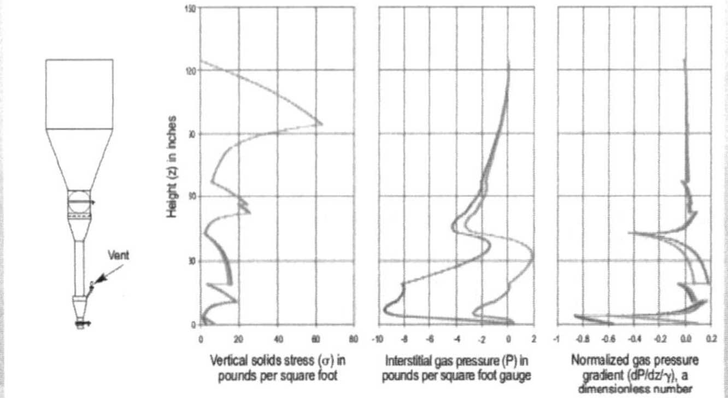

Trouble-Shooting: Segregation/CU Concerns

Trouble-shooting methods

1) Identify the flow pattern
2) Conduct flow properties and segregation tests
3) Analysis of stratified (systematic) blend and content uniformity data
   - Conduct ANOVA and assess stray values, trending, "hot spots", assay shift
4) Identify key equipment and process parameters
   - Bin and press parameters
   - Transfer chute parameters: dimensions, drop heights
   - Venting and valves; fill and transfer rates

Conduct segregation tests

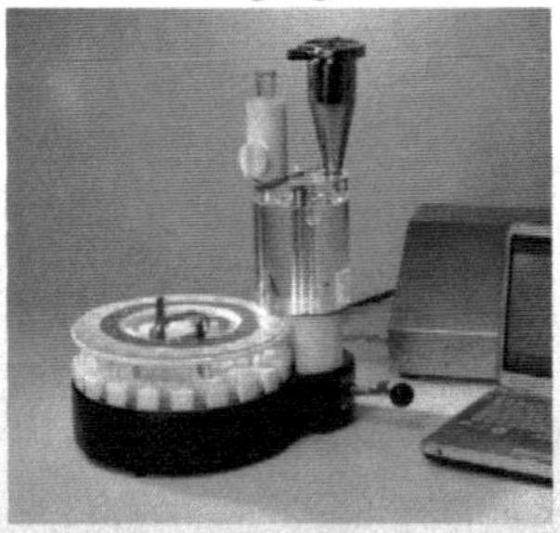

Analyze stratified CU data

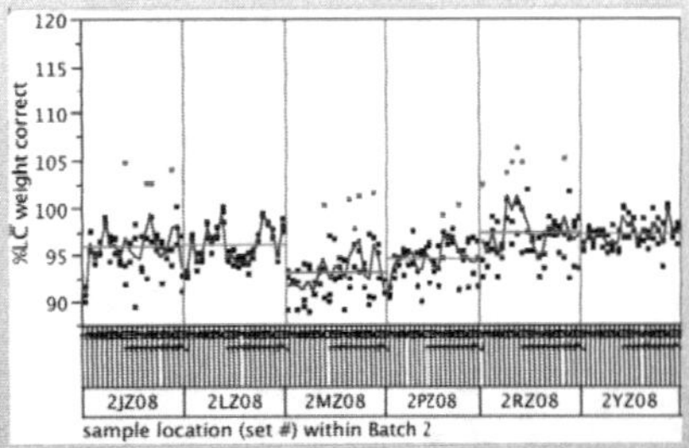

Characterize CU variation: E.g., "Trending"

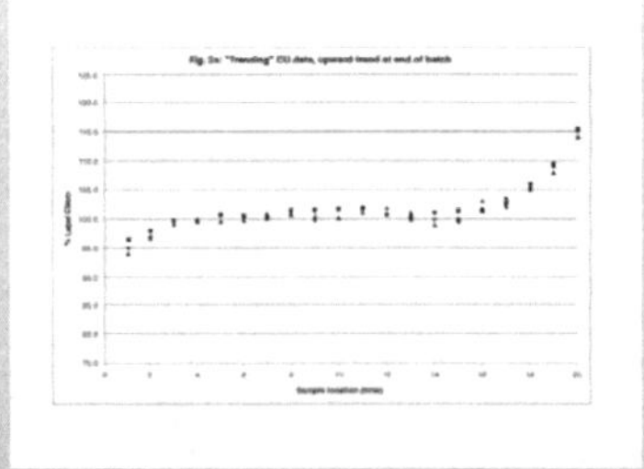

Trouble-Shooting: Segregation/CU Concerns:

Characterizing the type of CU variation

"Trending"

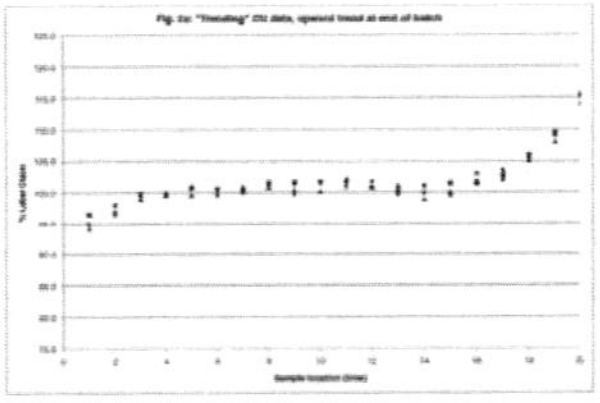

"Wandering"

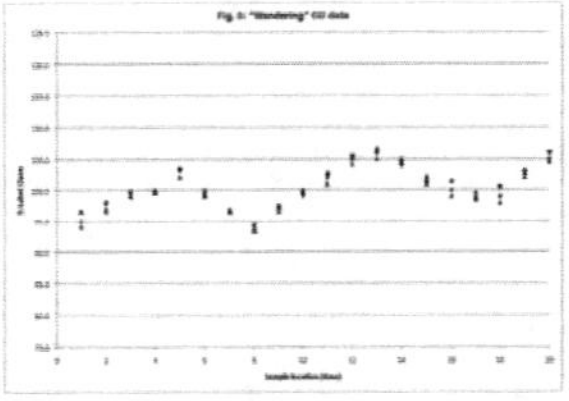

"Scatter"

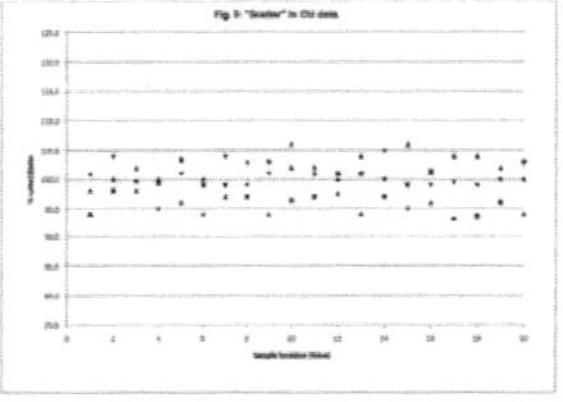

"Hot Spot"

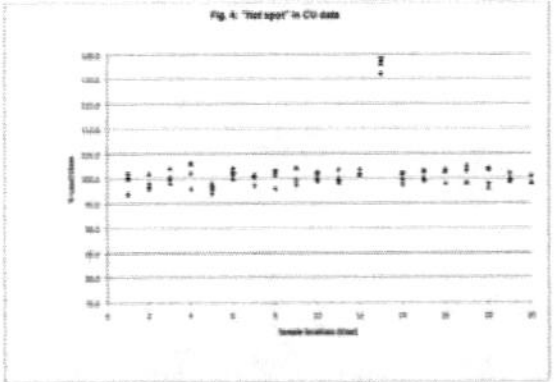

"Stray Value"

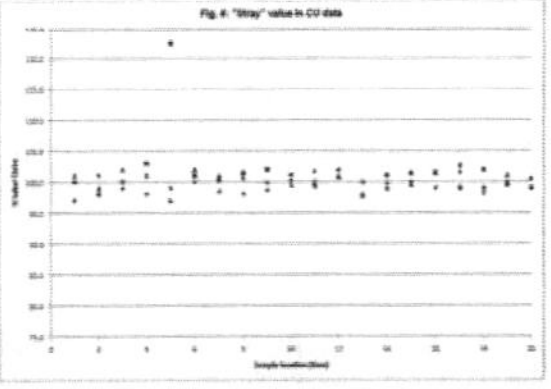

Note that different investigative methods and corrective actions may be required for the different CU issues

JENIKE & JOHANSON SCIENCE | ENGINEERING | DESIGN

19

Trouble-Shooting : Segregation/CU Concerns:

CU Trending vs. Flow Pattern

Note that trending may not always occur at the beginning or end of a batch

Mass flow

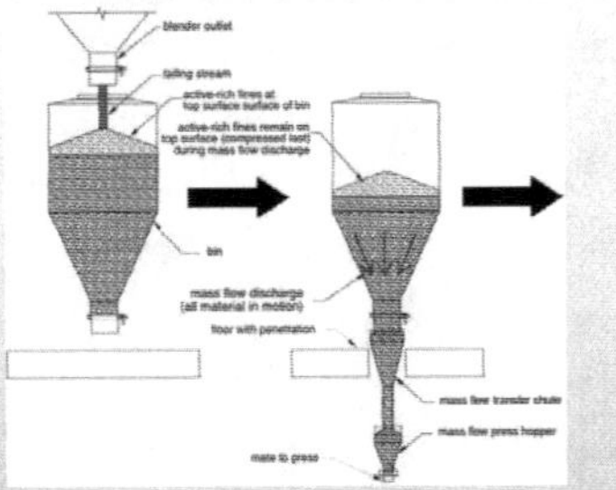

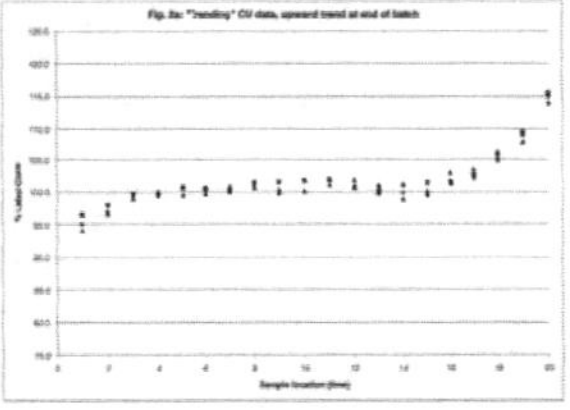

Funnel flow

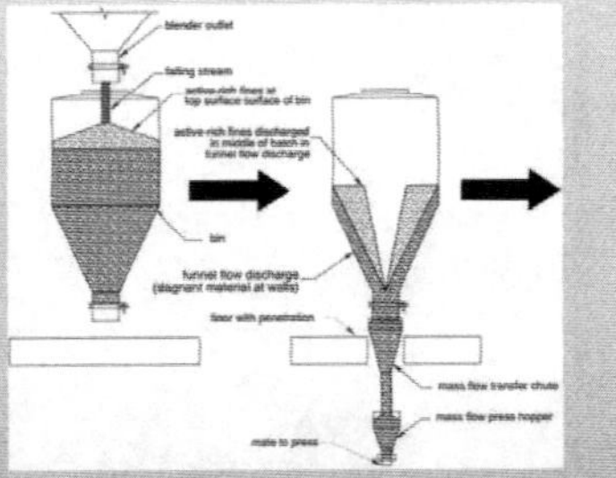

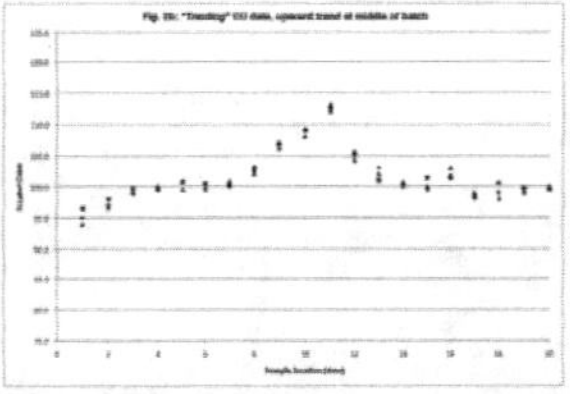

JENIKE & JOHANSON SCIENCE | ENGINEERING | DESIGN

20

Trouble-Shooting Segregation/CU Concerns:
Potential Corrective Actions

- *Potential* corrective actions to address CU concerns
  1) Change the material (particle size distribution, increase cohesion, granulate, ordered vs. random blend)
  2) Reduce transfer steps (e.g., in-bin blending)
  3) Design for mass flow (may not be effective in addressing fluidization)
  4) Control velocity gradients
  5) Control gravity transfers (blend-to-bin, bin-to-press) with additional valves and venting (see *following slide*)

Mass flow velocity gradients

Large gradient   Small gradient

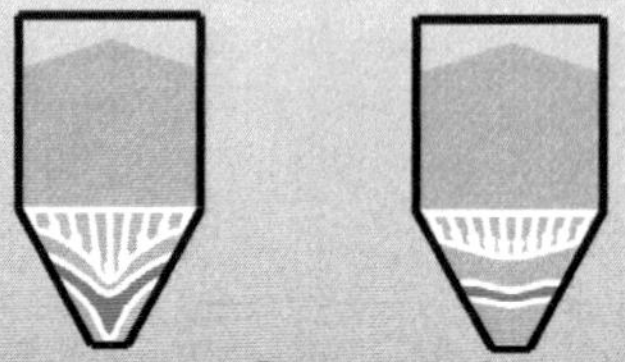

Close to design limit   Robust design

E.g., Anti-segregation BINSERT®

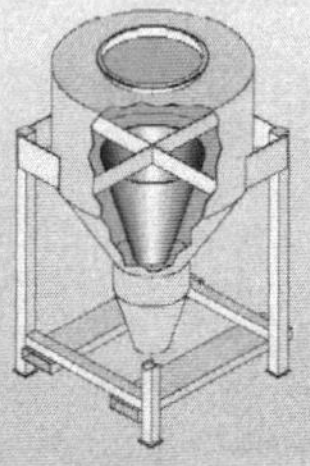

Trouble-Shooting: Segregation/CU Concerns:
Example of reducing CU variation

Transfer chute modification

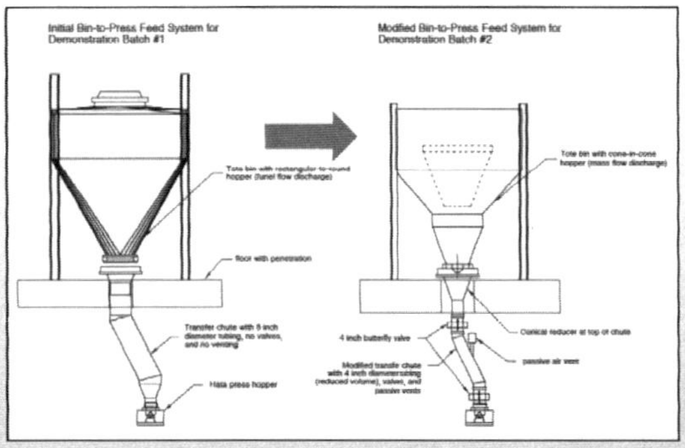

Reduction in CU trending

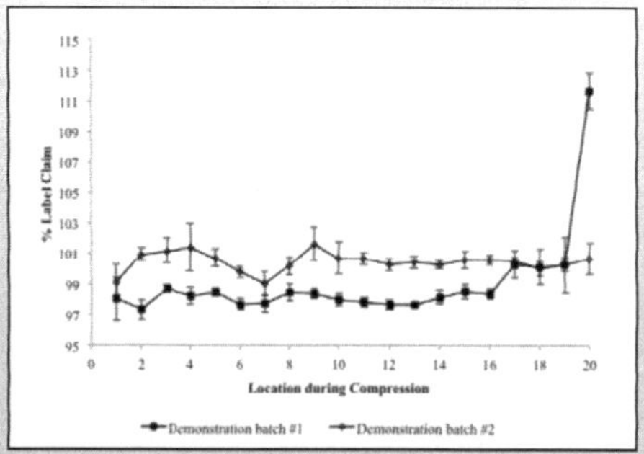

Ref.: "Addressing Segregation of a Low Dosage Direct Blend during Commercial Scale-Up"; Nipun Davar, Thomas Baxter, Pauly Kavalakatt, Sangita Ghosh, Herbert Schock, Pharmaceutical Technology, February, 2011

Conclusions

- Assess the key powder handling concerns prior to scale-up
  - Design equipment (bins, hoppers and chutes) for reliable flow at target production rates and reduce risks of segregation and CU variation
  - Key batch vs. continuous process concerns may be different
- Conduct powder characterization tests to assess risk
- Use powder test outputs as inputs to process & equipment design
- Utilize trouble-shooting methods during scale-up to address
  - Powder flow problems
  - Feed rate concerns
  - Segregation risks and CU concerns

The authors wish to acknowledge all their colleagues at Jenike & Johanson for their contributions

JENIKE & JOHANSON SCIENCE | ENGINEERING | DESIGN 23

429404 Establishing Content Uniformity Criteria throughout 3-Stage Validation: Estimate, Confirm, and Control

Abstract:

Two recent events have prompted the development of establishing content uniformity throughout a 3-stage process validation. The first was the withdrawal of the "Powder Blends and Finished Dosage Units - Stratified In-Process Dosage Unit Sampling and Assessment". The second was the re-emphasis that the Uniformity of Dosage Units (USU) / USP <905> is intended to be applied to product on the market, not as a release test. The goal is to guard against %RSD of 7.5%. During process validation Stage 1 and 2, an increased sample size is recommended to more accurately determine the process mean and standard deviation. The Stage 3 process validation will use the appropriate sample size based on the true estimate of the standard deviation from the larger sample size in Stage 2. Statistically based sampling plans and acceptance criteria is then developed for release to assure quality product on the market meets the USP <905>. This 3-Stage approach provides a path to estimate, confirm and control the product quality.

422206 Fda's Office of Process and Facilities: Facilitating Pharmaceutical Manufacturing Quality and Innovation

Abstract:

Christine Moore, Acting Director of the Office of Process and Facilities (OPF), will discuss how FDA CDER has re-organized its quality function into one super office—Office of Pharmaceutical Quality (OPQ) in mid-January 2015. One of the landmarks is that the newly established Office of Process and Facilities (OPF) is dedicated to and fully responsible for pharmaceutical manufacturing process review and facility inspection. The Office of Process and Facilities (OPF) ensures that quality is built into manufacturing processes and facilities over the product lifecycle. It's vision is to be the premiere organization for integrated manufacturing process review and inspection of pharmaceutical regulation, globally. Given the primary functions of newly established OPF and its potential huge impact to the pharmaceutical industry and pharmaceutical engineering practice, PD2M Forum is an appropriate avenue to share these changes.

440211 Systems Based Pharmaceutics Alliance: Research Opportunities for Academics

Abstract:

Officially launched in early 2014, the Systems Based Pharmaceutics industrial alliance seeks to close the gap between theoretical developments in process modeling and simulation; and the practical implementation of these technologies into the development and manufacture of new medicines.

" The SbP Alliance aims to develop a single model-based framework for drug substance manufacturing, drug product manufacturing, oral absorption and pharmacokinetics. The Alliance will seek to establish best practices in the application of model-based techniques which will be incorporated in validated commercial-quality computer software for the solution of industrially-important problems." (PSE Press Release, 20-Mar-2014)

During its two years of operation, the subject matter experts from the participant companies have implemented models that represent useful and practical decision making tools for the development of new drugs. The focus of the alliance for its Phase I (years 1 and 2), was the integration of models spanning the crystallization, filtration, wet and dry milling of API; and the resulting performance of the crystal in the drug product in models for dissolution, the ASD apparatus, content uniformity and oral absorption.

As the teams were implementing existing published models and approaches, several gaps were found to answer key dependencies to properly mimic dependencies and behaviors of interest in these system. This talk will present (primarily to the academic community) those areas of research that were identified as key areas of academic research where advances would make immediate impact to industrial needs.

Advancements in the Prediction and Control of Batch High Shear Granulation and Continuous Twin Screw Granulation

Author: Don Clancy
Title: Investigator - Modeling Lead
Department: Modeling Center of Excellence,
Division: R&D Product Development,
Company: GlaxoSmithKline
Contact E-mail: Donald.J.Clancy@gsk.com

1.1 Introduction to Modeling Batch High Shear Granulation

Useful predictive models for batch and continuous granulation utilize many of the same concepts, and it's constructive to understand batch high shear granulation prior to continuous twin screw granulation. Batch high shear granulation has been explored more extensively, and in some ways is less complex than continuous. The models that have been found to be useful to date for process control and design space exploration are a combination of first principles mechanistic linked with linear regression to predict an outcome such as granule particle size distribution, milled granule density, or tablet dissolution. The prediction in each case is going to be product specific, but the first principle nature of the mechanistic terms utilized in the linear regression models is shown to hold true across multiple drug products.

Most pharmaceutical high shear granulation processes granulate for a specified amount of time, or specify a maximum motor impeller power or torque before stopping the granulation process. Since most products aren't particularly sensitive to granulation endpoint, this simple control methodology is often adequate. Sometimes though, products can be particularly sensitive to changes in granulation extent. As an example, Drug Product "P" is a micronized hydrophobic API with 80% drug content in the granule. Relatively small changes in granule porosity and density have significant impacts on resulting tablet dissolution, and tablet breaking force. For this reason, the "Work" control model shown as Equation 1 that integrates both power and time was first developed and applied to drug product "P" and has controlled it's commercial manufacture.(Clancy, Method and Apparatus for Predicting Properties of Granulated Materials and Dosage Forms made Therefrom, 2010)

From a first principles perspective, "work done" is the instantaneous power applied to the system multiplied by the time the power was applied. Since power varies over time in granulation, the integral of power with respect to time needs to be considered. Typically the power measured in granulation is the power flow to the motor, and some power is lost due to the motor and power train inefficiencies. It's useful to subtract a baseline power (P_0) which is associated with the power usage needed by the motor and drive mechanism from the overall power used by the impeller, as shown in Equation 1.

$$\text{Work} = \int_0^{\text{Time}} \left(\text{Power}_{\text{Impeller}} - P_0\right) dt \qquad (1)$$

Where "$\text{Power}_{\text{Impeller}}$" is the measured power of the main impeller during granulation, and P_0 is the steady state baseline power measured prior to start of the granulation process with nothing in the bowl.

Graphically what Equation 1 is calculating is the net work done, which is the cumulative amount of power added above baseline, as the shaded in area shown in Figure 1.

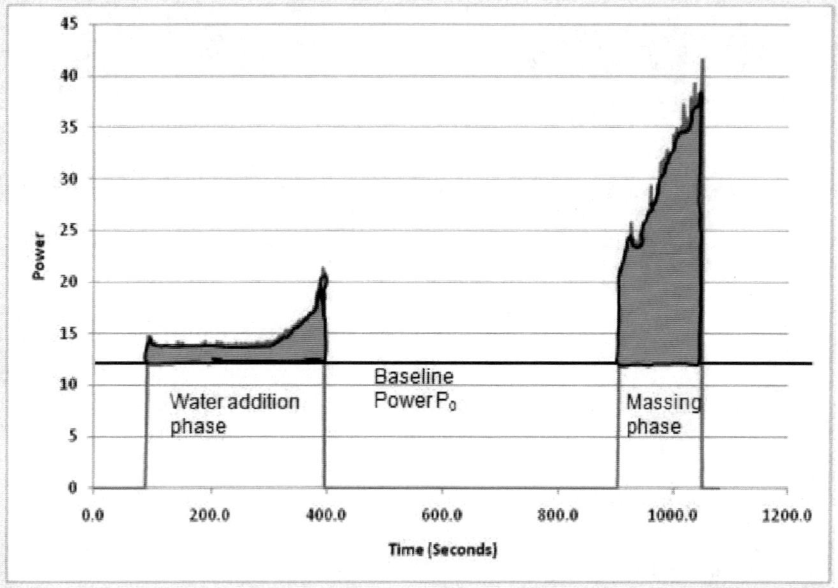

Figure 1: Calculation of Granulation Work Corresponds to Shaded Area

Calculation of a definite integral isn't generally possible or convenient, thus equation 1 is approximated using numeric integration and rapid time slicing. Equation 2 is the actual numeric calculation utilized to calculate the integral shown in Equation 1.

$$W_i = \left(P_i - P_0\right) \cdot \Delta t + W_{i-1} \tag{2}$$

W_i is the work calculated at the ith time slice, P_i is the impeller power at the ith time slice, $\triangle t$ is the time difference measured in seconds between "i" and "i-1" time slices. The output from above equation is integrated over time of the granulation once water input has started.

Work done per unit dry powder mass, water fraction added, and tablet thickness at a constant tablet mass can be used as parameters for empirical linear regression models to predict tablet dissolution. Examples of this are shown for three different drug products in Figure 2, Figure 3, and Figure 4.

The dissolution of tablets from drug product "P" can be modeled as a 4-dimensional surface of tablet dissolution as a function of work, water Fraction, and tablet thickness, as shown in Figure 2. The emperical model for dissolution in this case is a 10-term linear regression model as a function of work, water, tablet thickness and their cross-products with an R2 value of 92%. The data to form this model comes from a 10-batch response surface DoE as function of Work and Water, and split to 5 levels of compressed tablet thickness to make 50 sub-batches in a split plot experimental design. The raw data for this design is available in Appendix A, Table 1.

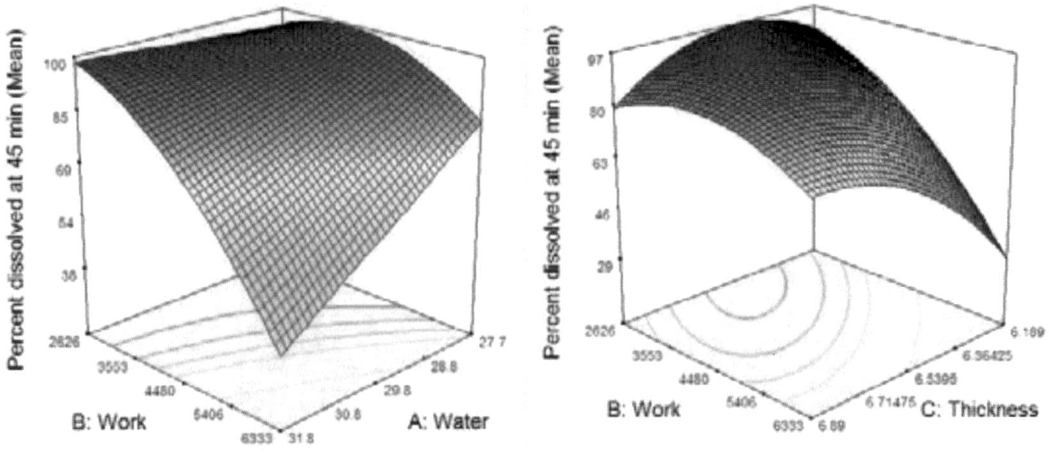

Figure 2: Percent Tablet Dissolved at 45-minutes for Drug Product "P" at Constant Target Tablet Thickness (Left) or Constant Target Granulation Water Fraction (Right)

If tablet thickness is held constant, granule dissolution and tablet hardness can be predicted as a 3-dimensional surface as function of granulation work and water fraction. An example surface for Drug Product "D" from a 9-batch response surface DoE is shown in Figure 3. In this case the tablet dissolution can be predicted from a simple 4-term empirical linear regression model considering work, work-squared, and water with an R2 value of 99%. The DoE data from this experiment is available in Appendix A, Table 2.

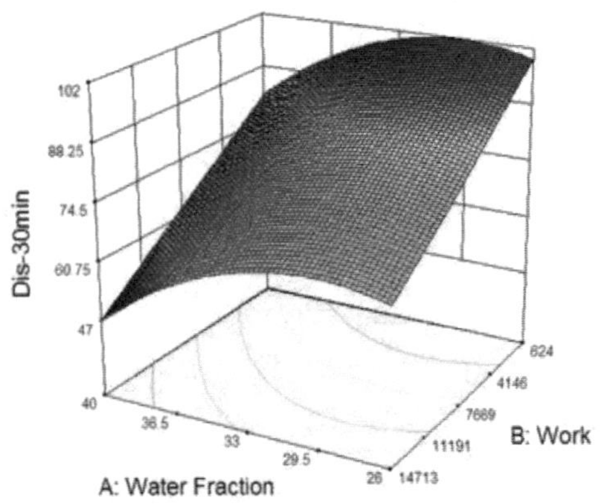

ANOVA for Response Surface Reduced Quadratic Model

Analysis of variance table [Partial sum of squares - Type III]

Source	Sum of Squares	df	Mean Square	F Value	p-value Prob > F	
Model	3160.72	3	1053.57	136.42	< 0.0001	significant
A-Water Fraction	177.52	1	177.52	22.99	0.0049	
B-Work	328.35	1	328.35	42.52	0.0013	
A²	130.80	1	130.80	16.94	0.0092	
Residual	38.61	5	7.72			
Cor Total	3199.33	8				

Std. Dev.	2.78	R-Squared	0.9879
Mean	78.56	Adj R-Squared	0.9807
C.V. %	3.54	Pred R-Squared	0.9664
PRESS	107.37	Adeq Precision	28.065

Figure 3: Percent Tablet Dissolved at 30-minutes for Drug Product "D" from 9-batch Response Surface Experiment at 65-L Scale

1.2 Modeling Batch High Shear Granulation by Sampling during Granulation

The concept has been explored of sampling from high shear granulation at varying time points and using the results from sampled batches to predict the behavior of whole batches on basis of Work done per unit mass. To accomplish this, the power input per unit mass needs to be integrated with respect to time to calculate a Work done per unit mass for each sample. The equation used for calculation is shown as Equation 3.

$$W_i = \left(\frac{P_i - P0}{mass_i} \right) \cdot \Delta t + W_{i-1} \tag{3}$$

Where "$mass_i$" is the mass of powder remaining in the bowl at the ith time point on a dry weight basis, and W_i is now scaled as the work done per unit of mass.

Use of sampling can have the useful outcome of reducing the material requirements to produce a granulation model by a factor of three to five. Sampling should be used with the caution that there is some power needed to restart the motor and impeller. For smaller or newer granulators this restart power appears to be less significant than with larger or older electric motors. In a typical sampling experiment 20% to 25% of the batch is removed at each of four or five time points, and work done per unit of dry powder mass is calculated at each time point.

An example of sampling three batches at constant water level at 300-L scale for drug product "P", compared to dissolution of a non-sampled batch is shown in Figure 4. Each of three batches was sampled at varying time points to make four samples per batch with each sample 25% of the batch. The results of the 12 sub-sampled batches at varying work/mass was used to calculate an optimum granulation endpoint that minimized the dissolution RSD while providing close to optimum mean dissolution. The results of the complete batch were within the standard error of prediction for dissolution mean and RSD for models built on the sampled batches.

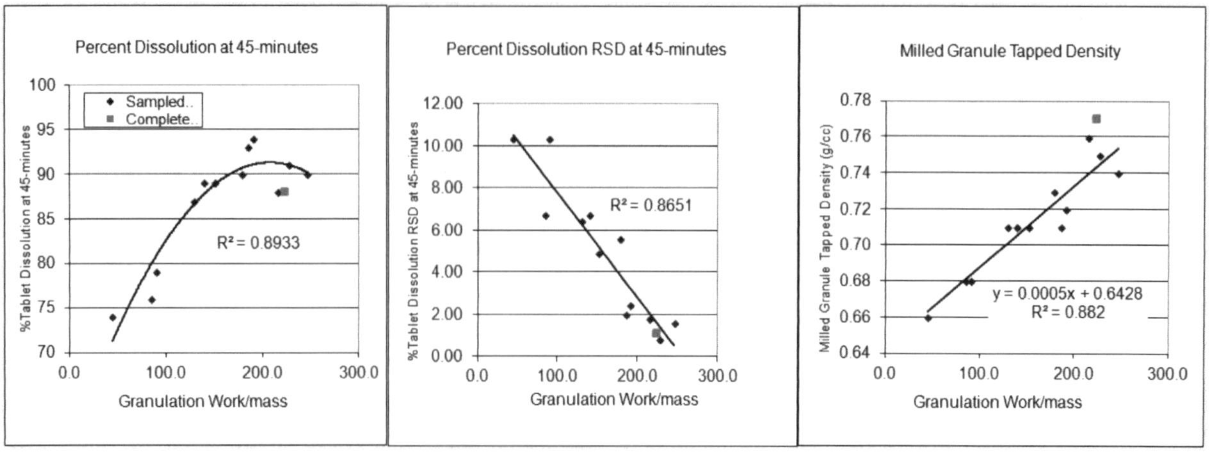

Figure 4: Granulation Work/Mass Models from Sampling Three Batches Four Times Each Compared to a Non-Sampled Batch

There is some evidence that models built from sampling will slightly overestimate the optimum granulation work/mass, for example the milled granule tapped density of the non-sampled batch shown in Figure 4 is slightly higher than anticipated from a model built from the sampled batches. This has been attributed to the power needed for startup and shutdown. For this particular 300-L granulator there is an initial power spike as the power is turned on, and some mathematical exploration was done that if this initial power spike were removed, the models improved. The initial power spike varies from equipment scale and motor, with some older motors having too large and variable of an initial power spike to allow meaningful sampling experimentation to occur. The reproducibility and size of the initial power spike and overall stability of the baseline power (P_0) should be looked at carefully if a sampled batch granulation model is desired. In some cases the initial startup power spikes can be reduced or eliminated by warming the motor up for 20 to 30 minutes prior to adding the granulation mixture to the bowl, and ensuring all moving components are lubricated.

Given the caveat about ensuring use of a granulator motor without a large initial power spike, sampling during granulation can be a useful strategy in R&D to allow modeling and assessment of how robust a formulation is to variation in granulation. Figure 5 shows an example of a 20-point experiment at 25-L scale for product "S" composed of four batches of varying water level each sampled five times. This sampled 20-point experiment was useful for the drug product team in defining the optimum work and water levels needed for subsequent production of clinical batches at 25-L scale. The data table from this experiment is available in Appendix A, Table 3.

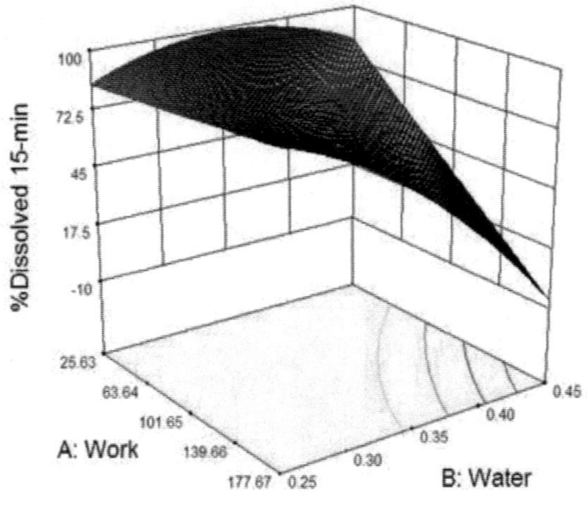

ANOVA for Response Surface Reduced Quadratic Model

Analysis of variance table [Partial sum of squares - Type III]

Source	Sum of Squares	df	Mean Square	F Value	p-value Prob > F
Model	3408.42	4	852.10	26.44	< 0.0001
A-Work	*1244.67*	*1*	*1244.67*	*38.62*	*< 0.0001*
B-Water	*1467.69*	*1*	*1467.69*	*45.54*	*< 0.0001*
AB	*626.50*	*1*	*626.50*	*19.44*	*0.0010*
B²	*528.56*	*1*	*528.56*	*16.40*	*0.0019*
Residual	354.52	11	32.23		
Cor Total	3762.94	15			

The Model F-value of 26.44 implies the model is significant. There is only a 0.01% chance that a "Model F-Value" this large could occur due to noise.

Std. Dev.	5.68	R-Squared	0.9058
Mean	79.06	Adj R-Squared	0.8715
C.V. %	7.18	Pred R-Squared	0.7680
PRESS	872.92	Adeq Precision	18.778

Figure 5: Response Surface of Drug Product "S" Dissolution at 15-minutes to Granulation Work/Mass and Water Fraction in Granulation from Sampled Experimentation at 25-L Scale

1.3 Impact of Raw Material Particle Size and Surface Area Changes on High Shear Granulation Modeling

As material is milled into finer particle size it requires more granulation work to achieve the same endpoint characteristics. A good example experiment was done at 65-L scale with drug product "P" that illustrates this relationship.

As material is milled to a finer size a variety of physical properties change the particle size distribution (PSD) becomes finer, specific surface area (SSA) increases, and bulk density decreases. A challenge is getting a good representative number describing the entire batch as there are typically significant sampling and replication variances for PSD and SSA. A novel technique was developed on a Freeman Powder Rheometer that provided the best signal to noise ratio of these techniques when comparing the RSD of replicate measurements and samples within batches.

A powder Rheometer is simply a very sensitive torque instrument mounted on a mixing blade. The mixing blade can be set to a specified RPM, then travels vertically with the rotating blade tracing a helix into and then out of a powder sample. The methodology established placed 30 grams of dry powder (corresponding to approximately 100 mL of volume) into the 200-mL sample apparatus. Tip speed was set to 300 mm/sec with a downward and upward helix angle of traverse of 5º. This rotation should fluidize the powder bed upwards as the blade traverses downwards through the bed. Ten traverses were done to "condition" the powder to remove affects of sample history. Twenty more replications were done, and the instantaneous work done by the impeller at 1-cm height from the bottom of the sample was then measured and called "Rheometer Work" measured in milliJoules (or milliNewton*meters) per 30 grams.

Rheometer work of the dry API powder was found to correlate to the particle size distribution X90 and the specific surface area of the API for drug product "P" as shown in Figure 6.

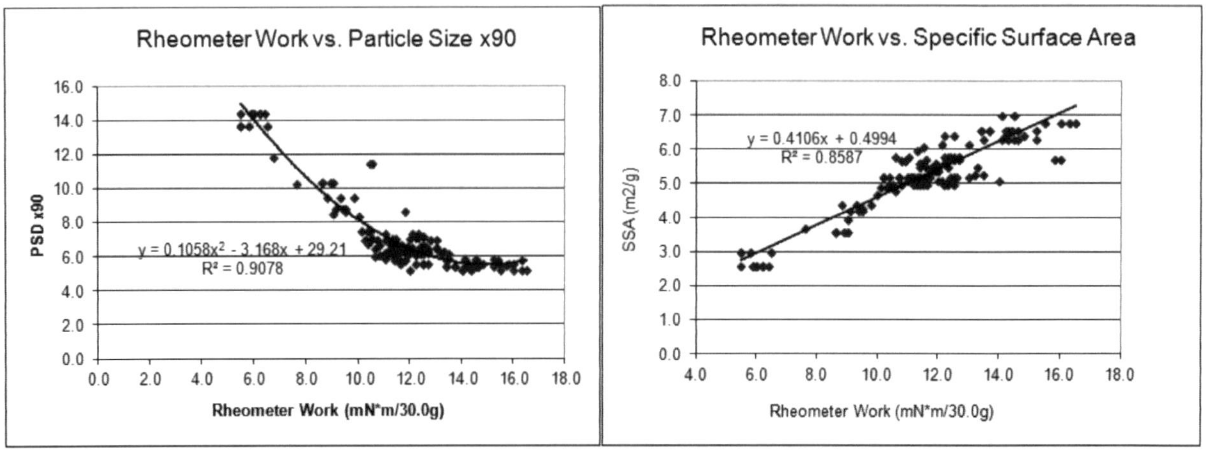

Figure 6: Relationship of API Surface area and Particle size X90 to Powder Rheometer Work

Specific surface area had an RSD of replicate measurements on a sample of 6%, PSD X90 had an RSD of replicate measurement of 4%, and the measurement of "Rheometer Work" had replicate measurement RSD of 2%. Sampling variance was also a significant concern, and the powder Rheometer uses 60 times as much sample as SSA and PSD testing, significantly reducing the impact of sampling variation.

To evaluate the impact of variation in API milling, a batch of drug product "P" API was micronized to three differing micronization work done of 60, 180, and 280 KJ/KG, resulting in 3 lots of API with particle size X90 of 13.2, 6.5, and 5.8 microns. These three lots of API were then tested on the Powder Rheometer, resulting in Rheometer Work values of 6.8, 11.0, and 12.2 mJ/30 grams.

The three lots of API were then granulated at 65-L scale with a constant water amount, and sampled at 4 different time points during granulation. For this drug product, the API is 80% of the granulation mixture, with remainder consisting of 5.9% PVP binder, 12.4% microcrystalline cellulose, and 1.7% sodium starch glycolate. The granulation Work per unit mass was calculated for each sample as described in the previous section. The DoE data from this experiment is available in Appendix A, Table 5.

If granulation work per unit mass is then used to correlate the resulting milled granule density, tablet dissolution, and tablet dissolution RSD, poor correlations are observed (Figure 7). Clearly there was a significant impact of the changes in API particle size distribution.

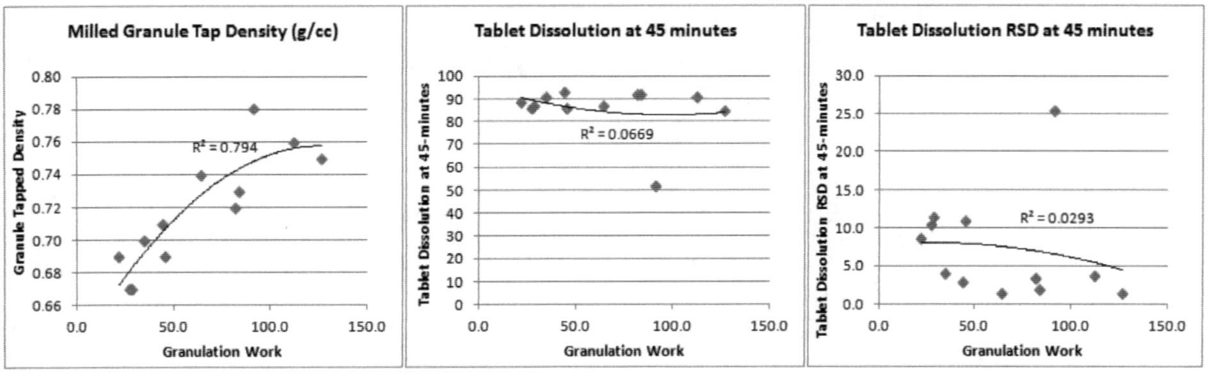

Figure 7: Poor Correlation of Granulation Work/mass to Output Granule and Tablet Characteristics

If the granulation work per unit mass is then normalized by dividing by the Powder Rheometer work done on 30 grams of dry sample, then the characteristic models for drug product "P" can be resolved. Milled granule tapped density, tablet dissolution, and tablet dissolution RSD can then be predicted by polynomial fits with goodness of fit ranging from 94% to 83% (Figure 8).

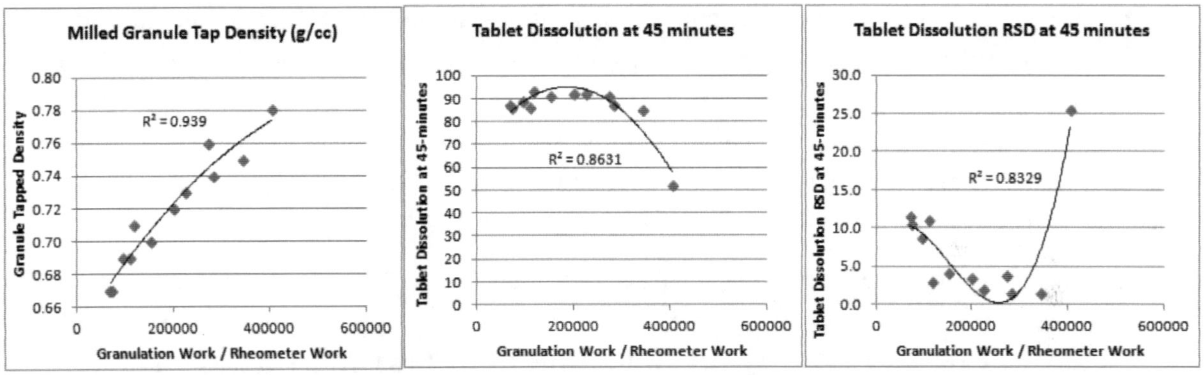

Figure 8: Correlation of (Granulation Work/mass)/Rheometer Work to Output Granule and Tablet Characteristics

There was a linear relationship noted between Rheometer work and the API specific surface area, thus the (Granulation Work / Rheometer Work) term could be expressed as a "Granulation Work per unit surface area" basis with similar results. From perspective of modeling the most precise measurement was chosen to be utilized, thus this experiment prompted work to qualify a Freeman Powder Rheometer for GMP purposes.

It was also possible to model this experiment as a response surface of granulation work per unit mass and API D90 value. This model had significantly lower goodness of fit for prediction of dissolution mean (74%), and problems with variance inflation in the estimation of model parameters. This response surface model is shown in Figure 9.

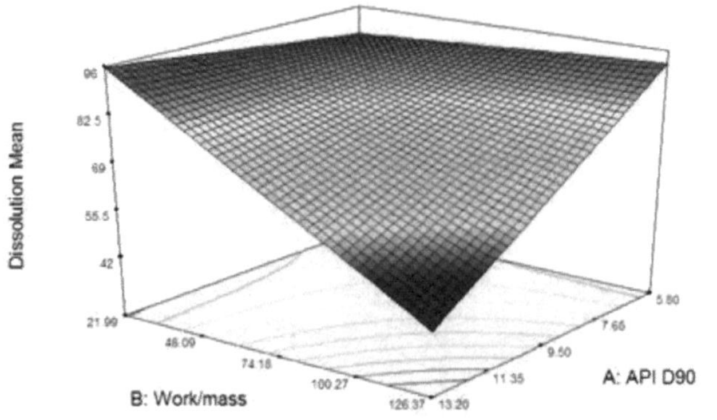

Std. Dev.	6.60	R-Squared	0.7400
Mean	85.92	Adj R-Squared	0.6425
C.V. %	7.68	Pred R-Squared	-0.3298
PRESS	1780.46	Adeq Precision	9.296

A negative "Pred R-Squared" implies that the overall mean is a better predictor of your response than the current model.

"Adeq Precision" measures the signal to noise ratio. A ratio greater than 4 is desirable. Your ratio of 9.296 indicates an adequate signal. This model can be used to navigate the design space.

Factor	Coefficient Estimate	df	Standard Error	95% CI Low	95% CI High	VIF
Intercept	-11525.43	1	3348.10	-19246.15	-3804.70	
A-API D90	-16808.12	1	4436.60	-27038.95	-6577.29	2.983E+006
B-Work/mass	-11622.30	1	3351.94	-19351.88	-3892.71	1.41
AB	-16820.35	1	4441.64	-27062.79	-6577.90	2.982E+006

Figure 9: Correlation of Granulation Work/mass and Particle Size D90 to Output Tablet Dissolution

Evidence suggests that major changes in raw material particle size and surface area have a significant impact on the granulation process. Finer powder of higher surface area is likely to require more granulation Work to achieve a similar endpoint. Since measurements of PSD and SSA have challenges of reproducibility and sampling variance, a more precise model can potentially be developed by examining how much work it takes for a Rheometer blade to traverse a dry sample while fluidizing the powder. This work found that dividing the granulation work by the Rheometer Work resulted in a dimensionless parameter that was predictive of endpoint variation.

1.4 Models Describing Scale-up and Equipment Transfer of Batch High Shear Granulation

Research has been done to understand how to transfer the granulation process between different granulators of varying impeller geometry or scale when "Work/mass" is being utilized as the control model. The main impeller blade angle and rotations per minute (RPM) have significant impacts on the amount of water needed to start the granulation process, and therefore affect the total amount of water and Work needed. Correlations were developed relating the Work and water requirements to main impeller Froude number or RPM, diameter, and blade angle. [8] With changes in impeller angle, a GSK designed 6-L granulator was made by FluidAir that mimics the performance of a 300-L granulator used in production.(Clancy, Campbell, Zhang, Gupta, & Oh, 2011)

Scale-up of high shear granulation is generally accomplished by ensuring similar geometry at each scale, and ensuring the main impeller tip speed is held constant across scale, typically at 5.5

meters/second. The main impeller tip speed has significant impacts on granule consolidation and breakage and needs to be fixed at a constant value. Low tip speed for example can result in runaway granule growth resulting in low density granules the size of golf balls. High tip speed can effectively mill the granules to point where little granule growth occurs though the granules that do form and survive will be small and highly dense.

It's impossible to fix all the important factors to consider when moving across scale. Fixing the impeller tip speed results in changes of g-forces as RPM and blade diameter are changed. Thus, the key to understanding scale-up of high shear granulation is to realize that the amount of water needed at each scale is not necessarily constant. One possible explanation is that the spread of water among surfaces of the mixture is affected by the g-forces being applied by the main impeller blade. Use of absorbent materials such as microcrystalline cellulose and super-disintegrants within the granulation mixture may amplify this affect as some of the water can be absorbed and will be released as these ingredients are subjected to g-forces.

An observation of impeller load behavior as water is added in granulation for drug product "P" is shown in Figure 10 for four different equipment scales. As the equipment scale was increased more water was need to invoke a power response where the impeller power started rising.

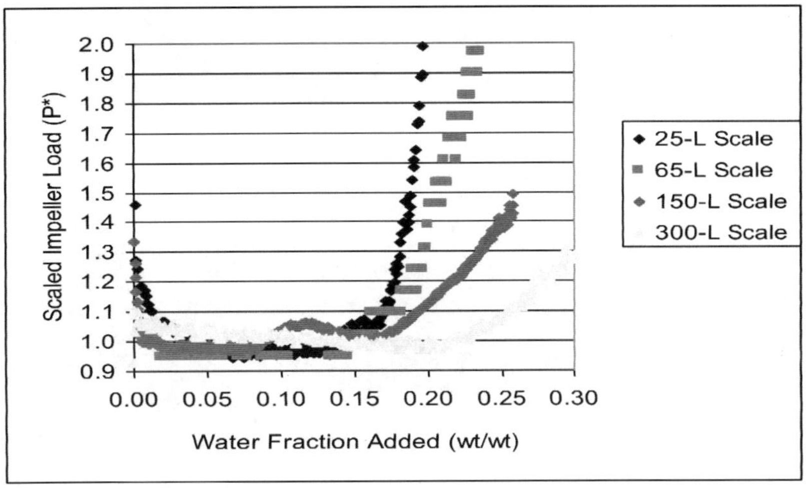

Figure 10: Change in Impeller Power During Water Addition as Granulation Scale is Changed

Initially, the impeller power decreased due to hysteresis effects and the time for the impeller motor to warm up then rises at a later point due to commencement of the granulation process. The amount of water added at the point of inflection in the impeller power indicates the beginning of densification in the granulation process and is called "Xsat" in this analysis.

X_{sat} can be found using one of two mathematical techniques. The first technique is to find the point where the impeller load mean value for 10 seconds is 3-sigma higher than the previous 30 seconds worth of data. A second way is to find the amount of water added at a point of minima of the following function.

$$\text{Objective\_Function} = (Power_{15sec} - P_0) / Mass_{H2O} \tag{4}$$

$Power_{15sec}$ is the 15-second backwards data averaging of the impeller power. 15-second backwards data averaging is used to smooth the impeller load data to prevent extraneous noise from providing a false minimum. The amount of backwards averaging needed will depend on the noise in the impeller power data. At 300-L scale, only 5-seconds of backwards averaging was used as the signal observed had a higher signal to noise ratio. The variable $Mass_{H2O}$ is the amount of water added. The two methodologies for Xsat determination agree with an R^2 of 90% for the data presented in this paper. The minima approach is chosen and utilized in this paper because it is simpler and more robust to calculate in a Programmable Logic Controller (PLC) than the statistical technique, and is not affected as strongly by signal noise, thus provides better models. An example of what taking the minimum of the objective function looks like during water addition, is shown in Figure 11.

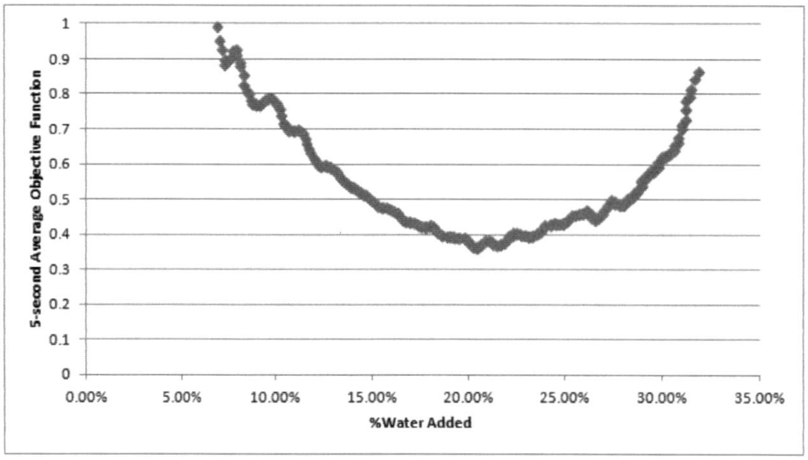

Figure 11: Calculation of an Xsat of 20% Water as the Minimum of Equation (4)

Observation was that an incremental approximately 2% more water was needed for each scale from 25-L, to 65-L, to 150-L, and 300-L in order to get an impeller power increase. Optimal granulation was consistently found using water amount of 8% to 10% in excess (X_s) of the point where impeller power starts to increase (Xsat), see Figure 12. The optimal water added at 300-L scale is 30%, or 9% in excess.

Granulator Scale (L)	Average Xsat Values at Tip Speed 5.5 m/sec	Optimal Water Requirement	Water Required in Excess of Xsat
25	14.7%	23%	8.3%
65	16.0%	26%	10.0%
150	19.4%	28%	8.6%
300	21.1%	30%	8.9%

Figure 12: Optimal water requirement at the various granulator scales to achieved Xat and suitable granules for Drug Product "P"

Centrifugal force is necessary for accelerating the powder bed into toroidal (roping) flow, which is the predominant flow pattern at the later stages in granulation. The centrifugal force also creates a compressive force pressing the powder bed toward the inner walls of the bowl. Thus Froude Number (*Fr*) (Equation 5) is a hypothesized scaling criteria for granulation, and can be thought of as a measure of the centrifugal forces being applied to the power.

$$Fr = \frac{4\pi^2 RN^2}{60^2 g} \qquad (5)$$

where *N* is the impeller rotational speed (rpm), *R* is the impeller radius (m), and g is the gravitational constant (9.81 m s^{-2}).

Xsat was noted to be linearly related to impeller Froude Number when granulators of dimensionally similar bowl (blade height/bowl height) are utilized at the same tip speed. In Figure 13, a plot of the average Xsat determined at each scale against Froude Number is presented. The linearity between Xsat and Froude Number is measured by the high correlation coefficient (R^2 = 0.97), which suggests that for dimensionally similar bowl at constant tip speed, possibly the Froude Number may be a reasonable predictor of Water or Work requirements. This may explain why Froude Number has been postulated as potential scaling factor; however, subsequent experimentation utilizing impellers of differing design and tip speeds suggests this relationship is limited to dimensionally similar blades at same tip speed.

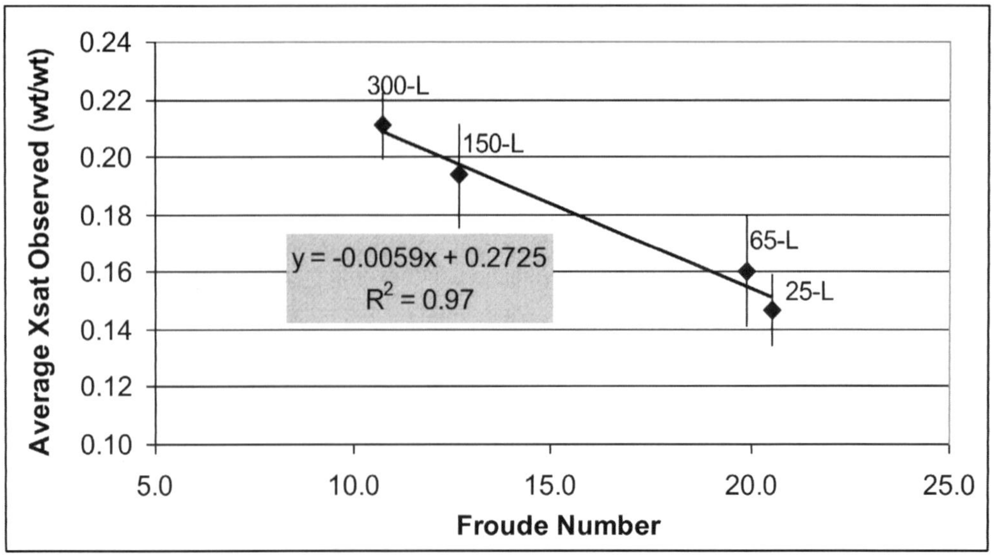

Figure 13: A plot of average Xsat values versus Froude Number for the various Fielder granulator scales at constant tip speed.

Subsequent experimentation found that Xsat can be manipulated by RPM at fixed tip speed (scale), and also by the blade angle. A lower angle blade at smaller scale and higher RPM can replicate the results of a higher angle blade at larger scale and lower RPM. Studies across multiple scales from 6-L, 25-L, 65-L, 150-L, and 300-L were then pursued, with details of the studies documented elsewhere.(Clancy, Campbell, Zhang, Gupta, & Oh, 2011)

If the desire is to mimic the behavior of a large scale granulator using small scale equipment it's useful to predict the correct blade angle. When a linear regression analysis was carried out an all the data generated from the studies conducted during the scale up of Drug Product "P" process impeller speed and blade angle gave the best fit model of Xsat with a 94.4% prediction coefficient. The 3-term model is as follows:

$$\text{Xsat} = 0.797 - 3.25\text{x}10^{-4}\,\text{xN} - 1.85\text{xsin}(\theta) + 1.53\text{xsin}(\theta)^2 \qquad (6)$$

Figure 14 shows the plot of actual Xsat against the predicted Xsat for the different granulator scales. No correlation to Froude number was observed once dissimilar blade angles and differing tip speeds were utilized in the study. Also, when the data is limited to geometrically similar blades to Fielder granulators (eliminating a flared tip 45-degee blade standard to FluidAir), the above correlation fits the data with a 99% coefficient. None of the other possible squared terms or cross products was significant at the 95%

confidence interval, suggestive that the inflection point changes linearly with impeller speed and nonlinearly with sine of the blade angle.

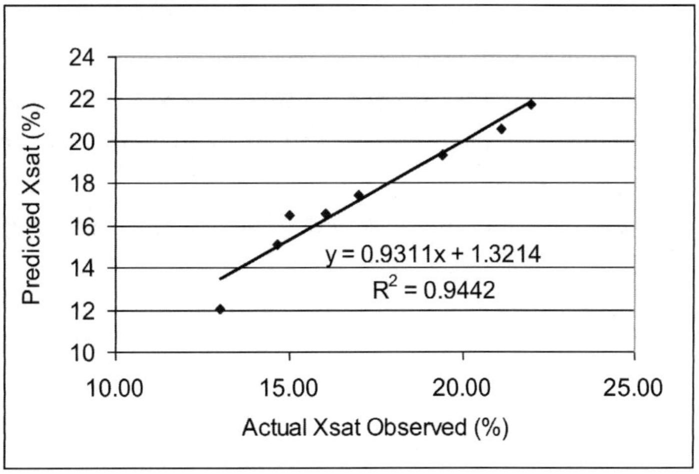

Figure 14: Actual Xsat Observed versus Predicted Xsat from impeller speed and blade angle model (Equation 6)

Another potential physical explanation for the trend observed in Figure 13 and Figure 14 comes from the observation that it appears to be important to maintain the same distribution of coverage of water spraying from the nozzle onto the powder surface passing below. If the geometry of the blade is held constant the powder bed at small scale passes below the spray nozzle many more times during water addition, thus achieving better spray distribution at small scale relative to large scale. The number of passes the powder takes below the spray nozzle for a given time period should be proportional to RPM x Sin^2(Theta). A plot of 1/(RPM x Sin^2(Theta)) creates a linear line of with nearly the same goodness of fit as the above fitted statistical trend, and can also be used to estimate the appropriate angle of the blade (Figure 15).

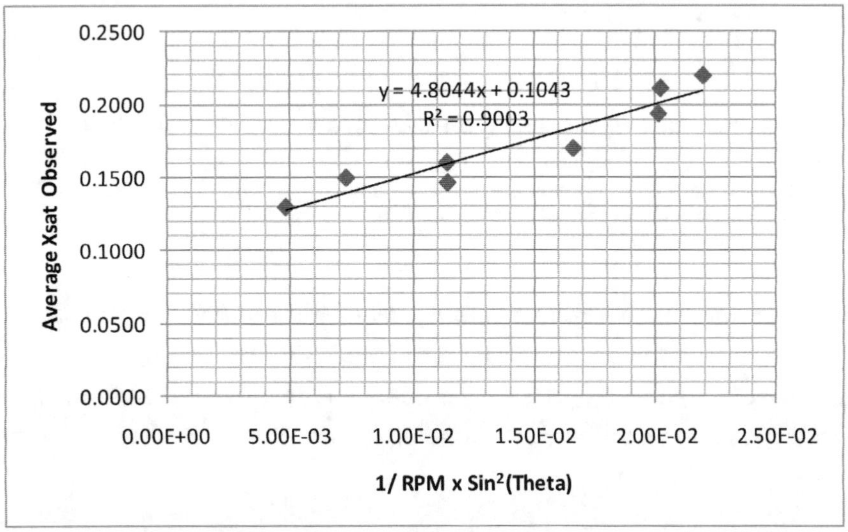

Figure 15: Actual Average Xsat Observed versus Scaling Factor of 1/(RPM x Sin^2(Theta))

The scaling factor of RPM x Sin^2(Theta) may already be designed into some Fielder granulators. The 25-L and 65-L granulator used in this study was a fixed-speed unit with two bowls built into the same piece of equipment. It is noted that the blade angle was varied and the impeller speed was fixed with result that RPM x Sin^2(Theta) was held constant between the two bowls. Similarly the 150-L and 300-L granulators had slight variation in blade angle such that this potential scaling factor was also held constant between those two machines; however, at a different value compared to the 25/65-L granulator. RPM x Sin^2(Theta) was the criteria utilized to design an experimental custom blade for the 6-L granulator utilized in this study. RPM x Sin^2(Theta) appears to be a less-perfect approximation compared to Equation 6, but may provide a close enough approximation for use with engineering the blade angle for use in scaling up of most products.

1.5 Evaluating the Significance of Work, Xsat and the Amount of Water Added Within Scale

Observation of the impeller power inflection point "Xsat" proved to be useful when explaining scaling differences, and also shows significant use in monitoring drug products controlled by the Work model at constant scale. As an example, for drug product "P" manufactured on a 300-L Fielder granulator, a total of 40 granulation batches were manufactured and presented as data in Table 4. The first ten batches were part of the experimental DoE described with the response surface shown in Figure 3. The next 30 batches were produced using the optimum granulation work and water amount found from the DoE. Dissolution profiles of the film coated tablets were used as an assessment of end product quality. The 40-batch data set with work endpoints and film coated tablets dissolution are tabulated in Appendix A, Table 4.

For this drug product held at constant tablet thickness and tablet weight, granulation work and granulation water amount are good predictors of dissolution performance. It is possible to use a quadratic model to define a region of proven acceptable quality such that if the Work, Water, and Thickness are controlled, acceptable dissolution performance will be achieved. Thus for this drug compound, Work, Water, and Thickness are utilized as effective control parameters to ensure passing dissolution.

Monitoring of 30 batches following the DoE showed there was variation in impeller load inflection point during water addition, and dissolution below 90% was noted to occur at water excess (Xs) values of less than 8% (Figure 16). This drug product has an undergranulation failure mode resulting in reduced dissolution mean and increased dissolution RSD at lower extents of granulation.

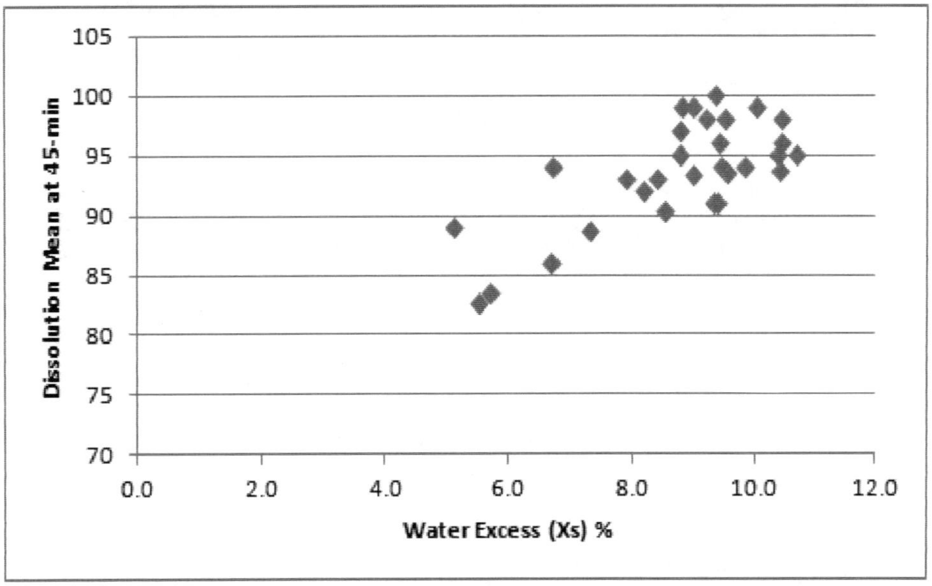

Figure 16: Variation in Water Excess and Tablet Dissolution for Drug Product "P" in 30 runs at Optimum Work and Water Setting at 300-L Scale

The general conclusion was that even at constant scale, there may be variation in the amount of water needed to start the granulation process (X_{sat}) which causes variation in the impeller load during water addition. If the drug product is sensitive to granule variation, the variation in excess water (Xs) can result in small to moderate changes in granule tapped density and drug dissolution. This is a risk factor that will need to be considered when establishing models and process control based on a granulation Work model, and might prompt need for a more complex model such as the SaWW model described in the next section. However it should be noted that even for Drug Product "P", the Work control model has been used for 700+ granulations to date of this writing with no failures below specification of 75% dissolution. Need for a more complex control model beyond "granulation work" such as the "SaWW" model discussed in the next section is only for cases where a finer degree of control is needed.

1.6 A Single Equation to Model Granulation – SaWW Model

The scale up development work and also monitoring granulation batches at 300-L scale suggested it's not the absolute amount of water added, but the amount of water added above the critical point where an increase is observed in the main impeller power. This led to development of the "Saturated Weighted Work" (SaWW) equation which can be used to predict granulation endpoint across equipment blade angle changes, water amount changes, Xs, or Work changes. The SaWW equation describes the granulation process using the idea that water in excess of a certain critical amount and work input can be integrated together to create a single predictive equation.

$$SaWW = \int_{0}^{t} \left(Power_{Impeller} - P_{0}\right) \cdot X_{S} \, dt$$

(7)

Where X_s is the amount of water above the saturation point, which is defined as follows:

$$X_S = (X_{H2O} - X_{SAT}) \qquad (8)$$

X_{H2O} is the fraction of water added at that point of time. X_{sat} is defined as the fraction of water needed at the point the impeller power begins to increase, indicating onset of granulation. X_{sat} changes with scale, and also with raw material physical property changes, and can be found using Equation 4 described in the previous section discussing scale-up.

The SaWW model, or a model that calculates the appropriate Work for a given water excess (X_S) could potentially be used as an improved control methodology when programmed into a PLC, but recommendation at this time is to use the Work model by first intent as most drug products are not as sensitive to granulation as drug product "P" and online measurement of Xsat may unnecessarily increase the complexity of the control code and control model.

SaWW, or Work and X_S were found to provide better model fits compared to Work and Water as predictive model terms for production and DoE data evaluated, and offers the advantage of reducing the number of factors needed to estimate to fit a linear regression model. As an example, for Drug Product "P", the 50-point DoE data available in Table 1 illustrated in Figure 2 as a complex 4 dimensional surface from a 10-factor linear regression model can be reduced to a 3-dimensional surface and a 6-term linear regression model with a similar goodness of fit (R^2=90%). The response surface of SaWW and compressed tablet thickness is shown in Figure 17. For the production monitoring data set in Table 4, the resulting tablet dissolution can be correlated to a single polynomial line as shown in Figure 18.

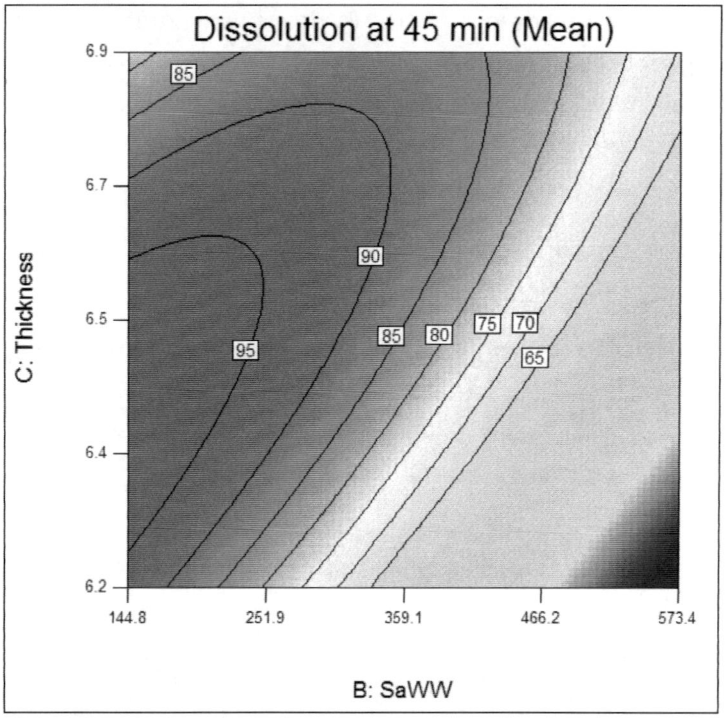

Figure 17: Drug Product "P" Data from 10-batch DoE with 5 compression levels to 50 observations, Dissolution Modeled as a single 3-dimensional surface using SaWW Model with R^2=90%

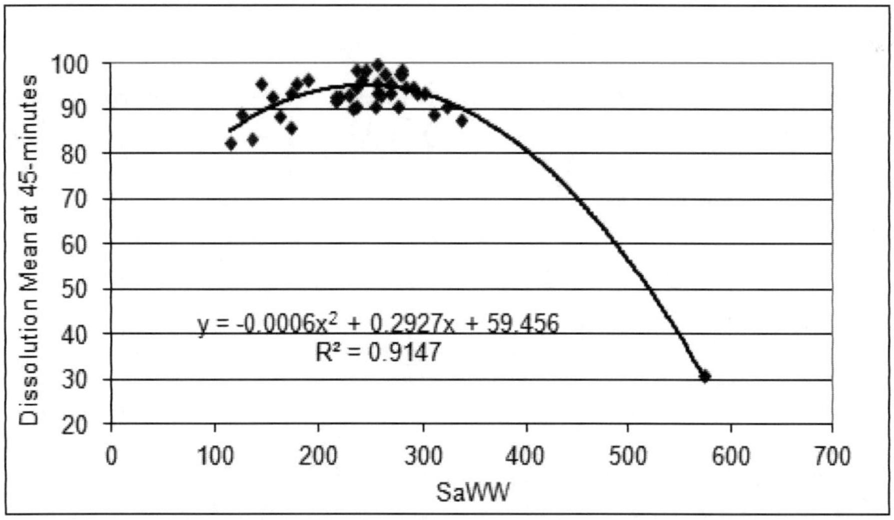

Figure 18: Drug Product "P" Data from 10-batch DoE at constant target tablet thickness, plus 30 subsequent batches, Dissolution Modeled as a single 2-dimensional plot using SaWW Model

The observation of the SaWW model describing the water and work interactions into a single parameter is not limited to just drug product "P", but has applied to all of them discussed in this chapter. For instance the SaWW model applied to Drug Products "D" (Figure 19 and Figure 20), finds the same observation of the prediction being reduced to a single characteristic line. This single line is different for each drug product, but the underlying mechanistic equation used for prediction is the same.

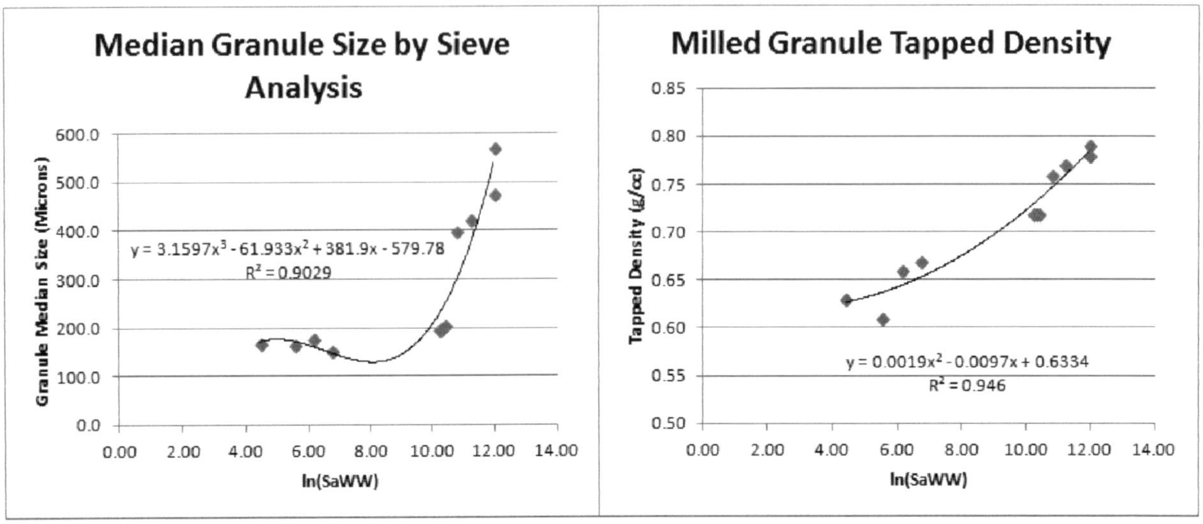

Figure 19: Drug Product "D", Response of Tapped Density, Granule Size to the natural log of SaWW

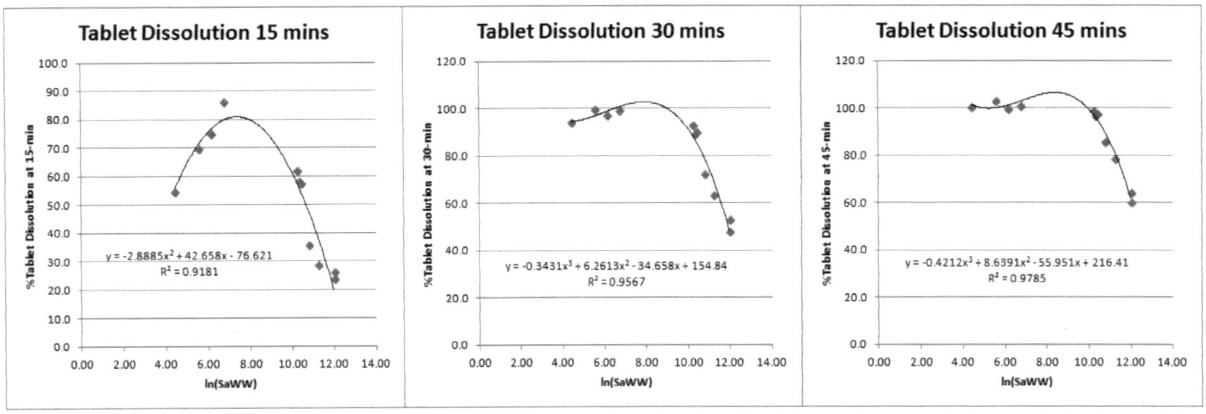

Figure 20: Drug Product "D", Response of Tablet Dissolution to the natural log of SaWW

Drug Product "S" (data Table 3) is modeled versus the SaWW model in Figure 21. This particular DoE examined water addition over a much larger range than typical, from 25% to 40% water. This allows an examination if the impact of water excess is truly linear over a wide range. The model equation used was equation 9 to test the significance of an exponent "N" on the water excess term. The model results with N=1 or N=3 is shown in Figure 21.

$$SaWW = \int_0^t \left(Power_{Impeller} - P_0\right) \cdot \left(X_S\right)^N \mathrm{d}t \tag{9}$$

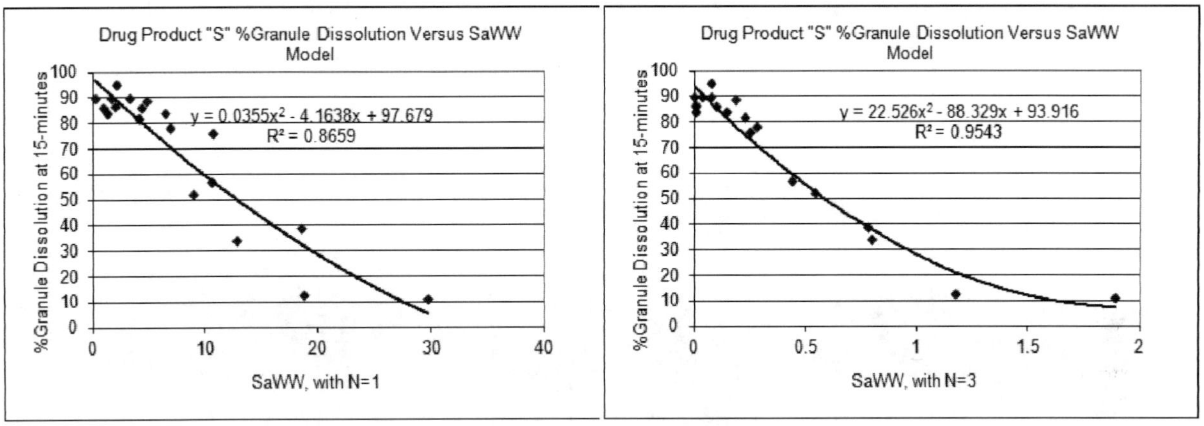

Figure 21: Drug Product "S" Data from 20-batch Sampled DoE at constant target tablet thickness, Dissolution Modeled as a single 2-dimensional plot using SaWW Model, N=1 on left, and N=3 on right.

The fits to granule dissolution data were notably improved using N=3 rather than N=1, indicating changes in water amount over a very broad range, may show a nonlinear impact of water excess, and Equation 9 is recommended.

The power of the SaWW model is the characteristic lines for a given product can be found with very few experiments. It only takes three to four observations to define a nonlinear line. Testing this line at multiple Work and Water levels then allows observation of reproducibility of the model prediction. Fewer linear regression terms are needed to fit the models, thus giving much higher capability to estimate the error distribution.

The strength and weakness of the SaWW model will lie in the capability to measure Xsat, the point of impeller inflection, as precisely as possible. If Xsat is significant and measured precisely, then the SaWW model will provide better model fits compared to models based on Work and Water amounts alone. As long as Xsat can be measured in a robust fashion, and is significant, then use of SaWW to control the process directly is likely to provide improved levels of process control compared to use of the Work equation.

1.7 Modeling Twin Screw Continuous Wet Granulation

Twin screw extrusion has been extensively studied and developed in the polymer industry since the 1950's. The first application to large scale OTC pharmaceutical granulation began in the 1970's and 1980's, primarily by GSK for antacid production. More recently, starting in 2002 twin screw extrusion came under more intensive study as an efficient and continuous means of replacing high shear wet granulation.(Keleb, Vermeire, Vervaet, & Remon, 2002) Use of "kneading" mixing elements (Figure 22) in the screw conveyer improves the granule flow, and is the primary source of granule consolidation.(Djuric, 2008)

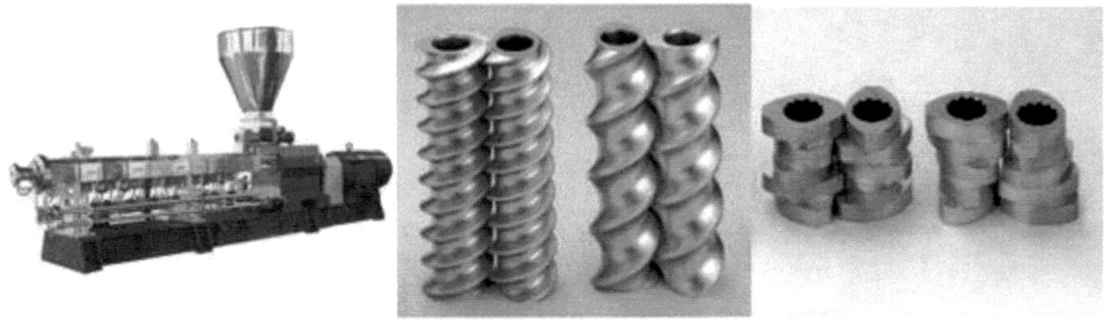

Figure 22: Twin Screw Extruder (left) with Conveying Elements (Center) and Kneading Elements (Right)

An excellent publically available data set that illustrates many of the principles for fitting a model to twin screw granulation data is from work done at Ghent University.(Vercruysse, et al., 2012) The experiment granulated a theophylline mixture in a 6-parameter DoE. The relevant data analyzed utilizing the first-principle equations described in this section is shown in Appendix A, Table 6.

The signal that can be best correlated from the granulation data is the observed granule yield, defined as material between 150 micron and 1400 microns in size. Too little granulation extent reduces yield due to increased fines, but too much granulation results in yield loss due to granules larger than 1400 microns in size. Regression analysis of the DoE parameters to predict granule yield finds the most significant terms are the number of kneading elements, and the barrel temperature. There's a possible cross product of throughput and screw speed that also appears to be significant. A regression model constructed from these parameters is shown in Figure 23 with impacts of factors shown in Figure 24.

Response Yield (Percent Granules between 150-1400 micron)

Actual by Predicted Plot

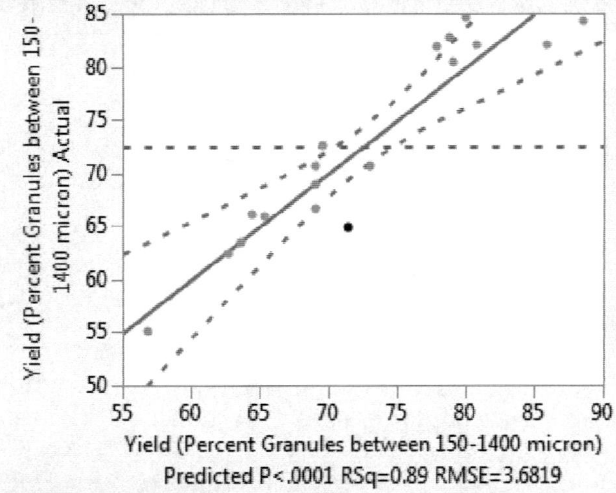

Summary of Fit

RSquare	0.885191
RSquare Adj	0.837354
Root Mean Square Error	3.681898
Mean of Response	72.52778
Observations (or Sum Wgts)	18

Parameter Estimates

Term	Estimate	Std Error	t Ratio	Prob>|t|
Intercept	44.971531	7.06944	6.36	<.0001*
Throughput (kg/hr)	-0.209037	0.127923	-1.63	0.1282
Screw Speed (RPM)	0.0080413	0.005482	1.47	0.1682
Kneading Elements	1.6275576	0.20184	8.06	<.0001*
Barrel Temp (°C)	0.4616674	0.128307	3.60	0.0037*
(Throughput (kg/hr)-17.9167)*(Screw Speed (RPM)-784.722)	0.0019671	0.000736	2.67	0.0203*

Figure 23: Regression Model of Twin Screw Theophelline DoE

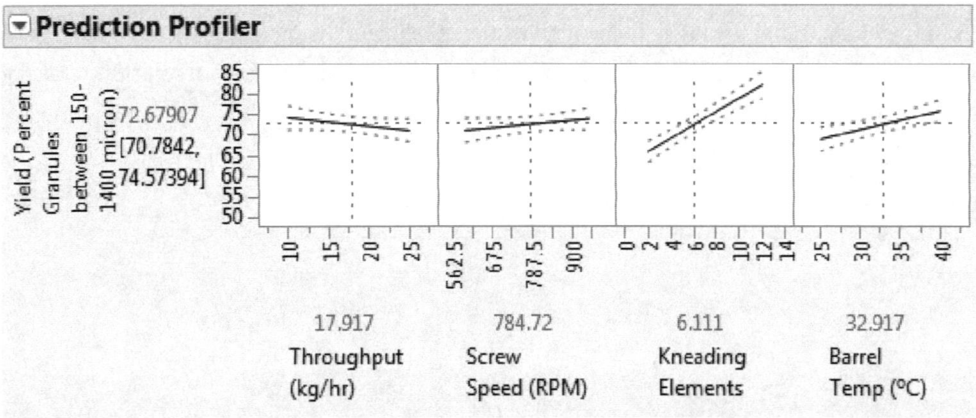

Figure 24: Factor Prediction Profiler for DoE Parameters for fit Granule Yield

Alternatively, it's possible to construct a model for twin screw granulation using first principles and the knowledge developed for high shear granulation, discussed earlier in this chapter. From first principles, the power input into the powder mixture is a function of torque and RPM as shown in equation 10.

$$Power = Torque \cdot \frac{RPM \cdot 2 \cdot \pi}{60} \tag{10}$$

High shear granulation was best modeled as power per unit dry powder mass, thus the above power term needs to be divided by the mass throughput measured in grams per second.

$$Power\_per\_mass = Torque \cdot \frac{RPM \cdot 2 \cdot \pi}{60 \cdot throughput} \tag{11}$$

There's an effect of temperature where higher temperature is noted to drive the granulation further. This effect is hypothesized to be due to changes in rheology. A mixture that requires more energy to overcome viscous forces or surface area interactions will require more power to reach an equivalent endpoint. Thus the impact of temperature is hypothesized to reduce the viscosity of the binding liquid thus allowing easier deformation and consolidation of the wet granules. The power per unit mass in the above equation is divided by the liquid viscosity at the granulation temperature resulting in Equation 12.

$$Power\_per\_Mass\_Viscosity = Torque \cdot \frac{RPM \cdot 2 \cdot \pi}{60 \cdot throughput \cdot \mu} \tag{12}$$

Plots of power per unit mass, and then power per unit mass divided by viscosity are shown in Figure 25. Use of viscosity to describe the temperature effect is noted to improve the correlation coefficient. The resulting model provides a better correlation coefficient (R^2=90%) than that achieved with linear regression, and only needs to estimate two fitted parameters.

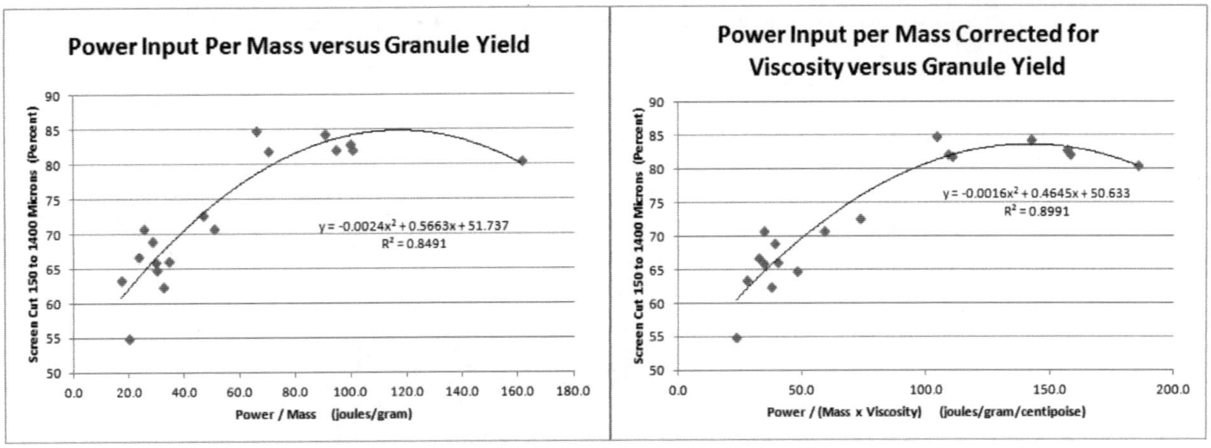

Figure 25: Prediction of Granule Yield from First Principle Equations

The same relationship to power is noted to be valid for both high shear granulation and twin screw granulation. If the twin screw granulation has motor power measured instead of torque then power per unit mass can be calculated as:

$$Power\_per\_mass = \frac{(P - P_0)}{throughput} \tag{13}$$

Where P is the measured power, and P_0 is the baseline power needed to turn the impeller without solids present. The resulting units from equation 13 are joules/gram, which is equivalent to the Work/Mass models described in the description of batch high shear granulation modeling.

1.8 The Impact of Feeder Variability on Twin Screw Wet Granulation

Continuous operations make the assumption that the flow into the equipment is continuous, and unfortunately powders that need to be granulated do not tend to flow in a continuous fashion. It is useful to consider the impact of granulation variation induced by variation in inlet powder flow to the twin screw granulator. If drug product tablet dissolution is sensitive to changes in granulation extent, the impact of feed variation is in-batch variation of granulation extent, resulting in increased in-batch variation in granule properties, and potentially reduced dissolution mean with higher RSD. In cases of very poorly flowing powders, the observation is the tablet dissolution properties achieved from twin screw granulation is not as good as those observed in high shear granulation. The situation can be improved by improving the powder flow of the mixture being granulated, through use of flow aids like fumed silica, or by limiting the region of operation of the powder hopper feeding the twin screw granulator.

Example hoppers used for feeding dry powder into twin screw granulation are manufactured by K-Tron, or Brahbender (Figure 26), and utilize either dual or single helical impellers (Figure 27).

Figure 26: K-Tron Hopper (Left) and Brahbender Hopper (Right)

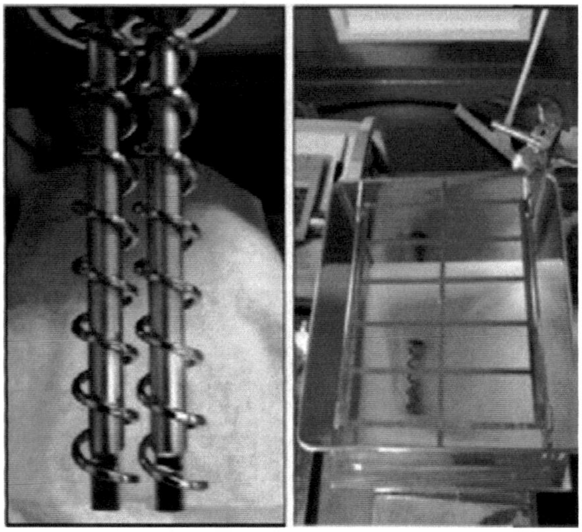

Figure 27: Dual Helical Impeller of K-Tron Hopper (Left) and Single Helix of the Brahbender Hopper (Right)

Figure 28 shows two different powders being fed from a K-Tron dual screw hopper. The freely flowing powder on left comes out as nearly a continuous stream, whereas the poorly flowing powder on right exits the hopper in cohesive packets.

Figure 28: Freely flowing powder from K-Tron Hopper (left), Very cohesive powder (right)

To examine the variability of powder flow from hoppers, ten different powders were analyzed using a 6-inch Brookfield PFT shear cell, with resulting density and flow index analytical data shown in Figure 29.

Rank #	Material Name	Flow Function Slope	Flow Function Intercept	Fill Bulk Density (kg/m^3)	Flow Index at 0.5 KPA	Flow Index at 1 KPA	Flow Index at 2.4 KPA	Flow Index at 5 KPA	Flow Index at 10 KPA	Type of Flow at 1 KPA
1	Manitol Pearlitol 200SD	0.03	0.002	484.4	0.034	0.032	0.031	0.0304	0.0302	Free Flowing
2	Avicel PH102	0.14	0.04	367.3	0.22	0.18	0.157	0.148	0.144	Easy Flowing
3	Avicel PH101	0.22	0.023	303	0.266	0.243	0.230	0.2246	0.2223	Easy Flowing
4	HPMC Pharmacoat 603	0.22	0.116	401.7	0.452	0.336	0.268	0.2432	0.2316	Cohesive
5	Improved 6% API Formulation	0.06	0.277	378.8	0.614	0.337	0.175	0.1154	0.0877	Cohesive
6	Improved 12% API Formulation	0.14	0.338	332.9	0.816	0.478	0.281	0.2076	0.1738	Cohesive
7	Manitol Pearlitol 160C	0.16	0.366	585.3	0.892	0.526	0.313	0.2332	0.1966	Very Cohesive
8	6% API Formulation	0.19	0.555	378.8	1.3	0.745	0.421	0.301	0.2455	Very Cohesive
9	12% API Formulation	0.23	0.561	332.9	1.352	0.791	0.464	0.3422	0.2861	Very Cohesive
10	100% API lot 76	0.16	0.835	134.9	1.83	0.995	0.508	0.327	0.2435	Very Cohesive

Figure 29: Shear Cell Analysis of 10 Powders Utilized in Hopper Study

Each powder was then loaded into the K-Tron hopper, and the feedrate setting was varied over values of 10%, 15%, 25%, 50%, 75%, and 99% of maximum RPM. After powder flow was established, data recording was enabled on a Mettler Toledo scale, collecting dynamic weight measurement at 8 measurements per second, with data averaging for 3 seconds per weight observation. The average feedrate for the trial was recorded, as well as the 3-second relative standard deviation (RSD) of the powder flowrate. Example data for Manitol 200SD is shown in Figure 30, and this process was repeated for the 9 other powders resulting in a total of 126 observations.

Name	Hopper Mass	Feed setting	Feed Rate (grams/sec)	Flow RSD% for 3 second time interval
Manitol 200SD	2.8	99%	5.472	1.35%
Manitol 200SD	2.55	75%	4.113	1.59%
Manitol 200SD	2.99	50%	2.587	2.00%
Manitol 200SD	3.2	25%	1.092	2.57%
Manitol 200SD	1.92	99%	5.404	2.89%
Manitol 200SD	1.53	75%	4.133	0.70%
Manitol 200SD	1.73	50%	2.623	2.54%
Manitol 200SD	1.26	25%	1.136	2.25%
Manitol 200SD	0.97	15%	0.541	5.59%
Manitol 200SD	2.147	15%	0.527	5.53%
Manitol 200SD	2.33	10%	0.215	14.93%

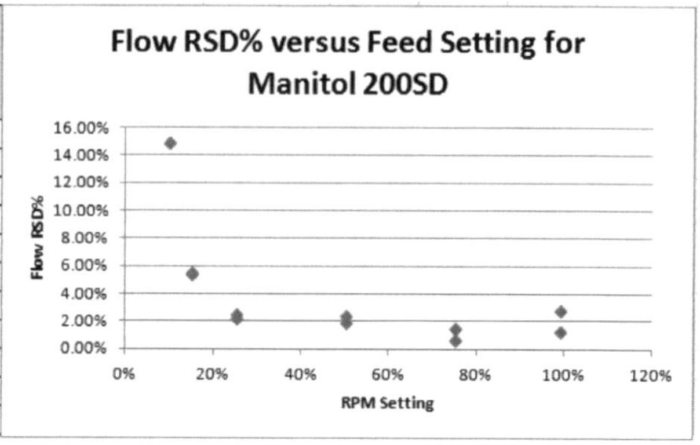

Figure 30: K-Tron Flow Data for Manitol 200 SD (left), Observed Relationship between RPM Setting and Flow RSD% (right)

At a fixed RPM, 25% for example, a correlation is observed between the shear cell flow index at 1 KPa, and the 3-second hopper flow RSD (Figure 31). Correlations were attempted at shear cell consolidation stresses ranging from 0.5 KPa to 10 KPa, with conclusion that the RSD data is best fit to shear cell data at 1 KPa consolidation pressure.

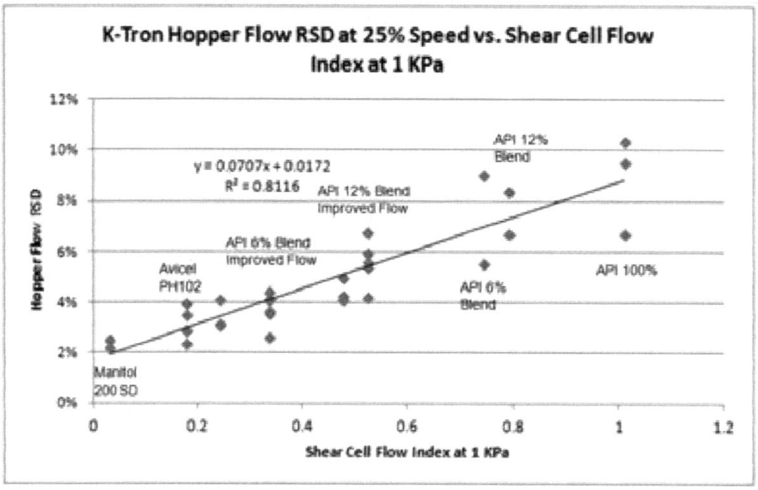

Figure 31: K-Tron Hopper Flow RSD at 25% Feed Setting versus Shear Cell Flow Index at 1-KPa

Modeling all the data collected on the K-Tron Hopper at all the RPM settings results in in a 5-term linear regression model built from 126 observations. The response surface is best fit to the reciprocal of feeder RPM, measured as seconds per feed screw revolution. The model for the K-Tron hopper, parameter estimates, and the resulting response surface predicting RSD fraction is shown in Figure 32.

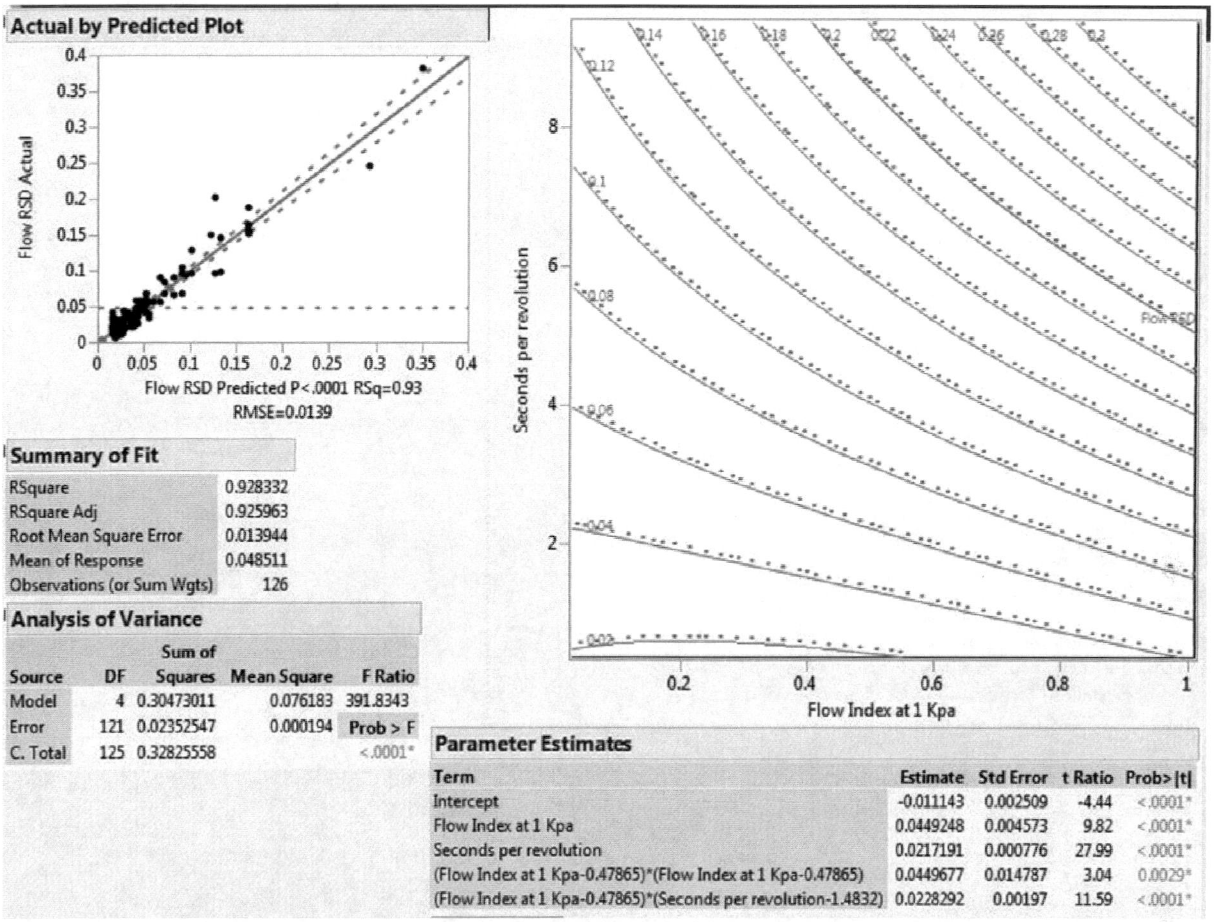

Summary of Fit

RSquare	0.928332
RSquare Adj	0.925963
Root Mean Square Error	0.013944
Mean of Response	0.048511
Observations (or Sum Wgts)	126

Analysis of Variance

Source	DF	Sum of Squares	Mean Square	F Ratio
Model	4	0.30473011	0.076183	391.8343
Error	121	0.02352547	0.000194	Prob > F
C. Total	125	0.32825558		<.0001*

Parameter Estimates

Term	Estimate	Std Error	t Ratio	Prob>|t|
Intercept	-0.011143	0.002509	-4.44	<.0001*
Flow Index at 1 Kpa	0.0449248	0.004573	9.82	<.0001*
Seconds per revolution	0.0217191	0.000776	27.99	<.0001*
(Flow Index at 1 Kpa-0.47865)*(Flow Index at 1 Kpa-0.47865)	0.0449677	0.014787	3.04	0.0029*
(Flow Index at 1 Kpa-0.47865)*(Seconds per revolution-1.4832)	0.0228292	0.00197	11.59	<.0001*

Figure 32: K-Tron Twin-Screw Hopper Powder Flow RSD Fraction Model, Response Surface to Powder Flow Index at 1-KPa, and Number of Seconds per Feed Screw Rotation

The K-Tron hopper is remarkably robust, and was able to successfully feed powder up to a flow index of 0.995 at 1-KPa, albeit with increasing levels of variation measured every 3-seconds. The disadvantage of the K-Tron in this particular configuration is that even for very good flowing powders, such as Manitol 200SD, there was still a significant relationship between hopper speed setting and the observed RSD of flow. It should be noted there is also an impact of screw pitch, a finer pitch impeller is able to have more discharge events per second thus lower RSD, but cannot handle as poorly flowing of powders.

The same experimental procedure was also applied to the Brahbender hopper, though fewer powders could be fed through this hopper. Powders above a flow index of 0.5 to 0.6 at 1-KPa have a probability of forming cohesive arching resulting in starving the impeller screw of powder. The Brahbender hopper was able to poorly feed a powder with a flow index of 0.75 at 1-KPa, but with numerous observations of cohesive arching and powder flow entirely stopping for seconds. Only six test powders could be fed through the Brahbender, resulting in 72 observations for the experiment. The resulting model to predict powder flow RSD fraction and response surface diagram is shown in Figure 33.

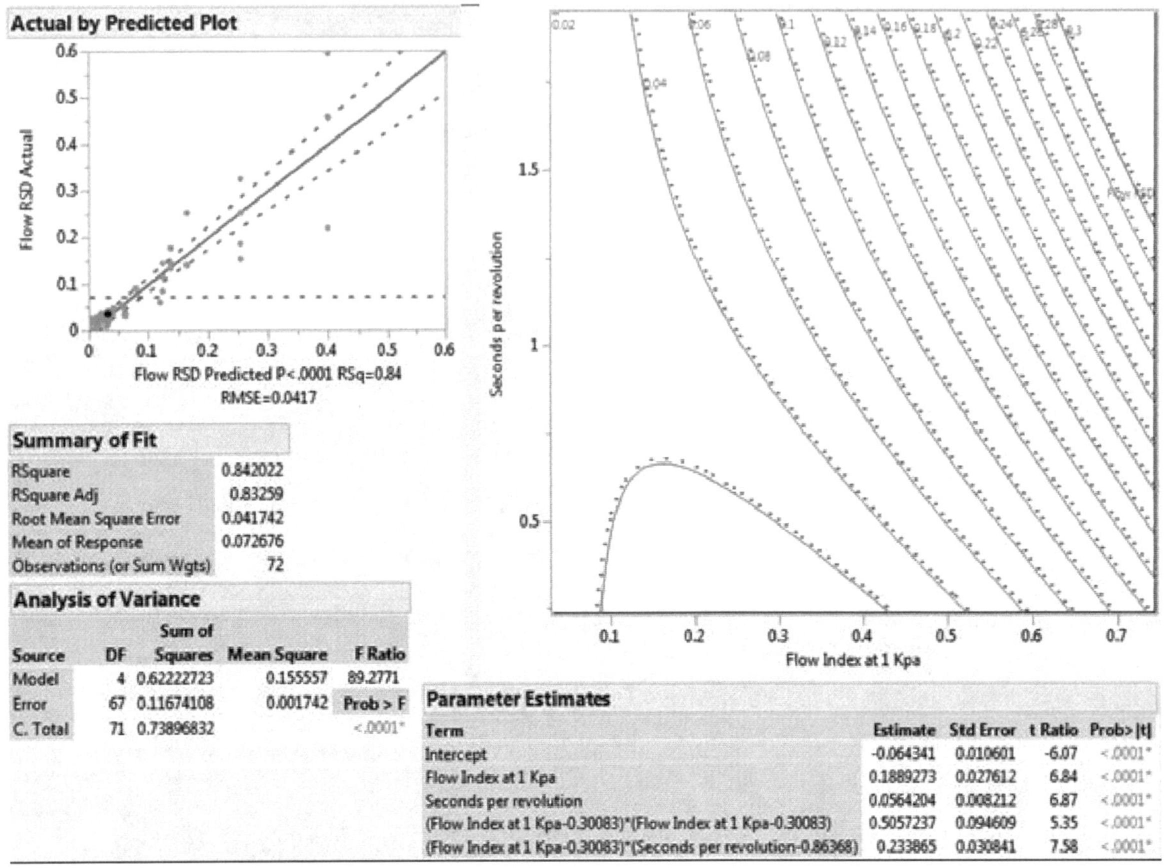

Summary of Fit

RSquare	0.842022
RSquare Adj	0.83259
Root Mean Square Error	0.041742
Mean of Response	0.072676
Observations (or Sum Wgts)	72

Analysis of Variance

Source	DF	Sum of Squares	Mean Square	F Ratio
Model	4	0.62222723	0.155557	89.2771
Error	67	0.11674108	0.001742	Prob > F
C. Total	71	0.73896832		<.0001*

Parameter Estimates

Term	Estimate	Std Error	t Ratio	Prob>|t|
Intercept	-0.064341	0.010601	-6.07	<.0001*
Flow Index at 1 Kpa	0.1889273	0.027612	6.84	<.0001*
Seconds per revolution	0.0564204	0.008212	6.87	<.0001*
(Flow Index at 1 Kpa-0.30083)*(Flow Index at 1 Kpa-0.30083)	0.5057237	0.094609	5.35	<.0001*
(Flow Index at 1 Kpa-0.30083)*(Seconds per revolution-0.86368)	0.233865	0.030841	7.58	<.0001*

Figure 33: Brahbender Hopper Powder Flow RSD Fraction Model, Response Surface to Powder Flow Index at 1-KPa, and Number of Seconds per Feed Screw Rotation

The significant advantage the Brahbender Hopper has is that for powders that flow well, up to a flow index of approximately 0.4, there is comparatively less impact of feed rate (RPM) variation compared to the K-Tron hopper, as illustrated by the response surface in Figure 33. This makes the Brahbender Hopper the preferred choice for feeding materials that flow well, such as excipient mixtures. The K-Tron should be chosen if the flow index is greater than 0.4 to 0.5 at 1-KPa consolidation stress.

To this point, the response surfaces shown are the RSD relative to a 3-second measurement of mass discharge. Three seconds is an arbitrary number which was decided upon based on the settling time of the scale. As the time period of measurement is reduced, the RSD will increase, so it's important to consider what the residence time is of the powder underneath the powder feed point and the water input point. The input ports where the water and powder enter the granulator require 3.66 revolutions of the impeller to traverse, resulting in a residence time of 0.5 and 1.5 seconds at 450 and 150 RPM respectively (Figure 34). As this residence time is decreased, more variable packets of powder then encounter the water addition nozzle, resulting in variable extents of moisture fraction, and thus more variable granulation.

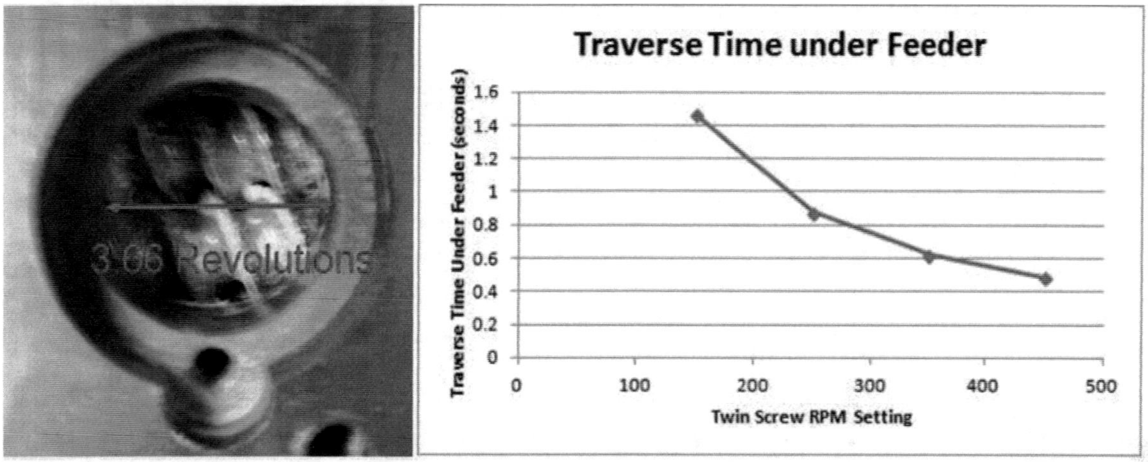

Figure 34: Impact of Twin Screw Granulator RPM on Residence Time Under Powder and Liquid Feed

The RSD should decrease the more powder discharge events can be averaged together from the hopper feeding the twin screw. If the measurement time is extended from 3 seconds to 12 seconds, a nonlinear trend is observed in the relative RSD that is measured. Theoretically this RSD should scale based upon the reciprocal of the square root of time, based on theory of how standard deviation of a population mean tends to vary with increasing number of observations. Figure 35 shows the impact of measurement time period and also the near-linear response if plotted versus the reciprocal of square root of time. This near-linear relationship can then be extrapolated to faster time periods, larger values of 1/(time^0.5), to get an estimate of the increase in RSD expected as the twin screw RPM setting is changed (Figure 36).

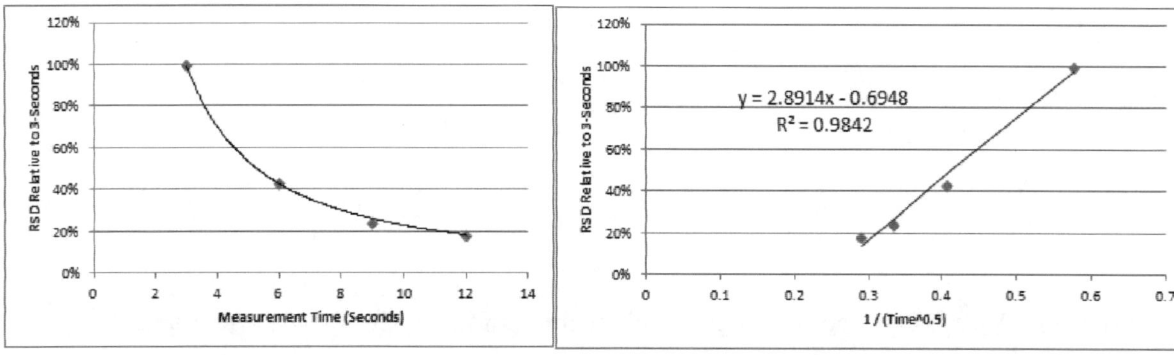

Figure 35: Impact of Change in Measurement Time on Calculated RSD Relative to 3-Seconds

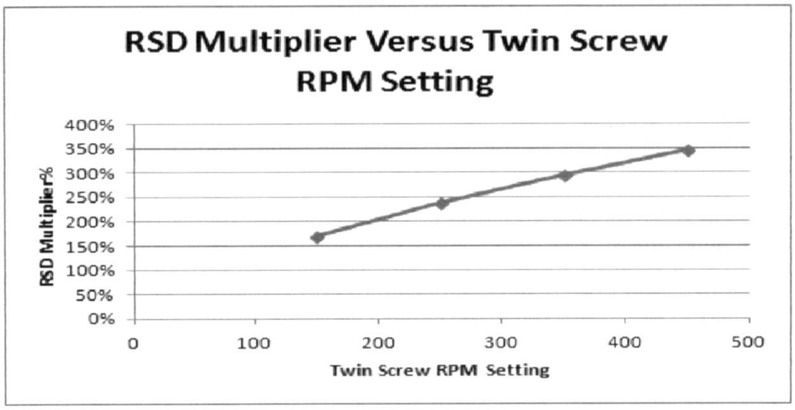

Figure 36: Impact of Twin Screw Granulator RPM Setting and RSD Multiplier Compared to 3-Second Data

The result of input powder variability is then going to be water wetting variability, and then variability in granulation extent in the resulting product granule. It's important to develop and include back-mixing zones that increase the powder residence time to seconds before and after the water addition zone to reduce the impact of this phenomenon for poorly flowing powders.

Figure 37 shows two case study predictions assuming impeller configuration is set up with no back mixing elements, for a powder with a flow index of 0.8 at 1-KPa. The first case, shown on left sets the K-Tron feed setting to 50% (1.08 revolutions per second) and a twin screw speed of 150 rpm. For a water target of 20%, 99% of the material would be expected to have wetting between 15% and 25%. Case two on the right shows it's possible to get much more variability by reducing the K-Tron feed setting to 20% while increasing the twin screw speed to 450 RPM. In this second scenario the 99% limits of wetting is between 3% and 37%, and significant variation in granule properties is expected.

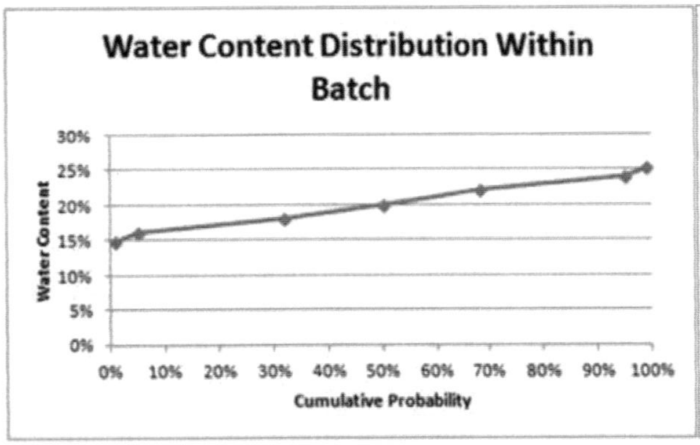

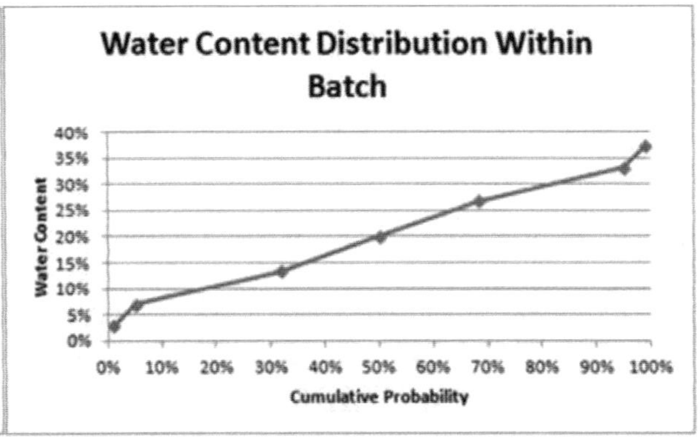

Figure 37: Model Predictions of Water Distribution for Case 1 (left) and Case 2 (right)

In reality the twin screw granulator would not be set up with no back mixing elements, and operated in such an extreme case of slow K-Tron feeder speed and rapid screw RPM. The situation is unlikely to be as extreme as this, and the impact of powder feed variation will be more subtle, more similar to the case study shown on the left in Figure 37. For drug products that are sensitive to granulation, it does illustrate the need to consider powder flow properties, it's impact on feed variability, the residence time distributions, the impact of back mixing, and the resulting granulation consistency. Models using the equations provided in this section can be used to estimate the worse-case water wetting consistency, before the addition of back mixing elements.

1.9 Conclusion

Batch high shear granulation and twin screw granulation can both be successfully modeled using equations relating to the granulation work done per unit mass of dry powder. Models based on work per unit mass can and have been used to successfully control granulation processes in real time, even for drug products fairly sensitive to granulation extent. If an even finer degree of control or potentially more precise modeling is desired or there is desire to predict water and work requirements across scale, then it's useful to measure the minimum amount of water needed to cause a significant change in impeller power during water addition, referred to as "Xsat" in this discussion. For high shear granulation, observation of Xsat as it varied with scale allow the generation of models that can predict how the water requirements will change with scale, and allows construction of granulation impeller blades designed to ensure similar water requirements across scale.

For high shear granulation, an equation (named SaWW) that integrates both granulation work and the amount of water in excess of Xsat is found to be useful in providing better models and reducing the number of dimensions that need to be considered when creating a linear regression response surface model. It appears likely the SaWW model shown as Equation 7 is a good fundamental equation for modeling most granulation processes over moderate changes in water excess, and will reduce the resulting predictions to a single characteristic line that requires fewer experiments to define. If large changes in water amount are explored (e.g. moving from 25% to 40% water addition), it's possible there may be significant nonlinearity in the impact of water excess, in which case Equation 9 should be explored.

Continuous twin screw granulation adds the complexity that the powder feeding the granulator does not necessarily feed in a continuous fashion, but rather in a variable fashion than can be predicted using the response surface models for each hopper type. A measure of powder flow using the Brookfield PFT shear cell flow index at 1-KPa was found to correlate to the resulting variability. Models for worst case water-wetting variation, assuming minimal back mixing are discussed. For drug products where dissolution is sensitive to variation in granulation extent, it is useful to consider optimizing the flow properties through API physical properties or through use of flow aids such as fumed silica, to reduce the variability in feed rate to the twin screw granulator. It's also useful to consider use of back mixing elements to reduce the severity of the feed variability.

1.10 Works Cited

Clancy, D. (2010). Method and Apparatus for Predicting Properties of Granulated Materials and Dosage Forms made Therefrom. *US Patent* , 20100318228.

Clancy, D., Campbell, G., Zhang, J., Gupta, M., & Oh, C. (2011). Closing the gap in series scale up of high shear wet granulation process using impeller power and blade design. *Powder Technology* , 184-192.

Djuric, D. (2008). *CONTINUOUS GRANULATION WITH A TWIN SCREW EXTRUDER.* Düsseldorf: Heinrich-Heine-Universität Düsseldorf.

Johannes von Orelli, H. L. (Volume 287, Issues 1–2, 9 December 2004). Search for technological reasons to develop a capsule or a tablet formulation with respect to wettability and dissolution. *International Journal of Pharmaceutics* , Pages 135–145.

Keleb, E., Vermeire, A., Vervaet, C., & Remon, J. (2002). Continuous twin screw extrusion for the wet granulation of lactose. *International Journal of Pharmaceutics* , 69-80.

Matthew J. Mollan Jr., P. a. (2004). Continuous Processing in Pharmaceutical Manufacturing. *Pharmaceutical Manufacturing* .

Trafton, A. (2012, March 12). Continuous drug manufacturing offers speed, lower costs. *MIT News* .

Vercruysse, J., Dias, C., Peeters, E., Fonteyne, M., Delaet, U., Assche, U., et al. (2012). Continuous twin screw granulation: Influence of process variables on granule and tablet quality. *European Journal of Pharmaceutics and Biopharmaceutics* , 205-211.

von Orelli, J., & Leuenberger, H. (2004). Search for technological reasons to develop a capsule or a tablet formulation with respect to wettability and dissolution. *International Journal of Pharmaceutics* , 135-145.

Wallack, T. (2007, September 28). Novartis to give MIT $65m to find new way to produce drugs. *The Boston Globe* .

Wilburn, K. (2010). *The Business Case for Continuous Manufacturing of Pharmaceuticals.* Massachusetts: Massachusetts Institute of Technology.

Appendix A: DoE and Repeat Run Data Tables

Granule Batch	Tablet Batch	Granulation Elapsed Time (seconds)	Power for 10-seconds at end of batch (%Motor Load)	Water Delivered (%mass/ mass)	Equation (2) Work (%Load* seconds)	Xsat	Equation (8) Xs(%)	Equation (7) SaWW (%Load*sec)	Tablet thickness (mm)	Milled Granule Tapped Density at 1250 taps (g/cc)	Percent Dissolved at 45-min	Dissolution RSD 45-min
1	A	481	41.9	29.9%	4268.1	20.2%	9.7%	322.730	6.21	0.73	65	22.7
1	B	481	41.9	29.9%	4268.1	20.2%	9.7%	322.730	6.34	0.73	88	7.2
1	C	481	41.9	29.9%	4268.1	20.2%	9.7%	322.730	6.51	0.73	91	1.1
1	D	481	41.9	29.9%	4268.1	20.2%	9.7%	322.730	6.67	0.73	89	1.7
1	E	481	41.9	29.9%	4268.1	20.2%	9.7%	322.730	6.86	0.73	87	0.7
2	A	484	43.1	31.8%	6364.2	20.2%	11.6%	573.387	6.21	0.81	5	41.4
2	B	484	43.1	31.8%	6364.2	20.2%	11.6%	573.387	6.36	0.81	13	86.4
2	C	484	43.1	31.8%	6364.2	20.2%	11.6%	573.387	6.51	0.81	31	49.6
2	D	484	43.1	31.8%	6364.2	20.2%	11.6%	573.387	6.68	0.81	69	24.6
2	E	484	43.1	31.8%	6364.2	20.2%	11.6%	573.387	6.85	0.81	77	2.4
3	A	442	47.3	31.8%	4125.6	21.9%	9.9%	309.117	6.23	0.77	70	22.8
3	B	442	47.3	31.8%	4125.6	21.9%	9.9%	309.117	6.35	0.77	89	1.3
3	C	442	47.3	31.8%	4125.6	21.9%	9.9%	309.117	6.53	0.77	89	2.2
3	D	442	47.3	31.8%	4125.6	21.9%	9.9%	309.117	6.70	0.77	89	2.4
3	E	442	47.3	31.8%	4125.6	21.9%	9.9%	309.117	6.86	0.77	82	2.6
4	A	501	30.0	27.7%	2796.6	20.5%	7.2%	144.777	6.21	0.73	88	5.7
4	B	501	30.0	27.7%	2796.6	20.5%	7.2%	144.777	6.36	0.73	96	1.2
4	C	501	30.0	27.7%	2796.6	20.5%	7.2%	144.777	6.55	0.73	96	1.3
4	D	501	30.0	27.7%	2796.6	20.5%	7.2%	144.777	6.71	0.73	94	1.5
4	E	501	30.0	27.7%	2796.6	20.5%	7.2%	144.777	6.89	0.73	90	5.3
5	A	664	33.9	27.7%	5587.9	20.1%	7.6%	301.534	6.20	0.72	69	36.3
5	B	664	33.9	27.7%	5587.9	20.1%	7.6%	301.534	6.36	0.72	94	1.6
5	C	664	33.9	27.7%	5587.9	20.1%	7.6%	301.534	6.50	0.72	94	1.5
5	D	664	33.9	27.7%	5587.9	20.1%	7.6%	301.534	6.69	0.72	93	2.8
5	E	664	33.9	27.7%	5587.9	20.1%	7.6%	301.534	6.87	0.72	80	8.7
6	A	413	33.5	31.8%	2647.0	22.1%	9.7%	178.592	6.19	0.73	92	3.1
6	B	413	33.5	31.8%	2647.0	22.1%	9.7%	178.592	6.34	0.73	96	0.8
6	C	413	33.5	31.8%	2647.0	22.1%	9.7%	178.592	6.50	0.73	96	0.6
6	D	413	33.5	31.8%	2647.0	22.1%	9.7%	178.592	6.69	0.73	93	2.4
6	E	413	33.5	31.8%	2647.0	22.1%	9.7%	178.592	6.87	0.73	84	9.7
7	A	533	41.0	29.9%	4934.7	20.7%	9.2%	337.302	6.20	0.73	60	41.5
7	B	533	41.0	29.9%	4934.7	20.7%	9.2%	337.302	6.34	0.73	90	0.9
7	C	533	41.0	29.9%	4934.7	20.7%	9.2%	337.302	6.52	0.73	88	1.5
7	D	533	41.0	29.9%	4934.7	20.7%	9.2%	337.302	6.68	0.73	89	2.2
7	E	533	41.0	29.9%	4934.7	20.7%	9.2%	337.302	6.87	0.73	84	3.3
8	A	413	30.2	30.1%	2626.3	21.0%	9.1%	156.056	6.20	0.70	89	7.2
8	B	413	30.2	30.1%	2626.3	21.0%	9.1%	156.056	6.37	0.70	94	2.5
8	C	413	30.2	30.1%	2626.3	21.0%	9.1%	156.056	6.52	0.70	93	1.4
8	D	413	30.2	30.1%	2626.3	21.0%	9.1%	156.056	6.71	0.70	87	7.7
8	E	413	30.2	30.1%	2626.3	21.0%	9.1%	156.056	6.89	0.70	86	10.5
9	A	544	29.5	28.0%	3612.5	20.5%	7.5%	190.096	6.21	0.70	90	10.9
9	B	544	29.5	28.0%	3612.5	20.5%	7.5%	190.096	6.36	0.70	97	0.9
9	C	544	29.5	28.0%	3612.5	20.5%	7.5%	190.096	6.53	0.70	97	0.8
9	D	544	29.5	28.0%	3612.5	20.5%	7.5%	190.096	6.70	0.70	93	4.8
9	E	544	29.5	28.0%	3612.5	20.5%	7.5%	190.096	6.88	0.70	83	6.2
10	A	456	39.3	30.1%	3712.7	20.9%	9.3%	275.916	6.20	0.74	80	8
10	B	456	39.3	30.1%	3712.7	20.9%	9.3%	275.916	6.36	0.74	91	3.8
10	C	456	39.3	30.1%	3712.7	20.9%	9.3%	275.916	6.52	0.74	91	0.6
10	D	456	39.3	30.1%	3712.7	20.9%	9.3%	275.916	6.68	0.74	89	1.9
10	E	456	39.3	30.1%	3712.7	20.9%	9.3%	275.916	6.86	0.74	81	10

Table 1: Drug Product "P" Granulation DoE at 300-L Scale, 10 granulation batch response surface design for Granulation Work and Water, Split Plot Design to 50 compression batches

	DoE Factors			Calculated DoE Factors					Dried Granule Properties			Tablet Properties		
Run ID #	Water Addition (% w/w)	Wet Massing time (min)	Water Addition Rate (g/min kg)	Work	Xsat	Xs	SaWW	ln(SaWW)	Tapped Bulk Density (g/ml)	Median Particle Size	Compaction Compressibility (%)	Tablet Dissolution 15 mins	Tablet Dissolution 30 mins	Tablet Dissolution 45 mins
1	40	0	29	5032.1	24.25	15.75	48207	10.78	0.76	396.6	20	35.7	72.0	86.0
2	33	2	24	5113.3	25.50	7.50	27550	10.22	0.72	194.0	18	62.0	93.0	99.0
3	40	0	19	7874.0	24.45	15.55	75101	11.23	0.77	421.3	21	28.7	63.3	79.0
4	26	0	29	624.2	23.21	2.79	257	5.55	0.61	165.1	18	69.7	99.3	103.0
5	33	2	24	4582.4	23.67	9.33	30046	10.31	0.72	199.2	18	58.0	89.0	97.0
6	26	0	19	1196.5	24.80	1.20	84	4.43	0.63	168.3	17	54.7	94.0	100.7
7	40	4	29	11870.8	22.99	17.01	158878	11.98	0.78	472.8	14	26.3	52.7	64.7
8	26	4	19	2424.8	25.26	0.74	844	6.74	0.67	152.0	17	86.0	99.0	101.0
9	26	4	29	1447.3	25.54	0.46	478	6.17	0.66	176.6	17	75.0	97.0	100.0
10	33	2	24	5170.4	24.08	8.92	32919	10.40	0.72	202.7	16	57.3	89.7	97.7
11	40	4	19	14713.2	25.26	14.74	159812	11.98	0.79	566.9	15	23.7	48.0	60.3

Table 2: Drug Product "D" 11 granulation batch Factorial DoE at 65-L Scale

Granulation Batch	Sample	Description	Water Content Added	Water Addition Time (seconds)	Granulation Time (seconds)	Impeller Load (kW)	Work (kj/kg)	SaWW (N=1) (kj/kg)	SaWW (N=3) (kj/kg)	%Granules Dissolved in 15-min
1	1	25% water sample 1	0.2540	320	360	0.808174	25.63106	0.219555	0.000556	90
1	2	25% water sample 2	0.2540	320	540	0.941444	39.00426	0.552105	0.001525	
1	3	25% water sample 3	0.2540	320	660	0.885915	49.55154	0.805422	0.002264	86
1	4	25% water sample 4	0.2540	320	960	0.808174	65.86184	1.174112	0.003339	84
1	5	25% water sample 5	0.2540	320	1240	0.808174	99.23494	1.869198	0.005366	87
2	1	35% water sample 1	0.3540	300	380	1.178368	31.57435	1.592449	0.033889	90
2	2	35% water sample 2	0.3540	300	540	1.178368	48.88172	3.153912	0.070921	90
2	3	35% water sample 3	0.3540	300	660	1.230196	60.52855	4.215445	0.096096	86
2	4	35% water sample 4	0.3540	300	960	1.256109	83.33616	6.310204	0.145776	84
2	5	35% water sample 5	0.3540	300	1200	1.282023	130.5264	10.6802	0.249414	76
3	1	40% water sample 1	0.4106	320	380	1.337552	31.02979	2.069927	0.074342	95
3	2	40% water sample 2	0.4106	320	540	1.415293	50.81866	4.694684	0.186448	89
3	3	40% water sample 3	0.4106	320	660	1.415293	67.07937	6.882321	0.279885	78
3	4	40% water sample 4	0.4106	320	960	1.522649	93.87543	10.53338	0.435825	57
3	5	40% water sample 5	0.4106	320	1200	1.548562	151.172	18.53245	0.777474	39
4	1	45% water sample 1	0.4546	300	380	1.759573	36.7678	4.000549	0.226612	82
4	2	45% water sample 2	0.4546	300	540	1.8114	63.06521	8.894825	0.543823	52
4	3	45% water sample 3	0.4546	300	660	1.785487	83.94737	12.77302	0.795179	34
4	4	45% water sample 4	0.4546	300	960	1.759573	116.4073	18.73007	1.181271	13
4	5	45% water sample 5	0.4546	300	1200	1.574476	177.6654	29.6435	1.888599	11

Table 3: Drug Product "S" Experiment at 65-L Scale, 4 granulation batches each sampled 5 times at Designated Time Points, Resulting in 20-sample DoE

Batch ID	Water Fraction	Granulation Work	Xsat	Xs	SaWW	Dissolution Mean at 45-min
1	29.93	4268	20.2%	9.7	322.7	91
2	31.81	6364	20.2%	11.6	573.4	31
3	31.81	4126	21.9%	9.9	309.1	89
4	27.74	2797	20.5%	7.2	144.8	96
5	27.74	5588	20.1%	7.6	301.5	94
6	31.80	2647	22.1%	9.7	178.6	96
7	29.93	4935	20.7%	9.2	337.3	88
8	30.14	2626	21.0%	9.1	156.1	93
9	27.95	3612	20.5%	7.5	190.1	97
10	30.14	3713	20.9%	9.3	275.9	91
11	30.15	3545	20.8%	9.4	236.9	91
12	30.15	3569	19.8%	10.3	290.7	95
13	30.15	3587	21.4%	8.7	236.3	95
14	30.14	3581	20.3%	9.8	268.4	94
15	30.15	3573	22.0%	8.1	215.7	92
16	30.15	3555	19.5%	10.7	282.4	95
17	30.15	3545	21.8%	8.4	216.6	93
18	30.15	3563	20.7%	9.4	255.3	94
19	30.15	3572	20.8%	9.4	255.4	96
20	30.15	3421	19.7%	10.4	269.2	96
21	30.15	3611	20.6%	9.5	278.6	98
22	30.15	3552	19.7%	10.4	277.1	98
23	30.15	3637	20.1%	10.0	279.0	99
24	30.15	3606	21.4%	8.8	235.6	99
25	30.15	3567	21.4%	8.8	241.5	97
26	30.15	3601	21.2%	9.0	245.0	99
27	30.15	3592	21.0%	9.2	262.6	98
28	30.14	3591	20.8%	9.3	256.3	100
29	30.14	3553	19.8%	10.4	292.7	94
30	30.14	3543	25.1%	5.1	124.5	89
31	30.14	3486	24.7%	5.5	114.3	83
32	30.15	3521	24.5%	5.7	135.7	84
33	30.14	3554	22.9%	7.3	162.2	89
34	30.14	3599	20.6%	9.5	228.6	94
35	30.16	3645	22.3%	7.9	219.3	93
36	30.15	3614	23.5%	6.7	173.9	94
37	30.15	3618	21.2%	9.0	259.7	93
38	30.15	3623	21.7%	8.5	231.7	90
39	30.16	3613	20.9%	9.3	254.0	91
40	30.15	3628	23.5%	6.6	173.4	86

Table 4: Data from Ten Experimental Granulation Batches for drug product "P" followed by 30 runs at Optimum Work and Water Setting at 300-L Scale

API Batch	Micronization Work (kj/kg)	API D90 (microns)	Rheometer Work (mJ/ 30grams)	Granulation Sample Number	Granulation Work (kj/kg)	Granulation Work/ Rheometer Work	Milled Granule Tap Density (g/cc)	Tablet Dissolution at 45-minutes	Tablet Dissolution RSD at 45-minutes
1	180	6.5	11.0	1	27.2	74117	0.67	86	10.6
1	180	6.5	11.0	2	44.0	120030	0.71	93	3.0
1	180	6.5	11.0	3	83.5	227782	0.73	92	1.9
1	180	6.5	11.0	4	126.4	344642	0.75	85	1.4
2	60	13.2	6.8	1	22.0	97018	0.69	89	8.7
2	60	13.2	6.8	2	34.8	153436	0.70	91	4.1
2	60	13.2	6.8	3	64.6	284842	0.74	87	1.4
2	60	13.2	6.8	4	91.5	403851	0.78	52	25.4
3	280	5.8	12.2	1	28.6	70446	0.67	87	11.5
3	280	5.8	12.2	2	45.2	111125	0.69	86	11.0
3	280	5.8	12.2	3	81.9	201398	0.72	92	3.4
3	280	5.8	12.2	4	112.1	275592	0.76	91	3.8

Table 5: Drug Product "P" Micronization and Granulation DoE at 65-L Scale Illustrating the Impacts of API raw material variation on Granulation

	DoE Parameters						Observed	Calculated Parameters			Granule Characteristics	
Run Number	Throughput (kg/hr)	Screw Speed (RPM)	Kneading Elements	Angle of Kneading	Barrel Temp (ºC)	Type Binder Addition	Screw Torque (N m)	Power/Mass (joule/gram)	Water Dynamic Viscosity (kg/m s)	Power/ (Mass* Viscosity)	Yield (Percent Granules between 150-1400 micron)	Milled Granule D50 (microns)
1	17.5	775	4	60	32.5	Wet	1.5	25.0	7.37E-04	34.0	70.7	442
2	25	600	2	90	25	Wet	2.2	19.9	8.68E-04	22.9	55	352
3	25	950	2	30	40	Dry	2.1	30.1	6.35E-04	47.3	64.8	399
4	10	600	12	30	40	Dry	4.0	90.5	6.35E-04	142.4	84.3	567
5	25	600	12	30	40	Wet	7.3	66.0	6.35E-04	103.9	84.7	583
6	25	950	2	30	25	Wet	2.4	34.4	8.68E-04	39.6	66.1	435
7	25	600	12	60	25	Dry	5.6	50.7	8.68E-04	58.3	70.7	445
8	25	950	12	30	25	Dry	6.6	94.5	8.68E-04	108.9	82.1	542
9	25	600	2	90	40	Dry	1.9	17.2	6.35E-04	27.1	63.4	405
10	10	600	2	30	25	Dry	1.3	29.4	8.68E-04	33.9	65.9	423
11	10	950	2	90	25	Dry	0.9	32.2	8.68E-04	37.1	62.4	394
12	25	950	6	90	40	Wet	4.9	70.2	6.35E-04	110.5	81.8	544
13	10	950	12	30	25	Wet	4.5	161.2	8.68E-04	185.6	80.5	536
14	17.5	775	4	60	32.5	Wet	1.7	28.4	7.37E-04	38.5	68.9	430
15	17.5	775	4	60	32.5	Wet	1.4	23.4	7.37E-04	31.7	66.7	418
16	10	950	2	30	40	Wet	1.3	46.6	6.35E-04	73.3	72.6	458
17	10	600	6	90	40	Wet	4.4	99.5	6.35E-04	156.6	82.8	542
18	10	950	12	60	40	Dry	2.8	100.3	6.35E-04	157.8	82.1	533

Table 6: Theophylline DoE on Twin Screw Granulator at Ghent University Recalculated Using First Principle Equation

Model based approach to control pinacol precipitation risk in telescoped Miyaura Borylation-Suzuki coupling process

Marimuthu Andiappan, Jeremy Merritt, Shankar Vaidyaraman

Small Molecule Design and Development (SMDD)

Eli Lilly and Company, Indianapolis, USA

AICHE Annual Meeting 2015

1

Outline

- Telescoped Miyaura Borylation-Suzuki coupling process
- Problem: Pinacol precipitation risk
- Distillation model, Empirical pinacol solubility model
- Use of Model
  - Predict & Avoid the risk of failure (Pinacol precipitation)
  - Design a robust process

2

Miyaura borylation/Suzuki cross-coupling reaction (C-C bond formation)

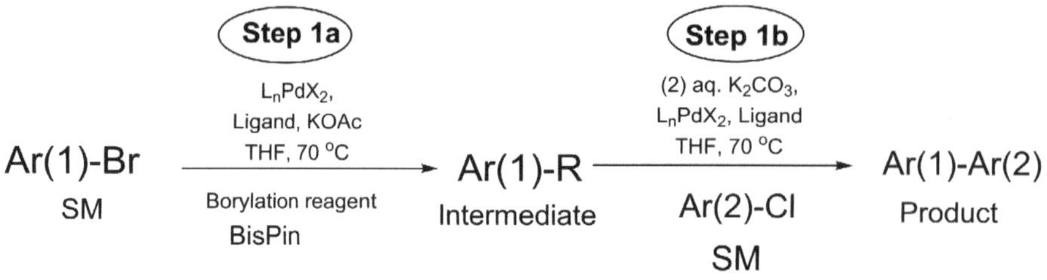

- Miyaura Borylation/Suzuki Cross-coupling reaction
  - 25% of all reactions performed in pharmaceutical industry consist of cross-coupling reactions
- Step 1a – Borylation
  - 1 equiv. Aryl bromide (LR), 1.4 equiv. borylation reagent - BisPin, 0.03 equiv. $L_nPd^{II}X_2$ catalyst and 0.03 equiv. ligand
- Step 1b – Suzuki reaction
  - 1.5 equiv. Aryl chloride and 0.02 equiv. $L_nPd^{II}X_2$ catalyst

3

Telescoped Miyaura borylation/Suzuki cross-coupling reaction: Big picture

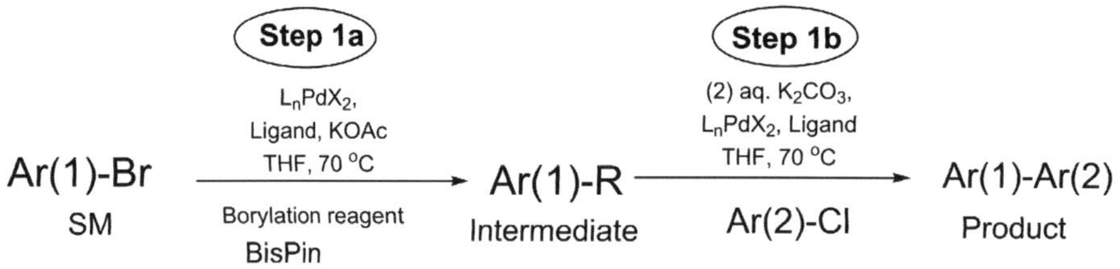

- Telescoped process (One pot, two step process): Minimize extra isolations associated with intermediate
  - Energy Efficient
  - High overall yield
  - Significant waste reduction

- E factor = Kg waste/Kg product
  - Oil refining : 0.1
  - Bulk Chemicals: 1-5
  - Fine chemicals: 5-50
  - Pharmaceuticals: 25-100

4

Problem: Byproduct precipitation in the final isolated wetcake

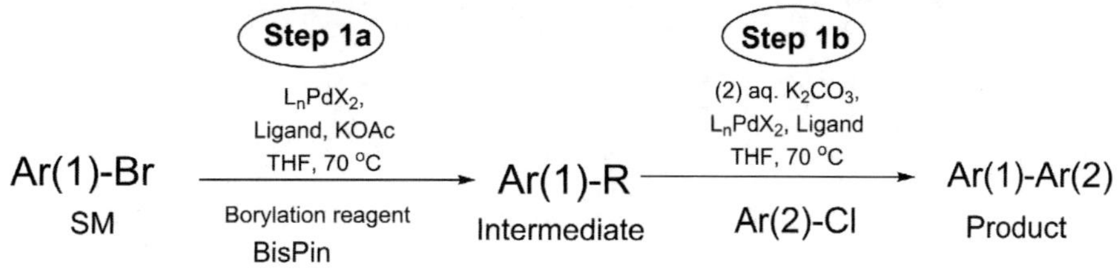

- THF process: 70 % Yield (Reactions 1a & 1b, Aqueous phase separation, Crystallization to 0°C, Filtration & Washing at 0°C, Vacuum drying at 50°C)

- Modified THF process: ~ 80% Yield (Organic contents: distilled from ~15 Vols. to ~6 Vols.)
  - Byproduct precipitation in wetcake

Vols. = Volume (litre) of mixture per Kg of limiting reagent (L/Kg)

5

Characterization of the wetcake – NMR, XRD, DSC

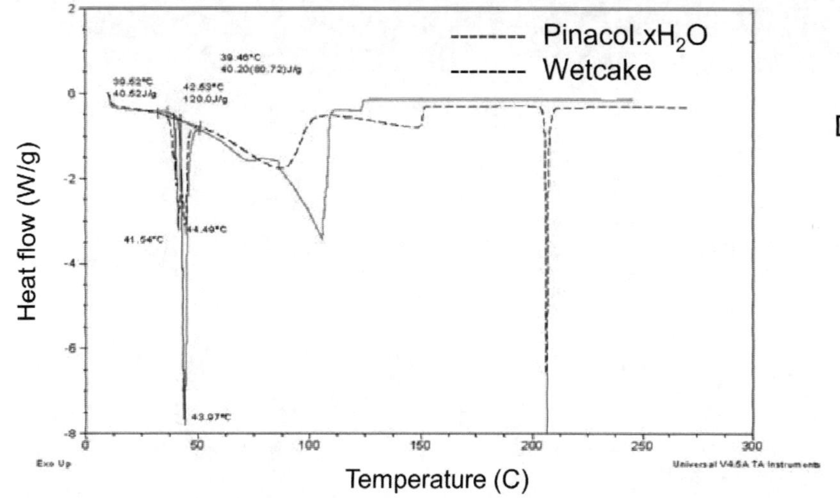

DSC

- Pinacol properties
  - Melting point: 40-43 °C
  - Different forms*: Hexahydrate, monohydrate, anhydrous

*Ref: X. Hao, S. Parkin and C. P. Brock *Acta Cryst.* (2005). B**61**, 689-699

6

Source of Pinacol: BisPin (Borylation reagent)

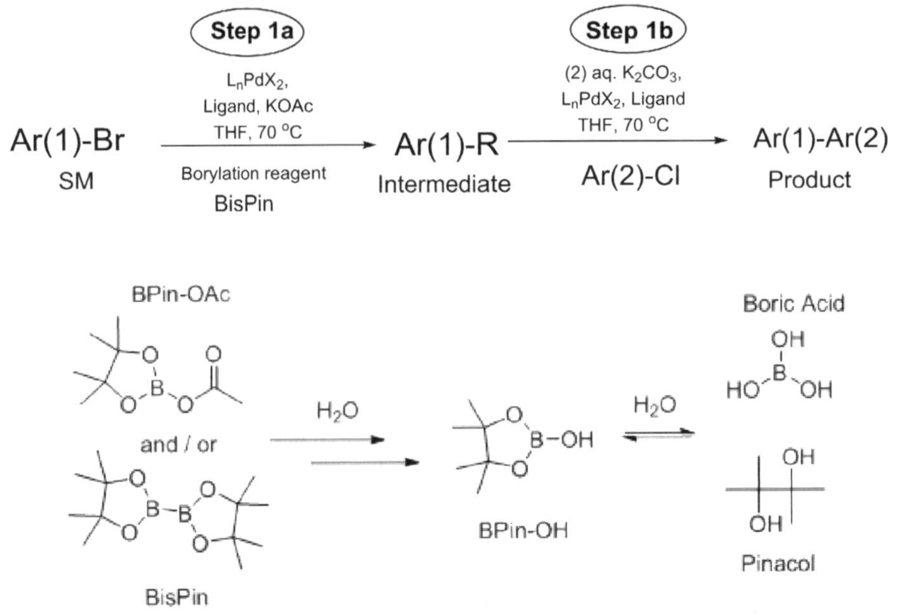

7

BisPin to Pinacol Mass Balance: 10 g batch

Component	Expected Pinacol Quantity (g)	Aqueous (% of mass balance)	Filtrate (% of mass balance)	Washes (% of mass balance)	Total Mass Balance
Pinacol	12.2g (100%)	1.78 g (14.57 %)	9.6 g (78.63 %)	0.39 g (3.18 %)	11.77 g 96.38%

- ~100% mass balance – BisPin to Pinacol

Pinacol precipitation problem: Operational risk

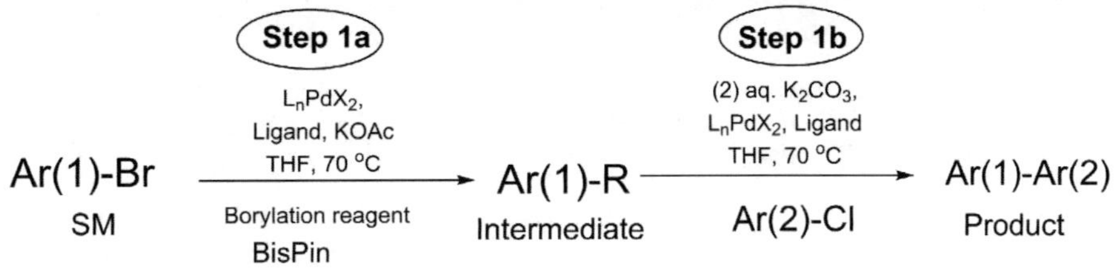

- THF process: ~ 80% Yield (with distillation, ~15 Vols. to ~6 Vols.)
  - Pinacol precipitation in wetcake (Pinacol M.pt: ~40°C, B.pt: ~170°C, Isolation temp: 0°C)
  - No pinacol in dried product: Evaporates at 50°C (vacuum)
  - Pinacol could precipitate in large scale vacuum drier or may not be able to dry – **Operational risk**

9

Pinacol Solubility Model

- Pinacol Solubility data in THF rich organic mixture: GC, XRD, water wt% by KF measurement

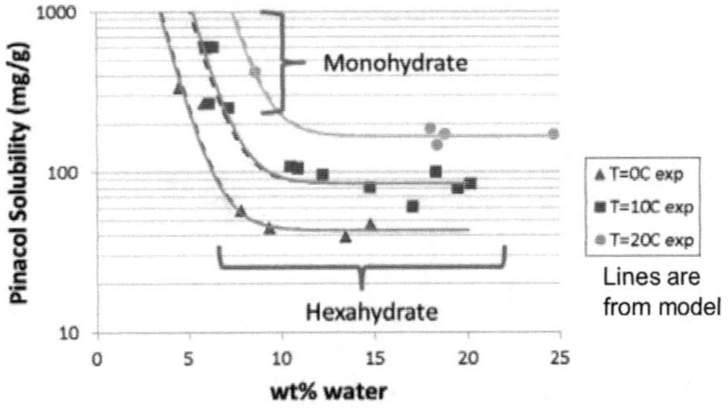

Pinacol (empirical) solubility model

$$\mathrm{Sol}\left(\frac{\mathrm{mg}}{\mathrm{g}}\right)[\mathrm{KF(wt\%)}, \mathrm{T(C)}] = \mathrm{A(T)} * \exp(\mathrm{KF_c(T)} - \mathrm{KF}) + \mathrm{S_0(T)}$$

*Ref:for pinacol forms: X. Hao, S. Parkin and C. P. Brock *Acta Cryst.* (2005). B**61**, 689-699:.

10

Distillation Model

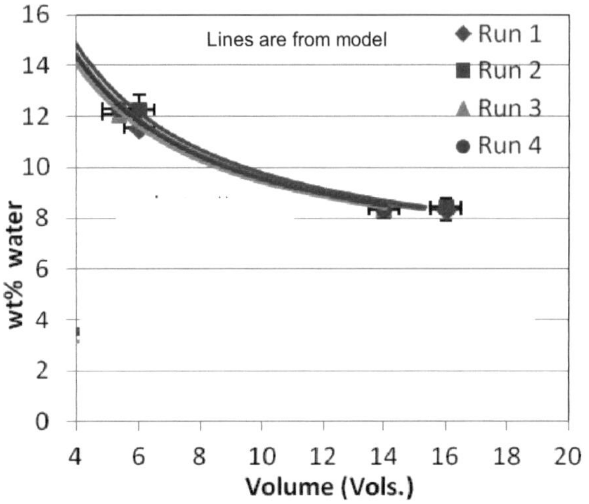

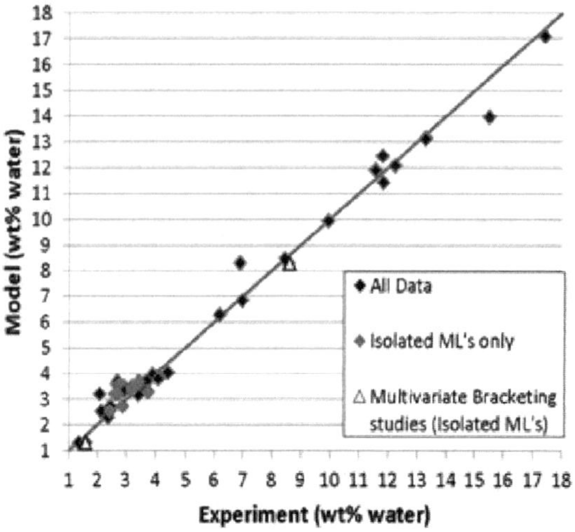

Overall fit for different set of data

- Distillation model: DynoChem tool, NRTL model

11

Use of (solubility+distillation) model

- Understand the risk of failure (Pinacol precipitation)
- Avoid the risk of failure (Pinacol precipitation)
- Identify PAR ranges for key process parameters

12

Why did pinacol precipitation happen when distilled? (~80% yield)

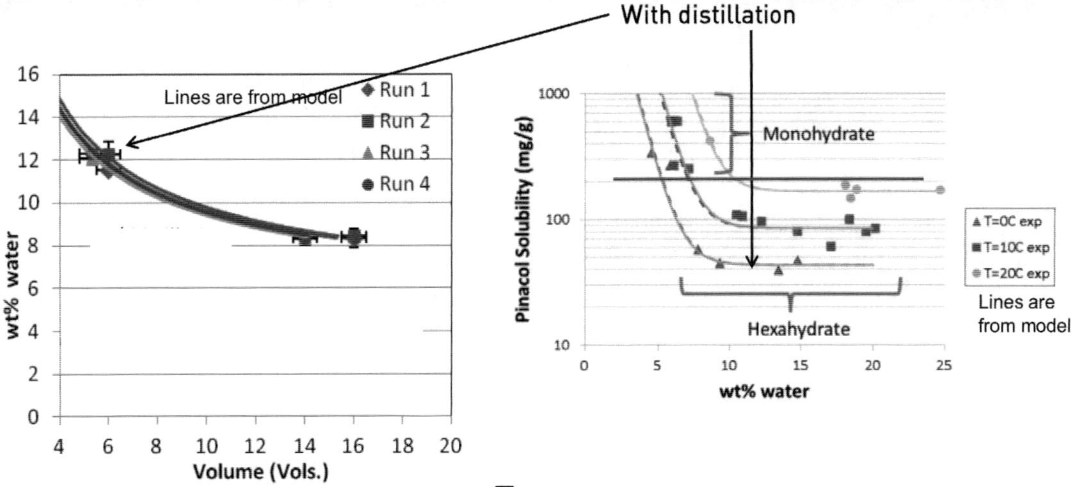

Distillation model ✚ Pinacol solubility model

- Pinacol concentration in the concentrated liquor (~6 vol., 1.4 equiv. BisPin) is ~200 mg/g (solubility is ~40 mg/g)

13

How to achieve KF below 5% in ~6 vols. of concentrated slurry?

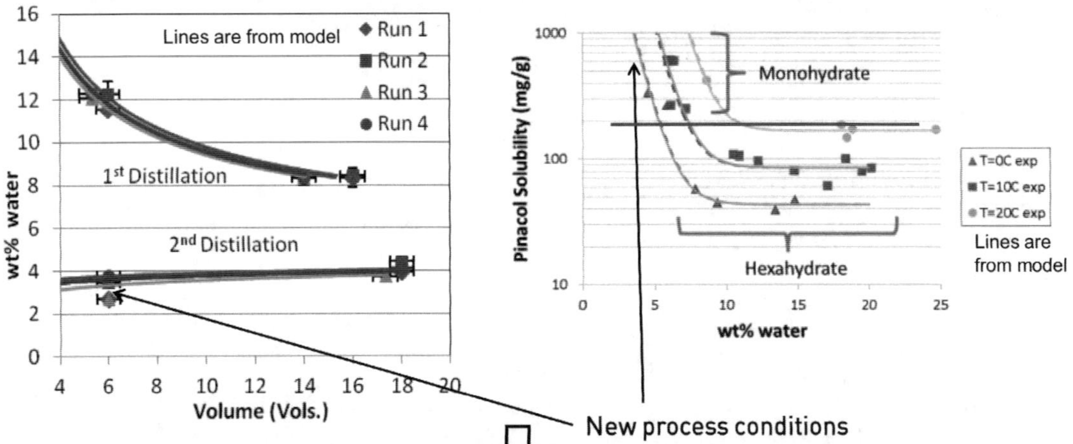

Distillation model ✚ Pinacol solubility model

- Water-THF azeotrope is ~4.9% water
- THF dilution and 2nd distillation to bring water wt% below 5 wt% in the concentrated liquor & Pinacol solubility higher than 200 mg/g
- New process : ~80% yield and no pinacol precipitation in lab & plant scales

14

Use of (solubility+distillation) model

- Understand the risk of failure (Pinacol precipitation)

- Avoid the risk of failure (Pinacol precipitation)

- Identify PAR (Proven Acceptable Ranges) for key process parameters

15

What are PAR (proven acceptable ranges) for key process parameters ?

- BisPin equiv.
  - Target= 1.2 equiv.
- 1st Distillation end volume
  - Target= 6 Vol.
- Fresh THF volume
  - Target= 12 Vol.
- 2nd distillation end volume
  - Target= 6 Vol.
- Isolation temperature
  - Target= 0 °C

16

Variability in key process parameters

- Routine operational variability (σ) for Lilly manufacturing equipments

Parameter	Routine variability (± σ)	Range Assessed for Criticality (± 6σ)
Charge Quantity (by scale); Solids	± 1%	± 6%
Charge Quantity (by mass flow meter); Solvents	± 2%	± 12%
Batch Quantity (by Radar Level Detector)	± 3%	± 18%
Temperature	± 2 °C	± 12 °C

Ref: Seibert, K. et al.; *The use of routine process capability for the determination of process parameter criticality in small-molecule API.* J. Pharm Innov. 2008; 3: 105

17

What are PAR (proven acceptable ranges) for key process parameters ?

- BisPin equiv.
  - Target= 1.2 equiv.
- 1st Distillation end volume
  - Target= 6 Vol.
- Fresh THF volume
  - Target= 12 Vol.
- 2nd distillation end volume
  - Target= 6 Vol.
- Isolation temperature
  - Target= 0 °C

18

Pinacol (Sol.-Conc./Conc.) in % at 0°C

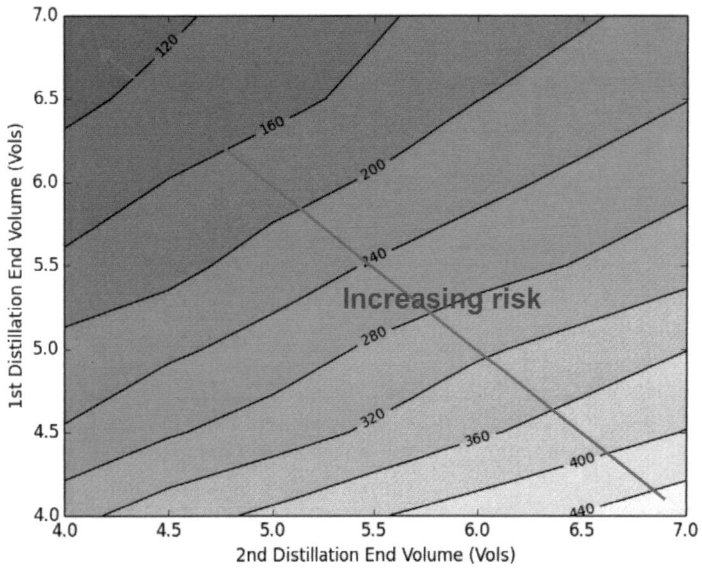

- Target
  - 1st dist. = 6 Vol.
  - 2nd dist. = 6 Vol.
- No pinacol precipitation risk at 0°C (targeted isolation temperature)

19

Pinacol (Sol.-Conc./Conc.) in % at -6°C

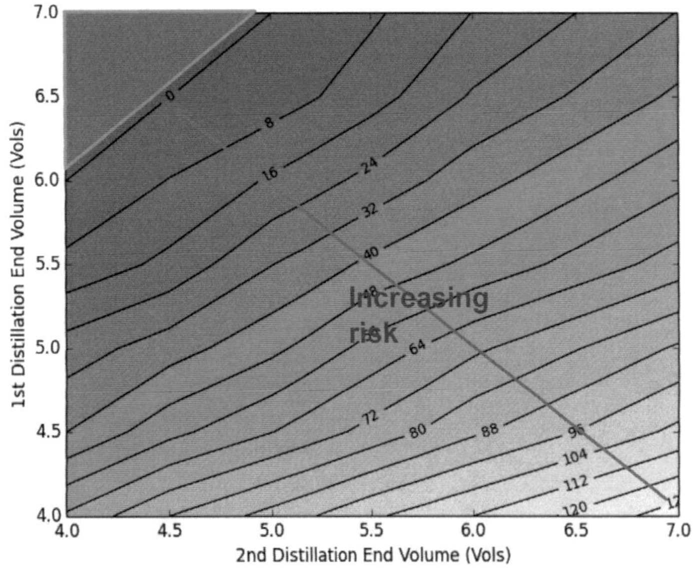

- Target
  - 1st dist. = 6 Vol.
  - 2nd dist. = 6 Vol.
- Pinacol precipitation risk at -6°C when under-distilled in 1st distillation and over-distilled in 2nd distillation

20

PAR ranges – Univariate

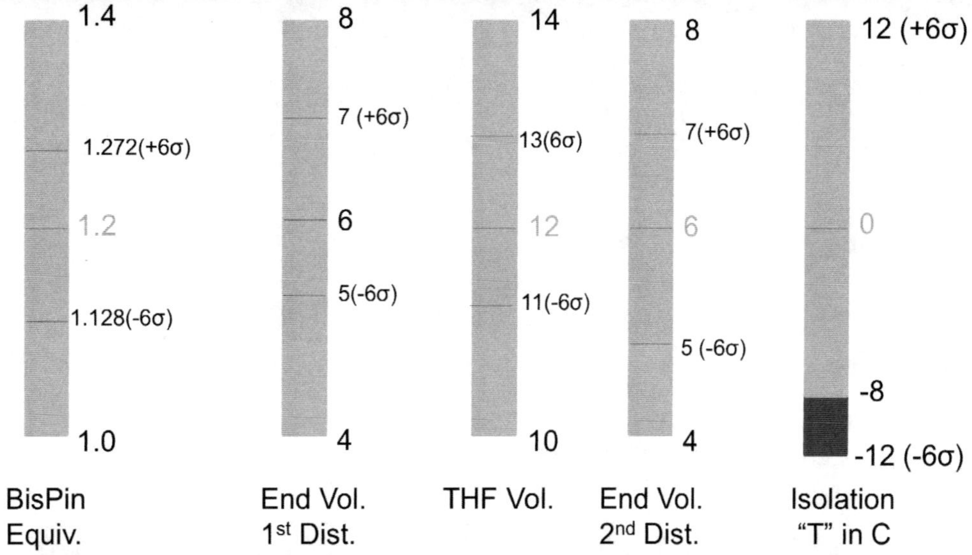

Yellow - The univariate range for a variable that results in no pinacol precipitation given that other variables are at their target
Red region - The univariate range for a variable that results in pinacol precipitation given that other variables are at their target

Ref. for visualization: Catagnoli, C. et al. Organic Process Research & Development, Vol. 14, No. 6, page 1407-1419, 2010.

21

PAR ranges: Multivariate (Green - All combinations)

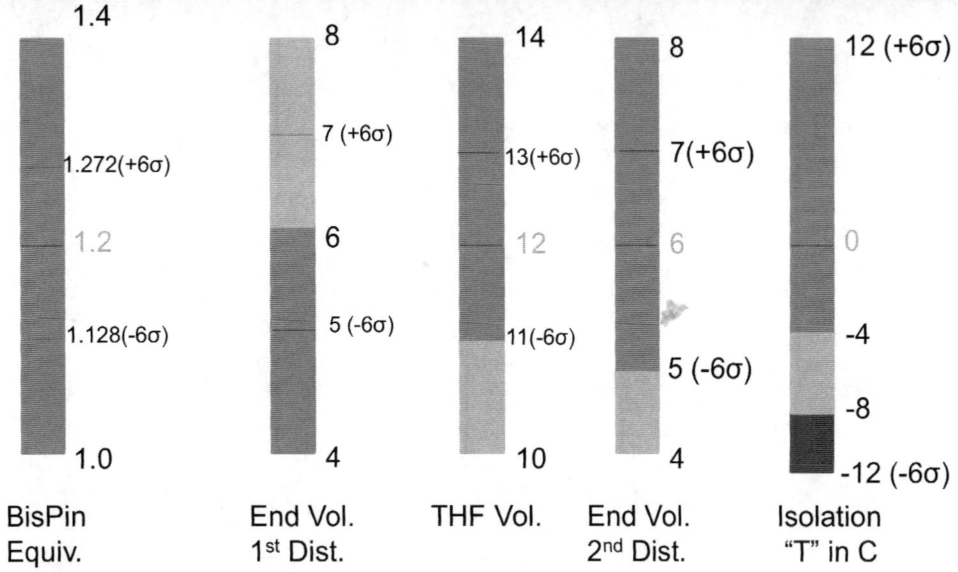

- Yellow+Green region - The univariate range for a variable that results in no pinacol precipitattion given that other variables are at their target
- Green region - Any combination of variable values in green region will result in no pinacol precipitation

22

PAR ranges: Multivariate (Green - All combinations)

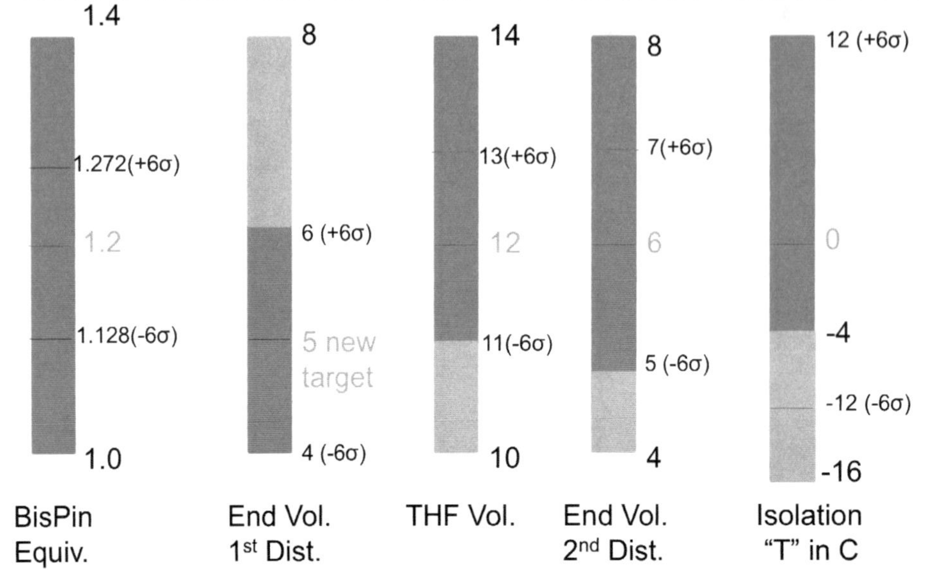

- Yellow+Green region - The univariate range for a variable that results in no pinacol precipitation given that other variables are at their target
- Green region - Any combination of variable values in green region will result in no pinacol precipitation

23

PAR ranges – Univariate

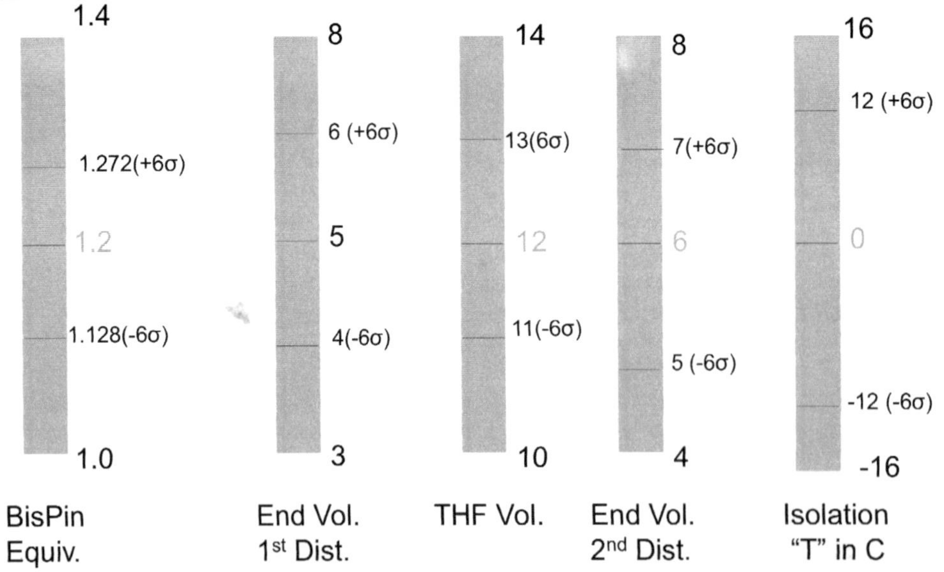

Yellow - The univariate range for a variable that results in no pinacol precipitation given that other variables are at their target

Red region - The univariate range for a variable that results in pinacol precipitation given that other variables are at their target

Ref. for visualization: Catagnoli, C. et al. Organic Process Research & Development, Vol. 14, No. 6, page 1407-1419, 2010.

24

Design of experiment to verify model prediction

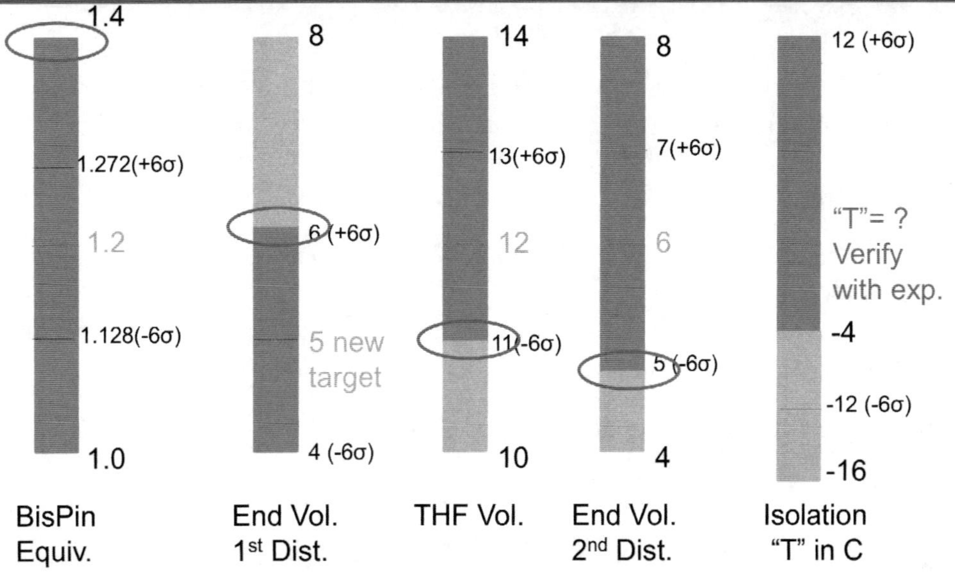

- Yellow+Green region - The univariate range for a variable that results in no pinacol precipitation given that other variables are at their target
- Green region - Any combination of variable values in green region will result in no pinacol precipitation

25

Model vs Experiment verification

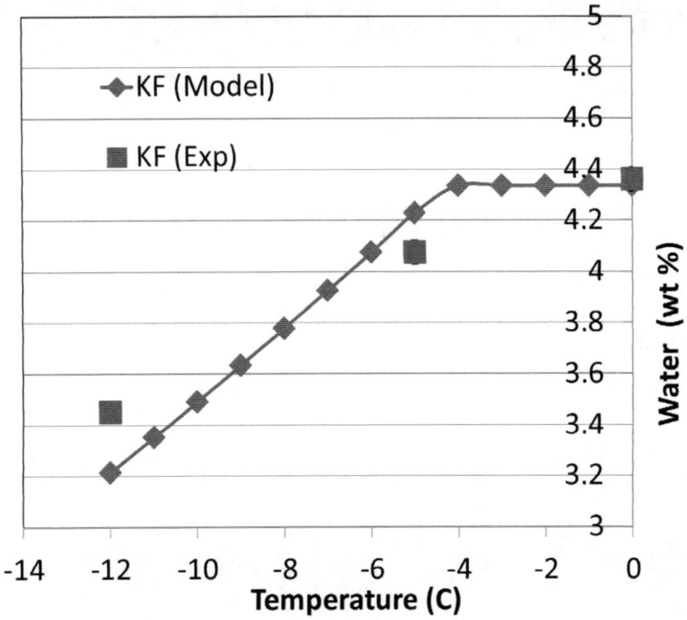

- No pinacol precipiation at 0°C
- Pinacol precipitation at -5°C and below

26

Summary

- Distillation+Solubility models in combination with experimental data to
  - Understand pinacol precipitation in the lab scale
  - Design new process with high yield and no pinacol precipitation
  - Predict the design space (PAR ranges)
  - Predict the operational flexibility
- This approach should be applicable to other Miyaura Borylation-Suzuki couplings to predict & avoid pinacol precipitation risk

27

Acknowledgements

- Team - Kevin Sullivan, Rachel Richey, Mark Pietz, Brad Held, Nathan Furr, Jen Kuehne-Willmore, Nick Magnus, Doug Kjell
- Stan Kolis
- Kevin Seibert
- Shekhar Viswanath
- Dave Varie

28

Thank You

Questions?

29

Option 2 – Make all Parameter ranges green

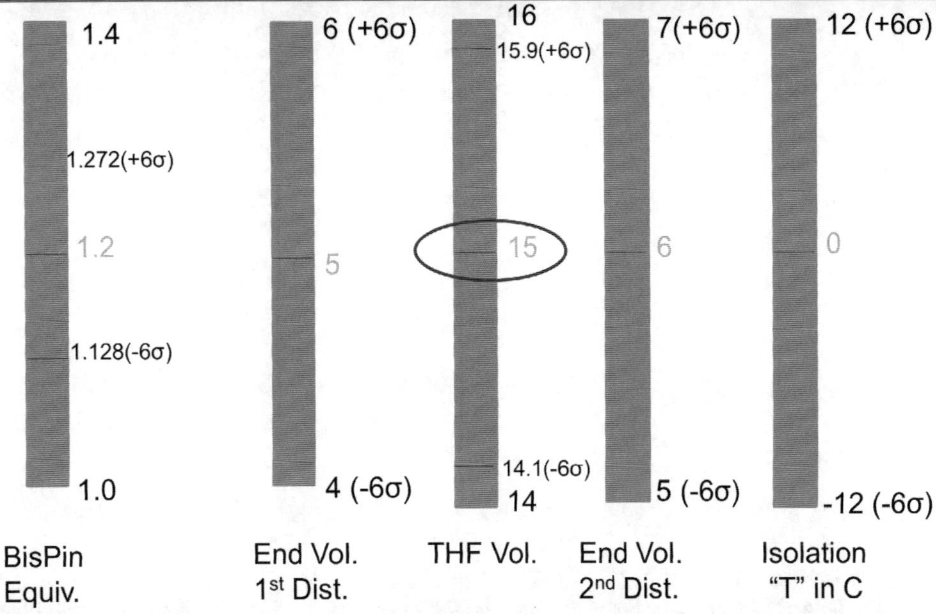

- The improvements (THF target from 12 to 15 Vol) help the yield and quality (no pinacol precipitation risk)
- There is a balance: Very low level of water leads to potency drop due to inorganics

Corrective action in the plant

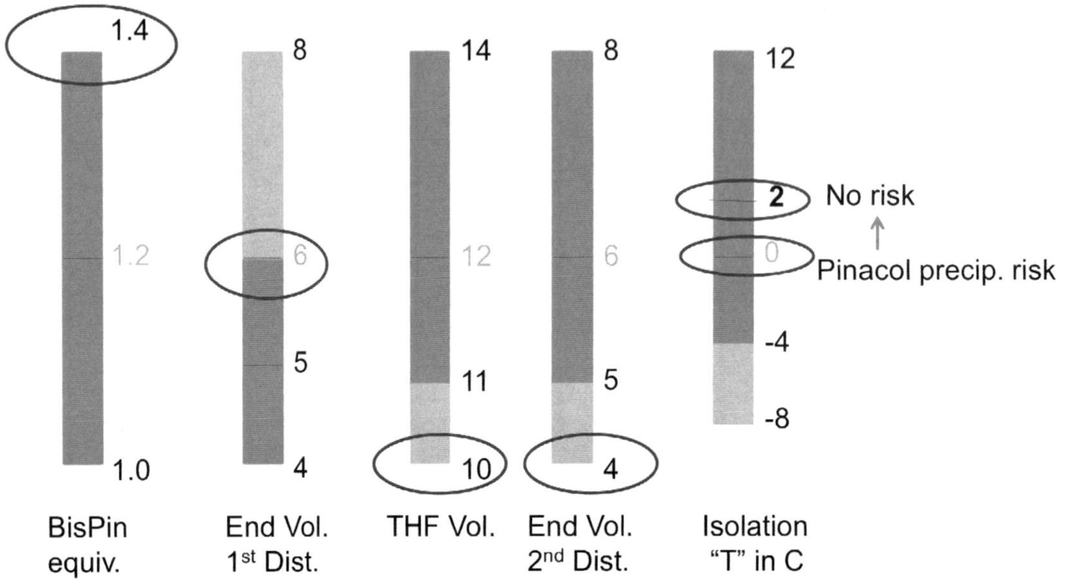

- Adjust the downstream set points when there are deviations in the upstream to eliminate pinacol precipitation in the wetcake.

31

PAR ranges (six sigma, actual numbers)

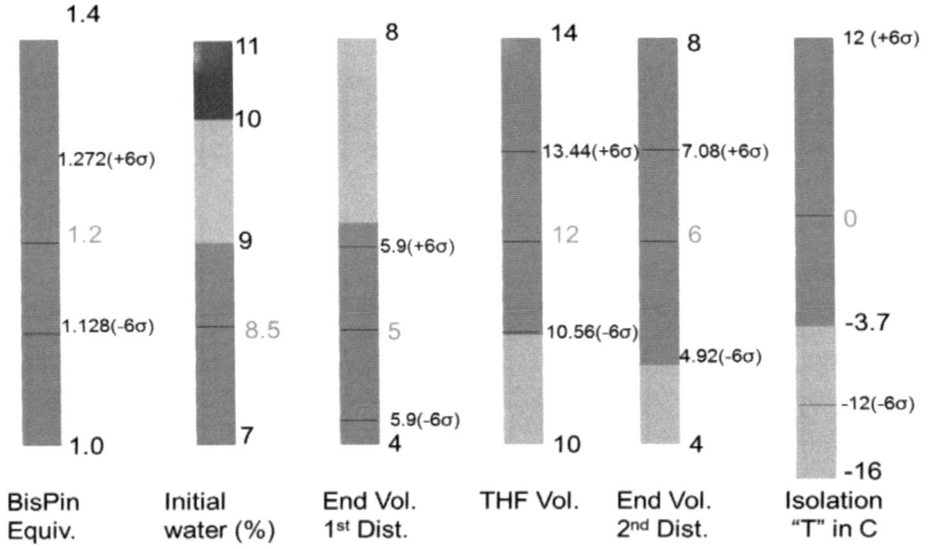

- Yellow+Green region - The univariate range for a variable that results in passing specs given that other variables are at their target
- Green region - Any combination of variable values in green region will result in passing specs

32

421537 Continuous Solid-Liquid Separation in Pharmaceutical Manufacturing

Abstract:

Continuous manufacturing processes offer many advantages for pharmaceutical production. Yet, continuous production is only slowly being adopted in the industry, main issue being a lack of ready-to-use continuous manufacturing equipment and missing process knowledge. In contrast to classical chemical production, small scale equipment would often be required for production of modern low-volume-high-value patient centred medication. In upstream production large efforts are already undertaken to close the existing technology gap [1]–[3]. Also, in secondary manufacturing many groups have formed, working on topics such as continuous wet granulation, -blending, -direct compaction or -thermal drying.[4], [5] However, at the interface between primary and secondary manufacturing, solid-liquid removal is a crucial step during pharmaceutical production.[6], [7] Nevertheless, little work has been reported regarding the development of continuous equipment for solid-liquid separation of pharmaceutical products, particularly to allow treatment of small process streams.

Commercially available continuous equipment for filtration and thermal drying was selected and thoroughly analyzed. Representative model APIs were chosen to evaluate the equipments aptitude to handle small process streams in a continuous pharmaceutical manufacturing environment.

The equipment's suitability for pharmaceutical production could be proven. Operation strategies for further enhancement of filtration performance were developed and successfully implemented. The achieved increase of permeate rates significantly sustains economic efficiency of the developed cross-flow filtration setup. To allow solvent removal below regulatory limits a spin-flash dryer was used, enabling treatment of highly viscous slurries under extremely robust process conditions within a given design space.

References:

[1] H. Zhang, R. Lakerveld, P. L. Heider, M. Tao, M. Su, C. J. Testa, A. N. Dantonio, P. I. Barton, R. D. Braatz, B. L. Trout, A. S. Myerson, K. F. Jensen, and J. M. B. Evans, "Application of continuous crystallization in an integrated continuous pharmaceutical pilot plant," *Cryst. Growth Des.*, vol. 14, pp. 2148–2157, 2014.

[2] M. O. Besenhard, A. Thurnberger, R. Hohl, E. Faulhammer, J. Rattenberger, and J. G. Khinast, "Continuous API-crystal coating via coacervation in a tubular reactor.," *Int. J. Pharm.*, vol. 475, no. 1–2, pp. 198–207, Nov. 2014.

[3] M. Sen, A. Rogers, R. Singh, A. Chaudhury, J. John, M. G. Ierapetritou, and R. Ramachandran, "Flowsheet optimization of an integrated continuous purification-processing pharmaceutical manufacturing operation," *Chem. Eng. Sci.*, vol. 102, pp. 56–66, Oct. 2013.

[4] J. Vercruysse, U. Delaet, I. Van Assche, P. Cappuyns, F. Arata, G. Caporicci, T. De Beer, J. P. Remon, and C. Vervaet, "Stability and repeatability of a continuous twin screw granulation and drying system.," *Eur. J. Pharm. Biopharm.*, vol. 85, no. 3 Pt B, pp. 1031–8, Nov. 2013.

[5] R. Singh, M. Ierapetritou, and R. Ramachandran, "An engineering study on the enhanced control and operation of continuous manufacturing of pharmaceutical tablets via roller compaction.," *Int. J. Pharm.*, vol. 438, no. 1–2, pp. 307–26, Nov. 2012.

[6] H. Zimmer and M. Drudel, "Optimized drying. Thermal processes. Mass spectrometric online gas analysis for improved process understanding, simplified maintenance, and shorter drying time," *CIT Plus*, vol. 10, no. 10, pp. 92–93, 2007.

[7] J. Burgbacher and J. Wiss, "Industrial Applications of Online Monitoring of Drying Processes of Drug Substances Using NIR," *Org. Process Res. Dev.*, vol. 12, no. 2, pp. 235–242, Feb. 2008.

421691 Model-Based Risk Mitigation Strategy for Late Stage Development of a Drug Substance Manufacturing Process

Abstract:

This work presents a strategy to identify options for the development of drug substance manufacturing processes through late stage while maximizing the use integration of computer based modeling and simulation technology and targeted laboratory experimentation. Our proposal considers the molecule specific product-associated risks and the process operational-associated risks as they systematically relate to and interact with each other across the entire GMP sequence. An input to this strategy is a new *unit operation model classification system* designed to provide estimates of modeler effort when a risk is studied and addressed. The classification system is employed to complete a process-wide risk , and a model-based development strategy is designed to mitigate (or eliminate) the multiple process risks. The output of this strategy is a set of proposed options for executing late stage development and serves as planning tool which provides quantitative predictions of the effort required for each proposed option. and the impact (measured as process risk reduction achieved) for each of the development options is predicted The various model-based development options are ultimately designed to produce the commercial-scale operating conditions and ranges as well as the necessary information package to write the CTD and successfully transfer the process to commercial manufacturing.

415050 Prediction of Feasible Operating Region for an Integrated Set of Unit Operations

Abstract:

Title: Prediction of Feasible Operating Region for an Integrated Set of Unit Operations

Authors: Carla Luciani, Salvador Garcia, Kevin Seibert, Thomas Wilson, Eric Crick

In this work, the prediction of feasible operating regions for a set of integrated unit operations will be discussed. Particularly, three unit operations will be investigated: a Suzuki coupling reaction, a liquid-liquid extraction, and metal removal in a series of continuous adsorption columns. The prediction of the feasible operating region is aided by a mechanistic model built.

425275 Technoeconomic Optimization for Green Continuous Pharmaceutical Manufacturing of Analgesics

Abstract:

Continuous Pharmaceutical Manufacturing (CPM) has recently emerged as a promising alternative to the currently dominant batch production methods which are expensive to develop and are characterised by low operational asset efficiency. Advances in continuous organic synthesis routes have induced demonstrations of full (end-to-end) continuous pharmaceutical production and comparative economic analyses explicitly illustrate CPM advantages (Schaber et al., 2011), thus attracting the strong and sustained interest of global pharmaceutical corporations (Jiménez-González et al., 2011) and authoritative regulatory bodies (Lee et al., 2015).

The foundation of every truly novel CPM process is the continuous synthesis chemistry, which has been demonstrated for numerous Active Pharmaceutical Ingredients (APIs), from common painkillers and blockbuster anti-cancer medications, to key therapeutic molecules against debilitating tropical diseases (Bogdan et al., 2009; Hopkin et al., 2013; Kopetzki et al., 2013). Landmark experimental studies have demonstrated viable end-to-end pilot-scale processes from raw feedstocks to final product formulations (Heider et al., 2014). Nevertheless, fully continuous separations must be fully elucidated and are critical toward maximising CPM potential at production scale (Baxendale et al., 2015). The importance of systematic process systems engineering methodologies is evident toward evaluating and demonstrating CPM benefits, both in primary/upstream (Jolliffe and Gerogiorgis, 2015) as well as in secondary/downstream (Sen et al., 2013) pharmaceutical processing.

The present paper addresses the problem of technical and economic optimisation of a conceptual upstream flowsheet for API production, with explicit consideration of reaction and separation unit operations. A nonlinear optimisation model is formulated to maximise the Net Present Value (NPV), subject to explicit mass and molar balance constraints for a reaction cascade comprising three plug flow reactors (PFRs), with explicit inclusion of revenue, fixed capital (CapEx) and operating expenditures (OpEx), plant operation lifetime and inflation rate. Chemical reaction kinetics have been determined from experimental data analysis of a previously published continuous organic synthesis pathway (Bogdan et al., 2009). Furthermore, feedstock price variation and heat transfer requirements are explicitly included in the model, to achieve a realistic estimation of operating costs.

As API recovery from the final separation affects revenue significantly, continuous final separation schemes and the impact of solvent selection on product recovery are explicitly addressed. The range of available options for continuous separation is systematically analysed via the inclusion of UNIFAC equations which correlate API solubility and product recovery: several possible green solvents (to ensure low toxicity and minimal environmental impact) and process operating conditions (temperature, solvent feed ratio) have been considered. Constraints include product purity and design sustainability (via waste limitation). Optimisation results indicate that the CPM plant design achieves profitability (positive NPV) within an acceptably short time. Product yields are in good agreement with published results, and an acceptably low environmental impact is envisaged on the basis of E-factor comparisons to current pharmaceutical practice metrics.

LITERATURE REFERENCES

1. Baxendale, I.R., Braatz, R.D., Hodnett, B.K. et al., 2015. Achieving continuous manufacturing: technologies and approaches for synthesis, workup, and isolation of drug substance. Continuous Manufacturing Symposium (May 20–21, 2014). *J. Pharm. Sci.* **104**: 781–791.

2. Bogdan, A.R., Poe, S.L., Kubis, D.C. et al., 2009. The continuous-flow synthesis of ibuprofen. *Angew. Chem. Int. Ed.* **48**: 8547–8550.

3. Heider, P.L., Born, S.C., Basak, S. et al., 2014. Development of a multi-step synthesis and workup sequence for an integrated, continuous manufacturing process of a pharmaceutical. *Org. Process Res. Dev.* **18**: 402–409.

4. Hopkin, M.D., Baxendale, I.R., Ley, S.V., 2013. An expeditious synthesis of imatinib and analogues utilising flow chemistry methods. *Org. Biomol. Chem.* **11**: 1822–1839.

5. Jiménez-González, C., Poechlauer, P., Broxterman, Q.B. et al., 2011. Key green engineering research areas for sustainable manufacturing: a perspective from pharmaceutical and fine chemicals manufacturers. *Org. Process Res. Dev.* **15**: 900–911.

6. Jolliffe, H.G., Gerogiorgis, D.I., 2015. Process modelling and simulation for continuous pharmaceutical manufacturing of ibuprofen. *Chem. Eng. Res. Des.* **97**: 175–191.

7. Kopetzki, D., Lévesque, F., Seeberger, P.H., 2013. A continuous-flow process for the synthesis of artemisinin. *Chem.-Eur. J.* **19**: 5450–5456.

8. Lee, S.L., O'Connor, T.F., Yang, X. et al., 2015. Modernizing pharmaceutical manufacturing: from batch to continuous production. *J. Pharm. Innov.* **10**: 1–9.

9. Schaber, S.D., Gerogiorgis, D.I., Ramachandran, R. et al., 2011. Economic analysis of integrated continuous and batch pharmaceutical manufacturing: a case study. *Ind. Eng. Chem. Res.* **50**: 10083–10092.

10. Sen, M., Rogers, A., Singh, R. et al., 2013. Flowsheet optimization of an integrated continuous purification-processing pharmaceutical manufacturing operation. *Chem. Eng. Sci.* **102**: 56–66.

409106 A Case Study of Data Analytics for the Manufacturing of a Monoclonal Antibody

Abstract:

The U.S. biotechnology sector has had double-digit growth rates in recent years (Imarc Group, 2012). In 2012, sales of biologics were approximately $63.6 billion, with monoclonal antibodies (mAbs) representing the largest fraction of this market with approximately 39% of sales (Aggarwal, 2014). Mathematical modeling of the manufacturing process is one possible way to both support the growing biologics market as well as decrease costs via improved control and understanding of process operations. Modeling can play an important role in understanding, controlling, and optimizing the process steps used in these processes (Tziampazis & Sambanis, 1994).

In this case study, analysis techniques are demonstrated in both bench-scale as well as process-scale datasets from the production of a monoclonal antibody product. Bench-scale data are analyzed using response surface methodology as well as regularization techniques. Process-scale data are analyzed using correlation analysis, latent variable methods, and regularization techniques. In all cases, this work shows that careful attention must be paid to model calibration and validation, particularly in biopharmaceutical applications where the amount of data is often small. This work also demonstrates that, where available, unit operations should not be isolated, but instead processed in sequence. Most importantly, this work demonstrates that there can be predictive value in modeling critical quality attributes using only a small, heterogeneous dataset. References

Aggarwal, S., 2014. What's fueling the biotech engine - 2012 to 2013. *Nature Biotechnology,* 32(1), pp. 32-39.

Imarc Group, 2012. *Global Biopharmaceutical Market Report and Forecast (2012-2017),* s.l.: International Market Analysis Research & Consulting Group.

Tziampazis, E. & Sambanis, A., 1994. Modeling of cell culture processes. *Cytotechnology,* Volume 14, pp. 191-204.

420144 Application of a Closed-Loop Analysis Approach to Achieve Process Robustness of a 25,000L Mammalian Cell Culture Process

Abstract:

Rapid advances in process data acquisition has enabled a large number of parameters to be monitored at any given time throughout the cell culture process. Modern large-scale cell culture facilities are fit with digital data-acquisition and aggregation systems that allow to scientists and engineers to analyze data. Although enormous amounts of data are collected through these systems in real-time, substantial effort is placed to extract meaningful information from the data. Recently, strategies have been developed to provide a more holistic view of the process, allowing analysts to extract useful process information that may not be captured via analysis of individual parameters. By unearthing these interactions, process improvement opportunities may arise that were not apparent prior to interrogating the data. Recently, multivariate analysis (MVA) has been used to observe and compare complex datasets for troubleshooting or characterization.

In this case study, a closed-loop process was utilized to implement process changes for increasing process yield and robustness. First, principles of MVA were applied to online and offline data from a 25,000L fed-batch CHO process to identify parameters that contribute to overall process variability. From this analysis, several parameters were identified as key contributors to variability in culture titer and glycosylation levels of the recombinant fusion protein produced from the culture. Results from this MVA were confirmed with appropriate experiments using a representative 5-L scale down model of the 25,000L bioreactor. Results from the scale-down model and MVA were used to implement at-scale process changes to reduce culture performance variability and increase overall process yield. Thus, the combined experimental and modeling approach enabled actionable changes to large-scale process, yielding increases in overall process yield.

428472 Clone and Media Selection in Mab Biosimilar Process Development

Abstract:

In the biopharmaceutical sector the topic of biosimilars remains a highly controversial issue. Until 2014, eleven different biosimilar active ingredients have been approved by the European Medicines Agency (EMA), while issuing a European marketing authorization to the first biosimilar monoclonal antibody (mAb) in 2013. Experience indicates that 7 to 8 years are required to launch a biosimilar on the market, at a cost of $100 – 250 million, while antibodies are likely to cost more. From the bioengineering perspective, this requires a target-directed process development approach.

This work presents a novel approach for the screening stage of the process development of a biosimilar. The objective of this stage is to evaluate and select both, cell line and media candidates, as well as media supplements and their levels. The procedure relies on the use of high-throughput experiments performed in deep well-plates. The results are analyzed by advanced multivariate data analysis tools, with the aim of maximizing the probability of reaching the bio-target in the following development stages. Regarding the manifold quality characteristics of the target molecule including its glycosylation and charge profiles, as well as aggregated and high and low molecular weight species, an approach combining principal component analysis (PCA) and decision trees is applied to provide selection criteria. The robustness of the approach is then tested on an external validation data set. Finally, the scalability of the derived process design space is discussed by a multivariate comparison to micro-bioreactor and lab scale experiments.

425441 Efficient Experiment Design and Model Generation to Assist Process Decision Making

Abstract:

Statistically designed experiments and empirical modeling techniques are commonly used in the pharmaceutical industry to define process ranges and enhance process knowledge. Parallel and automated experimental techniques are frequently used to execute the designs, as the experimental plan can be large. In some cases, parallel experimental techniques cannot be used due to mixing considerations or other setup specific issues.

This work presents a model-driven experimental strategy that was used to characterize a two reaction telescoped synthesis to product an active pharmaceutical intermediate. Additional complexity in this synthetic step includes the heterogeneity in the reactions and analytical stability issues. This process step required labor intensive experiments which had to be executed one at a time This work describes a process characterization strategy that started with an efficient D-optimal experimental design that was subsequently augmented based on findings from exploratory data analysis. The team used a model-driven approach, where analysis of the data was partially accomplished by the studying of the resulting statistical models and fitting characteristics. The resulting models were used to guide pilot plant parameter variation and were verified by the resulting on-scale data collection. The models were also used to perform robustness analysis of the system, where the expected process outcome variability was estimated by simulating expected process parameter variability. This experimental and modeling strategy provided actionable information with a modest number of experiments that allowed the project team to make decisions in a timely manner.

428862 Determination of Reaction Profiles Using Online FTIR Spectroscopy Coupled with Kinetics Model-Based Multivariate Curve Resolution

Abstract:

Fourier Transform Infrared (FTIR) spectroscopy is an important tool for the monitoring and controlling Active Pharmaceutical Ingredients (API) manufacturing processes. High spectral resolution and information content are particularly suitable for monitoring chemical reactions where differentiation of structurally similar molecules is required to maintain the quality of products. With a brief introduction to the setups of online FTIR in pharmaceutical reaction monitoring, this presentation will focus on a novel approach that uses the kinetics model-based multivariate curve resolution algorithm as a “calibration-free” tool for real-time analysis of batch reaction processes.

In this application, online FTIR was used to collect data from a hydrogenation process that makes an API (drug substance). The FTIR spectra are fed into the kinetics model-based multivariate curve resolution algorithm to generate the concentrations of product and starting material in real-time without using traditional calibration models. Experiments in the lab have shown that the results from kinetics model-based multivariate curve resolution were similar to those generated by Partial Least Square (PLS) prediction models in terms of analytical accuracy. This offers a “calibration-free” alternative to traditional multivariate (chemometrics) calibration for overcoming spectral overlapping in quantitative analysis. Because the process parameters are explicitly expressed in the kinetics model, their values can be changed easily to reflect the changes in the process condition. Therefore, unlike statistical calibration models that often require adjustment or complete rebuild once process conditions are changed, the kinetics model-based multivariate curve resolution algorithm is able to adjust to the new conditions and perform normally as before. This has been demonstrated in the scale up of the hydrogenation process to pilot plant, in which the batch surface area, hydrogen pressure and agitation were different from those in the lab experiments. The kinetics model-based multivariate curve resolution algorithm worked well without any adjustment. This study has demonstrated an alternative approach for meeting the challenges in supporting fast changing process development and scale up without using multivariate (chemometrics) calibration models that require lengthy adjustment or rebuild once process conditions are changed.

421708 Application of Multivariate Latent Variable Methods to Drug Product Development

Abstract:

This work presents three case studies of the use of Multivariate Latent Variable Methods (LVM) in pharmaceutical development: i) a root-cause investigation into the compression behavior of powder blend for oral dosage form manufacture, and further implications for commercial scale-up; ii) the study of the feasibility to transfer a roller compaction process from one roller compaction unit to another of a different brand and design; and iii) the use of LVM to develop a solvent replacement system based on solvatochromic properties of solvents and the further coupling with optimization technology for a D-Optimal selection of solvents for experimental screening.

429799 Improving Yield of Drug Product Manufacturing with Hybrid Approach

Abstract:

For remaining competitive in times of a growing public pressure on drug prices and the shrinking research productivity in the industry, drug product manufacturing must be performed increasingly efficient. The presented work is an expansion of Eberle et al. (2014), which comprises of a four-level approach for yield enhancements and is applied in an industrial case study to the manufacturing of injectable drug products, also referred to as Parenterals.

The four levels of the presented hybrid approach combine heuristic and statistical, data-based elements to take advantage of process specific information both from expert knowledge as well as process data. On the first level, the set of loss causing production processes at a unit under review are listed before identifying the processes with most improvement potential. Then, on the second level, expert knowledge is consulted for generating loss scenarios and assigning measurable parameters. Descriptive statistical modelling is performed on the third level before deriving enhancements or escalating process investigations on the fourth level.

As a case study, the presented approach was applied on an annual production output of a manufacturing facility for injectable drug products of Roche in Switzerland. The sampled output comprises of six products from liquid vials and pre-filled batches. Identified costs for lost product units on the first level revealed, that about 20 % of costs for non-commercialized products result from sampling for quality control activities. This loss cause was followed in magnitude by two different glass container defects, i.e. imperfectly formed and scratched vials. Scratched vials and their possible generation during filling processes were modelled with a variety of explanatory variables. However, the statistical models on these variables did not lead to the derivation of implementable information and, as a consequence of that, to an escalation of process investigations. An investigation of scratches with a vial replica that includes a pressure-sensitive outer surface was initiated; enabling the quantification of physical exposures on the glass body. Based on the resulting exposure information, processing steps were prioritised by their impact on glass scratches, hot spots were diagnosed and the statistical model was refined. In summary, the presented data-based approach structured the use of production data and supported efficiency enhancements of production processes.

References: Eberle L., Sugiyama H., Papadokonstantakis S., Graser A., Schmidt R., Hungerbühler K., 2014, Data-Based Tiered Approach for Improving Pharmaceutical Batch Production Processes, Proceedings of the 24th European Symposium on Computer Aided Process Engineering, Elsevier, Amsterdam, the Netherlands.

431431 Benefits of Quality By Design (QbD) for Consistent Product Quality

Abstract:

The pharmaceutical industry continues to embrace the principles outlined in ICH Q8(R2), Q9, Q10 and Q11 and to move from a Quality by Testing mentality to a Quality by Design paradigm. This talk will provide a high level overview of how the pharmaceutical industry can benefit by using QbD principles to develop robust manufacturing processes for small molecule drug products in order to consistently deliver quality drug products to the patient. Through examples, the talk will illustrate the QbD tools utilized by applicants during pharmaceutical development. It will also touch upon Integrated Quality Assessment (IQA) in the newly formed Office of Process and Facilities within the Office of Pharmaceutical Quality.

430873 Predicting Roller Compaction Process Capability

Abstract:

Roller compaction is a common processing option in the tableting process of small molecule pharmaceutical compounds. Because roller compaction can be described using a physical model of the process, it is possible to predict manufacturing behavior from laboratory experiments. However, it is desired to estimate the process robustness incorporating uncertainty at the defined manufacturing conditions. This presentation will describe an application of probabilistic modeling to leverage process models, lab data, manufacturing data, and uncertainty to best incorporate potential sources of process variability.

Using Markov Chain Monte Carlo, the process model can be regressed to laboratory data or pilot manufacturing data, resulting in an estimate of the possible outcomes of the measured process quality attributes, i.e. ribbon density. These outcomes are generated from a Bayesian network that integrates all the sources of variability (measurement error, model error, and input variability) associated with the process, predicting process capability (Cpk) and identifying any hidden deficiencies in the input data.

The framework quantifies the expected Cpk at the time of process qualification, increases the fundamental scientific knowledge of the process and through a sensitivity analysis, provides a rational mechanism to prioritize resources for future measurements.

415192 Impact of API Properties and Excipient Source on the Process and Performance of a Typical High Shear Wet Granulation Formulation

Abstract:

In a pharmaceutical formulation, some excipients, such as binder and disintegrant are used for their particular functionalities. While present only in small quantities, changes in these excipients can sometimes have significant impact on the processiblity and performance of a drug product. When a formulation contains high drug loading, properties of the API can also interplay with the process parameters to impact the properties of drug intermediate and drug product. In this study, we evaluated the effect of differences in binder, disintegrant, as well as API properties on the process and performance of a typical high drug loading high shear wet granulation formulation. The factors evaluated include:

1. Types of disintegrant (Sodium Starch Glycolate vs. Croscarmellose Sodium)
2. Source of the binder Hydroxylpropyl cellulose (Ashland Klucel vs Nisso)
3. API solubility and wettability
4. API particle size and surface area

The responses evaluated include:

1. Amount of granulation water needed for desirable granule properties
2. Granule growth ratio
3. Granule compactability
4. Tablet disintegration

Salicylic acid, aspirin, citric acid as well as some BMS compounds were used in this study, which covers a broad range of solubility and contact angle (0.002-590 mg/ml, 0-146°). It was found that the amount of granulation water needed for optimal granule properties is highly sensitive to the solubility and contact angle of the API. For highly soluble API, water amount needed for granulation is significantly reduced. Granule growth ratio has a significant dependence on the particle size/surface area of the API. Using the same granulation parameters, smaller API particle size results in higher granule growth ratio. For the low solubility API that was selected for the disintegrant type and binder source study, no significant difference was observed in the processibililty and disintegration of the drug product.

422303 Implementation of Advanced Multilayer Plant-Wide Control Architecture into a Direct Compaction Continuous Pharmaceutical Manufacturing Process

Abstract:

Continuous pharmaceutical manufacturing provides an appropriate platform where feed-forward and feedback control strategies, including necessary alarms, can be implemented to improve the product quality, minimizing product rejection and operating expenses. The precise control of the quality of the pharmaceutical product requires corrective actions on the process/raw material variability before they can influence the product quality. It is therefore highly desirable to implement a combined feed-forward/feed-back control system into a pharmaceutical manufacturing plant [1] where multivariate measurements of raw and intermediate materials, finished products, and process parameters are used to control quality. The feed-forward controller takes into account the effect of process disturbances proactively while the feedback control system ensures the end product quality consistently. The integrated feed-forward/feedback control system ensures minimum variability in the final product quality irrespective of process and raw material variations.

In this work, efficient multilayer plant-wide control architecture has been designed and implemented into a direct compaction continuous tablet manufacturing pilot-plant. The proposed control architecture consists of four layers. In the first layer, necessary sensors including PAT tools have been implemented for real time process monitoring. In second layer, feedback control loops have been implemented to ensure the critical quality attributes. In third layer, feed-forward control loops have been integrated with feed-back control loops to proactively mitigate the effects of process and raw materials variability and thereby to achieve more precise control of end product quality with minimum rejections. In final layer of the control architecture, the necessary alarms and signals have been implemented to control the plant operation more efficiently and reject the out of specs powder blends and tablets automatically.

The critical process parameters and quality attributes are monitored at each unit operation via a centralized control platform using advanced monitoring tools. For example, feeder flow rate and feeder screw rotational speed have been monitored at feeding operation. Powder blend composition, uniformity (RSD) and bulk density are monitored at the blending operation. Tablet weight, thickness, hardness, compression forces, ejection force, and operating parameters (e.g. fill depth, feed frame speed) are monitored at the tableting operation. NIR is applied first time to measure the powder bulk density in real time. As part of second layer of control architecture, supervisory feedback loops to control powder blend composition [2], main compression force, tablet weight and hardness are implemented. The tablet weight and hardness control loops have been decoupled so that both critical quality attributes can be controlled simultaneously. Main compression force and tablet weight are controlled through a cascade control arrangement. The necessary, feed-forward control loops are implemented so that the upstream measurements can be used to take downstream mitigation actions in real time. For example, measured powder bulk density was feed-forwarded in real time to the tablet press to proactively mitigate the effect of powder bulk density on tablet weight and hardness via changing the fill depth. Finally, the necessary alarms are implemented for unexpected events management including sending the warning/alarms signals to the operator, stopping the production if needed and rejecting powder blends and tablets out of specification. The four layers of control architecture have been implemented using industry standard control platform (e.g. DeltaV), PAT data management tool (synTQ), OPC communication protocol, fieldbus devises, chemometric tools and real time monitoring tools. The proposed control strategy is found to be better than implementations using only feedback control or without control strategy and therefore demonstrates potential to further improve pharmaceutical tablet manufacturing operations.

The objective of this presentation is two-fold. First to highlight the combined feed-forward/feedback based four layers of control architecture, and second to demonstrate its performance to control the direct compaction continuous tablet manufacturing process.

References

1. Singh, R., Muzzio, F., Ierapetritou, M., Ramachandran, R. **(2015)**. A combined feed-forward/feed-back control system for a QbD based continuous tablet manufacturing process. PROCESSES Journal, 3, 339-356.
2. Singh, R., Sahay, A., Karry, K. M., Muzzio, F., Ierapetritou, M., Ramachandran, R. (**2014**). Implementation of a hybrid MPC-PID control strategy using PAT tools into a direct compaction continuous pharmaceutical tablet manufacturing pilot-plant, International Journal of Pharmaceutics, 473, 38–54.

435447 Systematic Framework for the Design and Analysis of Control Systems for Continuous Pharmaceutical Manufacturing

Abstract:

In the pharmaceutical industry, fulfilling the Food and Drug Administration (FDA) requirements for product quality is one of the main concerns [1]. In order to assure product quality, batch operation is the most used manufacturing mode [2]. During batch production, the quality of the batch subject to operator action and quality is often controlled by statistical sampling and rejection of an entire batch if sampling indicates deviations from specifications. This leads to significant amounts of wasted material and significant overall cost [3]. The cost for manufacturing goods represent a large part of the products sales. For instance, the manufactured goods cost represent twenty-seven percent of the patented products sales, while for generic products the percentage is fifty percent [4]. This cost can be minimized by implementing improvements in the manufacturing process [5]. One of the possible improvements is to evolve from batch operations to continuous processing. Continuous manufacturing and on-line process monitoring have the potential to improve product quality and reduce manufacturing cost; the potential annual savings across the industry are projected to be as high as fifty billion dollars [3]. In 2002 the FDA initiated Process Analytical Technology (PAT). The goal of PAT is to augment the understanding and control the manufacturing process [6,7]. Through the implementation of QbD and PAT, the companies have to demonstrate understanding of how the operating conditions, process design and raw material variability affect the product quality [4]. Control design plays a significant role for continuous manufacturing implementation. It is by implementing appropriate process control strategies that poor quality product manufacturing can be mitigated or prevented [1]. By implementing Quality by Control (QbC), it is possible to reduce the operation cost, reduce the off-spec or segregated material and improve the safety environment.

In this contribution a generic framework is proposed for design and analysis of a hierarchical feedback control scheme for continuous manufacturing. The framework proposes three control levels that have to be considered in the control design. Level 0 includes the built-in control systems of the individual unit operations, in Level 1 PAT is used in simple feedback control schemes (property feedback control – PFC) to control critical quality attributes (CQAs), whereas level three uses dynamic models for prediction and advanced control schemes, such as model predictive control (MPC) for enhanced control performance. Control performance indicators suitable for the comparative analysis of various control schemes are also proposed. The potential impacts of using level 1 property feedback control (PFC) and level 2 model predictive control on product quality and process performance are analyzed using the proposed performance indicators. The proposed generic framework is exemplified using as a model system the feeder and blender unit operations. The control schemes are evaluated under realistic disturbance and measurement scenarios, at high and low active pharmaceutical ingredient (API) content.

References

[1] Ramachandran, R.; Arjunan, J.; Chaudhury, A.; Ierapetritou, M.G. Model-Based Control-Loop Performance of a Continuous Direct Compaction Process. *J. Pharm. Innov.* **2011**, 6, 249-263.

[2] Singh, R.; Boukouvala, F.; Jayjock, E.; Ramachandran, R.; Ierapetritou, M.; Muzzio, F. Flexible Multipurpose Continuous Processing of Pharmaceutical Tablet Manufacturing Process. *PharmPro Magazine*, **2012**, 27, 22–25.

[3] Reklaitis G. The Pharmaceutical Manufacturing Institute. **2013**, (White Paper) Purdue University, West Lafayette, IN.

[4] Rogers, A. J.; Hashemi, A.; Ierapetritou, M. Modeling of Particulate Processes for the Continuous Manufacture of Solid-Based Pharmaceutical Dosage Forms. *Processes*. **2013**, 1, 67-127.

[5] Rogers, A. J.; Inamdar, C.; Ierapetritou, M.G. An Integrated Approach to Simulation of Pharmaceutical Processes for Solid Drug Manufacture. *Industrial & Engineering Chemistry Research*. **2013**, 53, 5128-5147.

[6] US Food and Drug Administration. *Guidance for Industry, Q8 Pharmaceutical Development*; Food and Drug Administration, 2006

[7] Lee S. L., O'Connor T. F., Yang X., Cruz C. N., Chatterjee S., Madurawee R. D., Moore C. M. V., Yu L. X. Modernizing Pharmaceutical Manufacturing: from Batch to Continuous *Production. J Pharm Innov.* **2015**

413370 An Optimization Framework to Combine Design Space Maximization with Design of Experiments

Abstract:

While pharmaceutical quality by design enables the specification of a design space in terms of process inputs, the space of process outputs may be more relevant to safety and product quality. Thus, while design of experiments techniques are often used to explore the effect of process parameters, it is often desirable during development, and for the purpose of process understanding, to maximize the explored space in terms of process outputs or quality attributes. However, timeline and resource constraints limit the number of experimental trials available to properly characterize both spaces, prompting the need for systematic design of experiments. An optimization framework based on the sometimes conflicting objectives of maximizing the size of the output or quality attribute design space and maximizing the D-optimality of the input parameters is proposed and applied to a spray-coating process. This example shows that optimizing only one of the two objectives yields unacceptable results in the neglected objective. Two strategies are explored. First, a multi-objective optimization approach is presented, enabling visualization of possible tradeoffs as a Pareto-optimal front—a set of solutions where no objective can be further improved without degrading another objective value. This allows the decision maker to choose among a set of efficient designs. It also provides insights into the potential benefits of increasing (or decreasing) the number of experiments. Second, a bilevel formulation is also presented which enables process engineers to quickly identify good compromise designs without the need to map out multiple points along the Pareto-optimal front.

421707 Control Strategy for Manufacturing of Dry Powder Inhalation Capsules: A Regulatory Perspective

Abstract:

Dry Powder Inhalation (DPI) is a leading drug delivery technology for the treatment of chronic pulmonary diseases, such as asthma and chronic obstructive pulmonary disease (COPD). DPI capsules provide a portable, simple-to-use, and cost-effective way to deliver inhalation medication. However, manufacturing of low-dose DPI capsules with consistent delivered dose and fine particle dose is not an easy task due to the complex interactions between drug formulation, delivery device design, and patient usage. Quality by design (QbD) provides an effective tool to systematically evaluate, understand, and refine the formulation and manufacturing process to ensure predefined product quality to meet the needs of the patient. In this presentation, the manufacturing process of low-dose DPI capsules will be used as an example to illustrate how to implement QbD principles to enhance product and process understanding and to implement effective process controls to ensure reproducible commercial manufacturing for consistent product quality. The purpose of this talk is to provide a regulatory perspective with regard to a QbD-based control strategy for the manufacturing process aspects of low-dose DPI capsules.

414742 Efficient Calibration for in-Situ Measuring L-Glutamic Acid Solution Concentration Via ATR-FTIR

Abstract:

In this presentation, an efficient calibration method is proposed for in-situ measuring the concentration of L-glutamic acid (LGA) solution via ATR-FTIR. We clarify that using the unsaturated zone spectra as commonly practiced for establishing a concentration calibration model may lead to evidently biased measurement of the solution concentration. By comparison on using the spectra of unsaturated and metastable zones for calibrating the concentration of LGA aqueous solution, it is demonstrated that the use of metastable zone spectra to establish a calibration model can guarantee prediction accuracy for the solution concentration via real-time ATR-FTIR monitoring. Experiments under different concentrations and temperatures are performed to verify the efficiency of the proposed calibration method. In addition, it is shown that the proposed calibration model can also be used for nucleation detection, therefore facilitating real-time monitoring of batch cooling crystallization in practice.

415213 Optimization of a Reactive Crystallization Using a Pharmaceutical Engineering Approach

Abstract:

Reactive crystallizations are complex and difficult to control due to the number of competing parameters influencing the chemical transformation and particle growth. Here we will discuss a salt-to-salt transformation in the presence of multiple reactive chemical species. The crystallization is further burdened by the potential to create a poorly soluble free form that poses a risk to both product quality and yield. The presentation will highlight the use of FBRM in combination with other analytical techniques to identify and elucidate the side reactions responsible for observed variability in product potency. Detailed understanding of the upstream composition and the seed point proved vital to delivering a robust process. To offer reasonably broad parameter ranges to manufacturing and further mitigate risk, the refined process incorporates a crude crystallization focused upon yield and a re-crystallization that assures high quality. A fundamental understanding of the interactions of the chemical species and the root causes of variability were needed to deliver a suitable process for commercial manufacturing.

417528 Monitoring Co-Crystal Purity with on-Line Raman Spectrascopy

Abstract:

Pharmaceutical co-crystals are defined as stoichiometric molecular complexes that contain an active pharmaceutical ingredient (API) and a coformer in crystal lattice via noncovalent interactions, predominantly hydrogen bonds. Compared with APIs themselves, pharmaceutical co-crystals have advantages in solid state properties, such as solubility, melting point, chemical interaction, stability and bioavailability. Solution crystallization is an important method to produce co-crystals. However, crystals of constituent components, either API or co-former, may crystallize out concomitantly due to operational fluctuations in starting composition or temperature, compromising the purity of co-crystals. Therefore, it is desirable to detect crystals of single components when they appear during co-crystallization, whereby corrective operation can be taken to ensure co-crystal purity. In this study, a methodology is developed to monitor co-crystal purity using Raman spectroscopy with caffeine-glutaric acid-acetonitrile as the model system. Caffeine and glutaric acid can form co-crystal from solution in acetonitrile at the appropriate conditions. Raman spectra were collected in three classes of suspensions with a solid mixture of caffeine crystals and co-crystals, pure co-crystals, and a mixture of glutaric acid crystals and co-crystals, respectively at different temperatures. Principal component analysis (PCA) was performed for these spectra and the resulting first few principal components were put to discriminant analysis. It was found that Raman spectroscopy is able to distinguish the three classes of suspensions clearly.

Key Words, co-crystal, PAT, Raman spectroscopy, crystallization, principal component analysis

425352 Automated Multi-Loop Direct Nucleation Control Using Focused Beam Reflectance Measurement in Multistage Continuous Mixed Suspension Mixed Product Removal Crystallizer

Abstract:

Batch crystallization is an economical and widely used separation and purification unit operation especially in pharmaceutical industry. It has been extensively studied in literature and very well understood due to the development of various process analytical technology (PAT) tools like focused beam reflectance measurement (FRBM), particle video microscope (PVM), UV/Vis spectroscopy, IR spectroscopy and Raman spectroscopy. In addition, some of these tools have been successfully applied in feedback control frameworks in order to control the batch crystallization process in real time. Direct nucleation control (DNC) is a methodology that uses the chord count information from in situ FBRM to actuate adaptive heating or cooling rate to have controlled dissolution in the batch cooling crystallization process. It has been proved being able to produce much larger and more uniform crystal as compared with simple temperature control (T-control) approach in batch process.[1-2] Supersaturation control (SSC) and active polymorphic feedback control (APFC) apply UV/Vis or/and Raman spectroscopy in similar feedback control loops to maintain controlled supersaturation level or polymorphic transformation during batch cooling process.[3-4] These model-free feedback control approaches are proved very robust to automatically reject disturbances or to adjust the process when the required target product quality is changed. They can provide automated fast recipe design and only require simple mathematical tools without any modeling effort or much system knowledge.[5]

Continuous Crystallization is of great interest in recent decades due to consistent product quality, high process and equipment efficiency and reduced operation cost compared with batch crystallization. Mixed suspension mixed product removal (MSMPR) crystallizer is one of the most important continuous crystallization systems due to its simple operation and capability for long residence time. There are some published papers that investigate the recipe design for continuous MSMPR crystallization systems via modelling or experiment.[6-9] However, according to the authors' best knowledge, currently there are no published works that study the application of previously mentioned feedback control approaches (i.e. DNC, SSC, APFC) in any type of continuous crystallization systems.

In this work, a multi-loop DNC approach is experimentally investigated in a multistage MSMPR cooling crystallization process in order to automatically control the nucleation rate at each MSMPR stage. The feed solution is free of crystals. Each MSMPR stage is equipped with a heating circulator, a temperature sensor and a FBRM probe. A Labview based software Crystallization Monitoring and Control (CryMOCO) is installed on the supervisory computer to collect the chord count information from each individual FBRM probe. The information is then compared with the desired set-point of the corresponding MSMPR stage in CryMOCO, which remotely controls each temperature circulator and actuates a predefined adaptive heating or cooling rate in the corresponding MSMPR stage. In each DNC loop, the total chord count is targeted as the controlled variable and the set-point since it indicates information about nucleation, generation of fine particles as well as the average particle size. For example, in a certain stage the adaptive heating rate is actuated automatically once the measured FBRM total count is larger than the set-point of the corresponding stage, and vice versa. Paracetamol and ethanol are used as the model compound and solvent, respectively. The proposed multi-loop DNC approach is found being able to reach desired steady state and produce crystals of consistent desired size distribution after a few heating/cooling cycles with or without disturbances. The proposed approach has the following advantages: (1) It can automatically help avoid nucleation in the second stage and after when fine particles are not desired in the final product. This is achieved by designing a proper set-point for each MSMPR stage so that limited amount of nucleation or breakage is allowed. (2) It can automatically achieve optimal operating recipes to produce desired crystals of various sizes (i.e. either small or big). (3) It can automatically reject disturbances such as accidental seeding or nucleation. It is faster than simply washing out without any control. (4) It can reduce startup duration time and startup waste as compared with simple T-control that uses constant temperatures. Simulation work is also carried out in the group to validate and explain some of the experimental observations, and therefore to optimize the control performance.

References

[1] Saleemi, A.; Rielly, C.; Nagy, Z. K. Automated direct nucleation control for in situ dynamic fines removal in batch cooling crystallization. *CrystEngComm* **2012**, 14(6), 2196-2203.

[2] Abu Bakar, M. R.; Nagy, Z. K.; Saleemi, A. N.; Rielly, C. D. The impact of direct nucleation control on crystal size distribution in pharmaceutical crystallization processes. *Cryst. Growth Des.* **2009**, 9(3), 1378-1384.

[3] Saleemi, A. N.; Rielly, C. D.; Nagy, Z. K. Comparative investigation of supersaturation and automated direct nucleation control of crystal size distribution using ATR-UV/vis spectroscopy and FBRM. *Cryst. Grwoth Des.* **2012**, 12(4), 1792-1807.

[4] Simone, E.; Saleemi, A. N.; Tonnon, N.; Nagy, Z. K. Active polymorphic feedback control of crystallization processes using a combined Raman and ATR-UV/Vis spectroscopy approach. *Cryst. Growth Des.* **2014**, 14(4), 1839-1850.

[5] Nagy, Z. K.; Fujiwara, M.; Braatz, R. D. Modelling and control of combined cooling and antisolvent crystallization processes. *J Process Control* **2008**, 18, 856-864.

[6] Zhang, H.; Quon, J.; Alvarez, A. J.; Evans, J.; Myerson, A. S.; Trout, B. Development of continuous anti-solvent/cooling crystallization process using cascaded mixed suspension, mixed product removal crystallizers. *Org. Process Res. Dev.* **2012**, 16, 915-924.

[7] Vetter, T.; Burcham, C. L.; Doherty, M. F. Regions of attainable particle size in continuous and batch crystallization processes. *Chem. Eng. Sci.* **2014**, 106, 167-180.

[8] Yang, Y.; Nagy, Z. K. Advanced control approaches for combined cooling/antisolvent crystallization in continuous mixed suspension mixed product removal cascade crystallizers. *Chem. Eng. Sci.* **2015**, 127, 362-373.

[9] Yang, Y.; Nagy, Z. K. Combined cooling and antisolvent crystallization in continuous mixed suspension mixed product removal cascade: steady state and startup optimization. *Ind. Eng. Chem. Res.* **2015**, DOI: 10.1021/ie5034254.

423519 Digital Phase Diagram of K2SO4-KOH-H2O and Crystal Growth of K2SO4 in the Strong Alkali Solution

Abstract:

Potassium sulfate has always been an important and superior potash fertilizer. In this paper, we investigate the crystallization process of potassium sulfate from the strong alkali solution that is involved in the clean production process of raw alunite mineral. The research focuses on the digital phase diagram for a ternary system of K_2SO_4-KOH-H_2O and the crystal growth of K_2SO_4 in the strong alkali solution.

Firstly, the phase equilibrium data for the ternary system of K_2SO_4-KOH-H_2O are measured from 40°C to 80°C, where the contents of potassium and sulfur in the aqueous solutions are detected by ICP"COES and the composition of solid phases are tested by X-ray diffraction, respectively. Based on the measured phase equilibrium data, origin is used to present the 3D colormap and ternary contour drawings, as shown in Figure 1. With the interpolation and fitting results obtained from the software, the saturation temperature of a given K_2SO_4-KOH-H_2O content can be predicted from the smooth surface of the ternary phase diagram. It is found that the K_2SO_4 solubility decreases with the increase of the KOH concentration in the aqueous solution, and the K_2SO_4 solubility becomes smaller in the strong alkali solution which is beneficial to separate K_2SO_4 from the KOH solution.

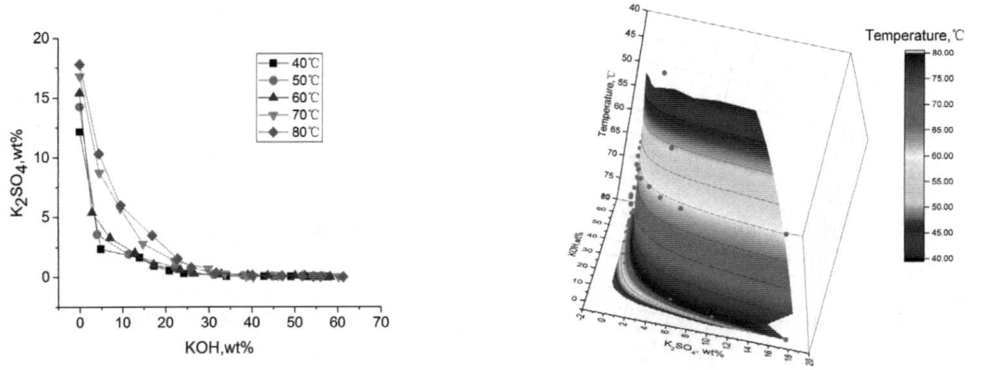

Figure 1. Solubility of K_2SO_4 in KOH solution and digital phase diagram of K_2SO_4-KOH-H_2O at different temperatures

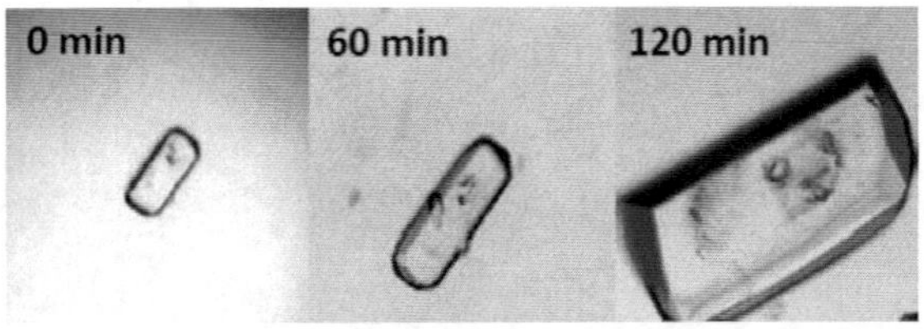

Figure 2. Crystal growth of K_2SO_4 in the strong KOH solution measured using a reflection microscopic technique at 313K

The supersolubility and metastable zone width of K_2SO_4 in the concentrated KOH solution are investigated by focused beam reflectance measurement (FBRM) and ultrasonic sensor online monitoring for different operating conditions, which will provide the important information for the design and optimization of K_2SO_4 crystallization process. For the system of K_2SO_4-KOH-H_2O, the FBRM and the ultrasonic sensor can provide a high sensitivity for nucleation detection, which detects changes related to the amount of grown nuclei and changes in solution concentration, respectively. Furthermore, the crystal growth rates of K_2SO_4 in the strong KOH solution are measured using a reflection microscopic technique, where the saturated solution is poured into a thermo-controlled microscopic cell with a volume of 5 mL and a diameter of 3.5cm. Experiments are conducted by decreasing the solution to different temperatures to create different supersaturation levels, and then the crystal growth rates of K_2SO_4 are calculated by measuring the changes in the dimensions of the crystals with the analysis Image Processing Program (Olympus). Figure 2 shows the crystal growth process of potassium sulfate in the strong KOH solution measured under the microscope at 313K. The crystal shape of K_2SO_4 is a typical rectangular shape, which is convenient for easy identification and measurements of specific crystallographic faces. The influence of impurities from the alkaline solution on the crystal shape will be investigated using the reflection microscopic technique. Moreover, the two-dimensional population balance equation is proposed for the calculation of the crystal growth rate of K_2SO_4 with an acceptable accuracy. From this study of crystal growth, a given content of solution can be designed with a desired growth mechanism in order to effectively separate K_2SO_4 from the strong KOH solution by crystallization.

435128 Controlling Crystal Size, Shape and Purity during Crystallization By Using Additives

Abstract:

Crystallization which is a purification and separation unit operation plays an important role in the industrial production of pharmaceuticals. The properties of the produced crystals such as purity or size and shape distribution have strong impact on the downstream processes as well as on the end usage. It is known, that chemicals are generally contaminated by different amounts of impurities, which can come from different sources, such as from upstream processes or chemical side reactions [1]. In most of the industrial processes, the effect of impurities may result e.g. in needle shape crystals which impeding manufacturability of the solid dosage form. While crystallization is often used as a purification process, trace amount of impurities generally exist in the product. While in some cases impurities of various amounts are acceptable for certain highly toxic impurities the acceptable limits are very low and difficult to achieve using a single crystallization step. For example genotoxic impurities carry significant health risk and strict regulations are applied on the concentration of these materials [6,8,9]. Impurities, even if it is in trace amounts, can influence the crystal properties like crystal shape. In this case, impurities are called as crystal growth modifiers (CGMs) [2-6]. In this work, a method is presented, which reduces the possible undesired effect of the CGMs during crystallization by using a control additive (CA). The control method is based on a chemical reaction between the control additive and the growth modifier, which results in a non-adsorptive reaction product. Thus, the concentration of CGM in the solution and thus in the solid phase is the function of the additive concentration in the solution. It is also known that the concentration of the CGMs in the solution affects the rate of the change of crystal growth.

This work deals with the investigation and control of batch cooling crystallization in the presence of crystal growth modifiers. The objectives of the control method are to maintain the purity of the crystal product and also to control the crystal size and shape distribution. A mathematical model including nucleation and growth of two characteristic crystal sizes as well as impurity adsorption and chemical reaction model is developed in order to investigate and control the purity and the crystal shape. The growth modifiers influence directly the growth kinetics. This inhibitory effect on the crystal surfaces leads to changed crystal size and shape distribution [2-6]. On the other hand, growth modifiers are built in the crystal structure according to surface adsorption mechanism and so CGMs also change the purity of the crystal product [7]. The proposed purity control method is based on chemical reaction between the undesired impurity and the control additives which provides an alternative transformation route for the compound against the surface adsorption. Hence, the chemical reaction inhibits the adsorption rate by lowering the concentration in the solution.

Utilizing the benefits of the competitive adsorption and reaction mechanisms, model based control for crystal shape and purity is developed. The simulator is suitable to calculate the optimal concentration of control additive in order to achieve desired crystal shape and/or reducing the impurity concentration below the desired limits. The results demonstrate the efficiency of the proposed approach for shape and purity control.

Acknowledgement: Financial support provided by the European Research Council grant no.[280106-CrySys] is acknowledged.

References:

[1] N., Variankaval, A.S., Cote, M.F., Docherty, 2008. From form to function: Crystallization of active pharmaceutical ingredients. AIChE Journal, 54(7), 1682-1688.

[2] A., Majumder, Z.K., Nagy, 2013. Prediction and control of crystal shape distribution in the presence of crystal growth modifiers. Chemical Engineering Science, 101, 598-602.

[3] R.J., Davey, 1976. The effect of impurity adsorption ion the kinetics of crystal growth from solution. Journal of Crystal Growth, 34, 109-119.

[4] N., Kubota, J.W., Mullin, 1995. A kinetic model for crystal growth from aqueous solution in the presence of impurity. Journal of Crystal Growth, 152(3), 203-208.

[5] N., Kubota, S., Sasaki, N., Doki, N., Minamikawa, M., Yokota, 2004. Absorption of an Al(III) impurity onto the (100) face of a growing KDP crystal in supersaturated solution. Crystal Growth and Design, 4(3), 533-537.

[6] A., Borsos, A., Majumder, Z. K., Nagy, 2014. Model development and experimental validation for crystal shape control by using tailored mixtures of growth modifiers. Computer Aided Chemical Engineering, 33, 781-786.

[7] K., Maeda, R., Tabuchi, Y., Asakuma, K., Fukui, 2006. Distribution of metallic ions in single KDP crystal grown from aqueous solution. Cryst. Res. Technol., 41(10), 955-960.

[8] A., Garcia-Arieta, 214. Interactions between active pharmaceutical ingredients and excipients affecting bioavailability: impact on bioequivalence. European Journal of Pharmaceutical Sciences, 65, 89-97.

[9] S., Gorog, 2006. The importance and the challenges of impurity profiling in modern pharmaceutical analysis. Trends in Analytical Chemistry, 25 (8), 755-757.

413094 IQ Consortium Enabling Technologies for Pharma

Abstract:

The International Consortium for Innovation and Quality in Pharmaceutical Development (IQ) is a technically-focused organization of pharmaceutical and biotechnology companies with a mission of advancing science-based and scientifically-driven standards and regulations for pharmaceutical products worldwide. Since 2013, it has established an 'Enabling Technologies Working Group' to identify opportunities among member companies in the pre-competitive space to enable cost-effective and relevant solutions to improve efficiency and quality in the R&D process with respect to API process development. The working group has identified projects of mutual interest, and subject-matter experts and champions to lead technical area sub-teams with the goal of creating proposals to progress on these projects. This presentation provides an update on the progress of the overall 'Enabling Technologies' effort with focus on modeling, crystallization, drying, solubility, and automation.

419660 Precompetitive Collaboration for Pharmaceutical Crystallization Processing

Abstract:

In 2014, the IQ Enabling Technologies Working Group conducted a survey of participating pharmaceutical companies to identify areas of common interest. Among these, crystallization development was the topic that generated the most interest, with 11 of 12 members of the Working Group expressing a desire to participate in this sub-team which kicked off at the beginning of 2015.

One of the first activities the team performed was to brainstorm a list of potential areas for collaboration, and then each company ranked their relative interest in each of the topics. Topics ranged from best practices for salt/polymorph selection to lab equipment for process development to modeling to sharing experiences related to product filings.

The highest scoring areas were related to sharing of best practices. The team agreed to focus initially on knowledge sharing, with the recognition that team members might be able to help each other close certain gaps and improve each company's specific practices. The knowledge sharing would lead into performing a collective gap analysis in order to identify prioritized projects that will require funding and external collaboration. The intent is for these prioritized projects to kick off in 2016, likely under the banner of a new construct being proposed as a spin-off from the IQ consortium.

This presentation will focus on both the outcomes of the collaboration thus far and, perhaps more importantly, on the approaches the team has taken to advance its objectives in the face of numerous challenges.

Allotrope Foundation
Building an Open Framework for Laboratory Data

Collaborating to Deliver an Innovative Framework for Laboratory Data Based on Open Standards

James M. Vergis
Pre-Competive Collaborations in Pharma Industry
AIChE Annual Meeting
November 9, 2015

"Every decision we make about our products is based on data. To deliver safe and efficacious medicines to the patient, we must be confident in the quality and integrity of our data. Our integration project with Allotrope Foundation will deliver standard metadata and data stored in a standard format, addressing the root cause of our data challenges and revolutionizing the way we generate, access and use data to make decisions and deliver our pipeline of medicines."

Dr. Annette Doherty
GSK Senior Vice President
Product Development & Supply

ALLOTROPE PRESS RELEASE
SEPTEMBER 30, 2015

Why are we here today?

- Provide an update to our peers in the industry and relay the message "**the Allotrope Framework is a reality**"
- Demonstrate how vendors and Allotrope members are collaborating to leverage the Framework through integration projects
- Invite you to participate in 2016

Allotrope Foundation

Member Companies	• Subject Matter Experts • Project Funding
Secretariat DrinkerBiddle	• Project Management • Legal & Logistical Support
Professional Software Firm OSTHUS success with R&D	• Framework Development • Technical Leadership
Partner Network	• Requirements & Specifications • Contributions, PoC Applications

Allotrope Foundation

Member Companies

AbbVie, Amgen, Baxter, Bayer, Biogen, Boehringer Ingelheim, Bristol-Myers Squibb, Eli Lilly, Genentech/Roche, GlaxoSmithKline, Merck & Co., Pfizer

Secretariat — DrinkerBiddle

- Project Management
- Legal & Logistical Support

Professional Software Firm — OSTHUS success with R&D

- Framework Development
- Technical Leadership

Partner Network

ACD/Labs, Agilent, Biovia, BSSN, IDBS, Mestrelab Research, LEAP Technologies, Mettler Toledo, Persistent Systems, Riffyn, Sartorius, Shimadzu, Thermo Scientific, Waters, Erasmus Univ. Med Center, University of Southampton

What is the problem we are trying to solve?

Comparison of Music vs. Scientific Data

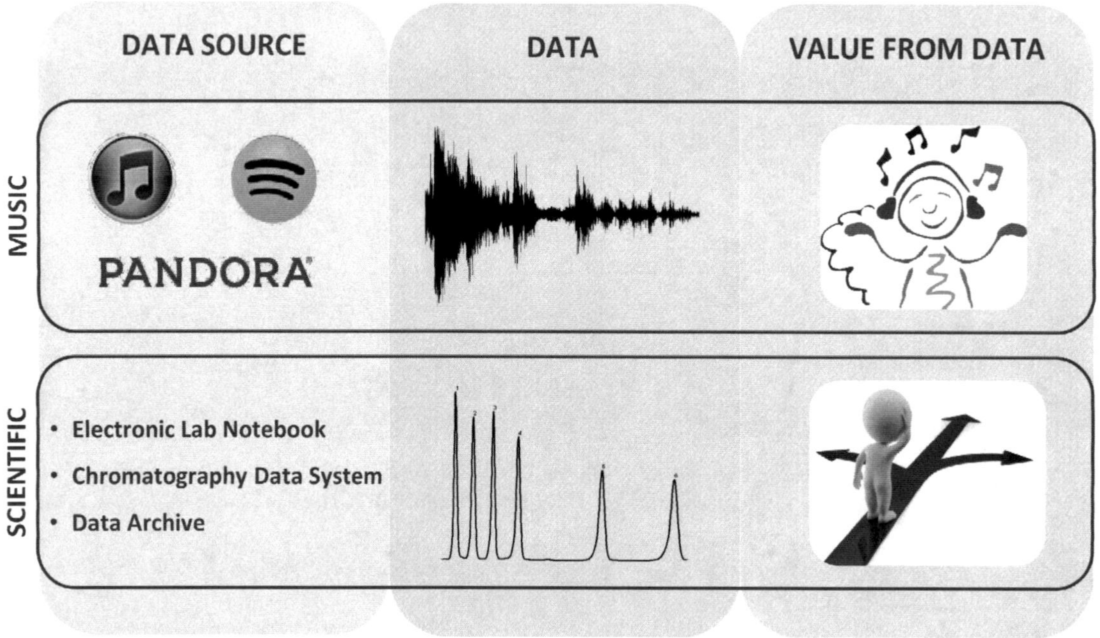

What is the problem we are trying to solve?

Comparison of Music vs. Scientific Data

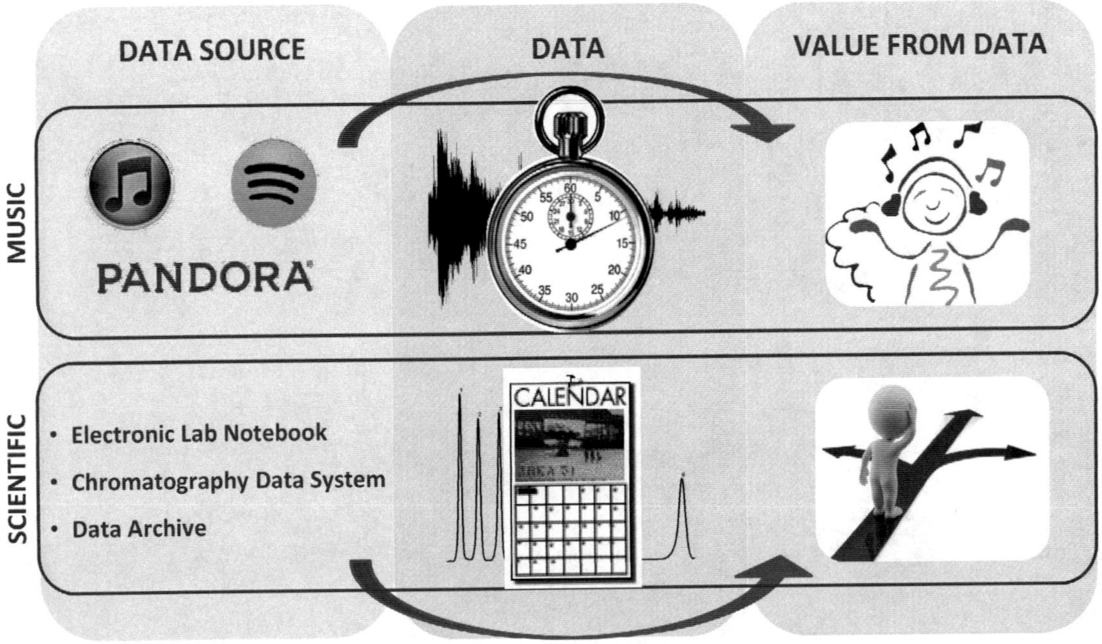

Why is Data for Music so Much Better?

Think about music...

Music is typically stored in a small number of *standard, non-proprietary* formats...

MP3 AUDIO

...with contextual *metadata* that are *complete, consistent & correct*

Artist	**Genre**
Album	**Date**
Song	**Artwork**

...enabling the user to *find, share* and *enjoy* it years later from any device *easily*!

Think about scientific data...

Scientific data is typically stored in a wide variety of *non-standard, proprietary* formats...

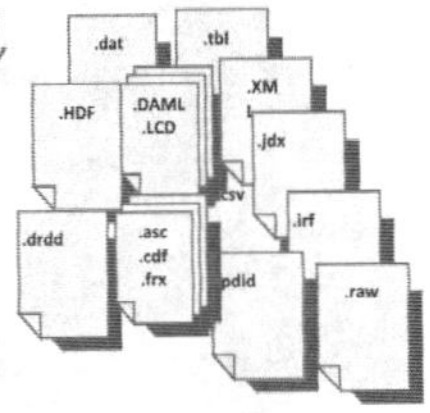

...with contextual *metadata* that are *missing, inconsistent & incorrect*

Material	**Process**
Equipment	**Result**

...making it *costly, difficult or practically impossible* to find and get value from it.

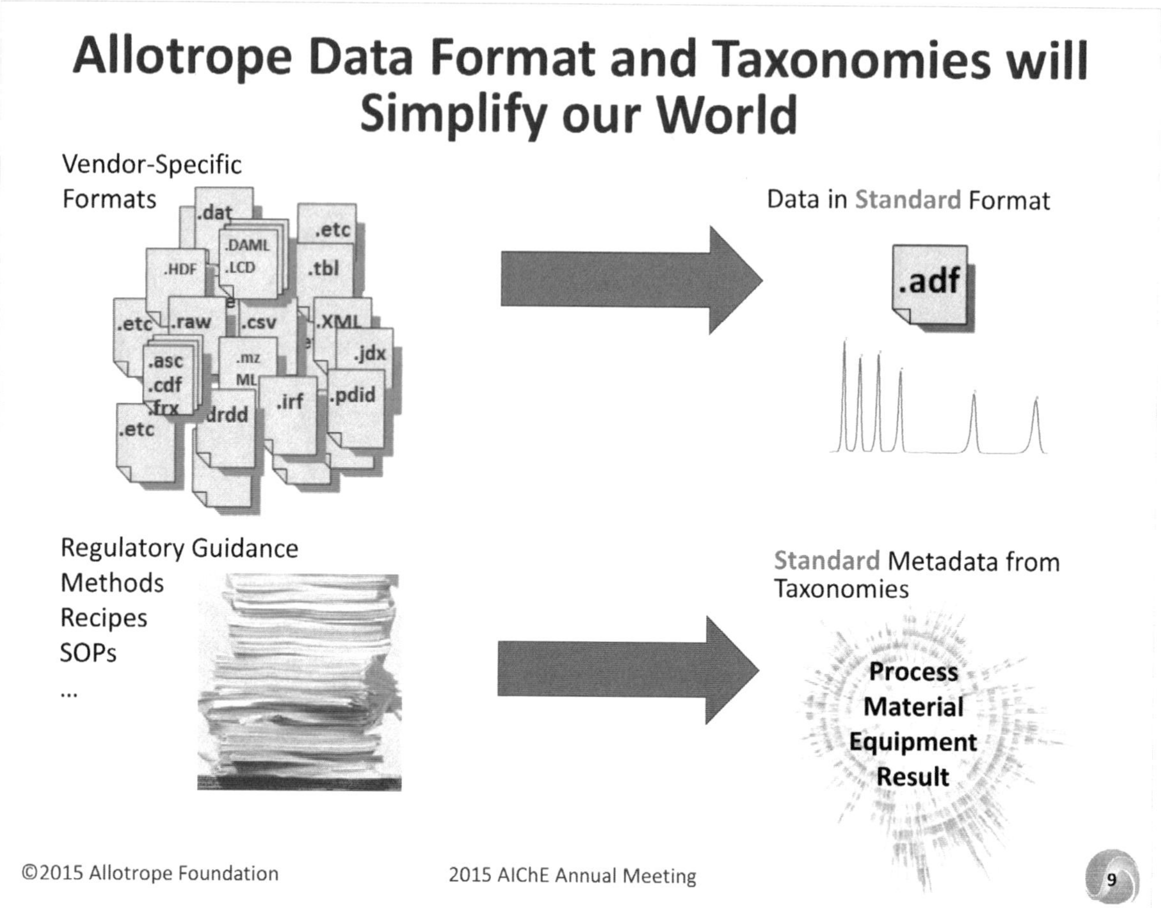
Allotrope Data Format and Taxonomies will Simplify our World
Vendor-Specific Formats
.dat
.DAML
.LCD
.etc
.HDF
.tbl
.etc
.raw
.csv
.XML
.jdx
.asc
.cdf
.frx
.mz
ML
.irf
.pdid
.etc
drdd
Data in Standard Format
.adf
Regulatory Guidance
Methods
Recipes
SOPs
...
Standard Metadata from Taxonomies
Process
Material
Equipment
Result
©2015 Allotrope Foundation
2015 AIChE Annual Meeting
9

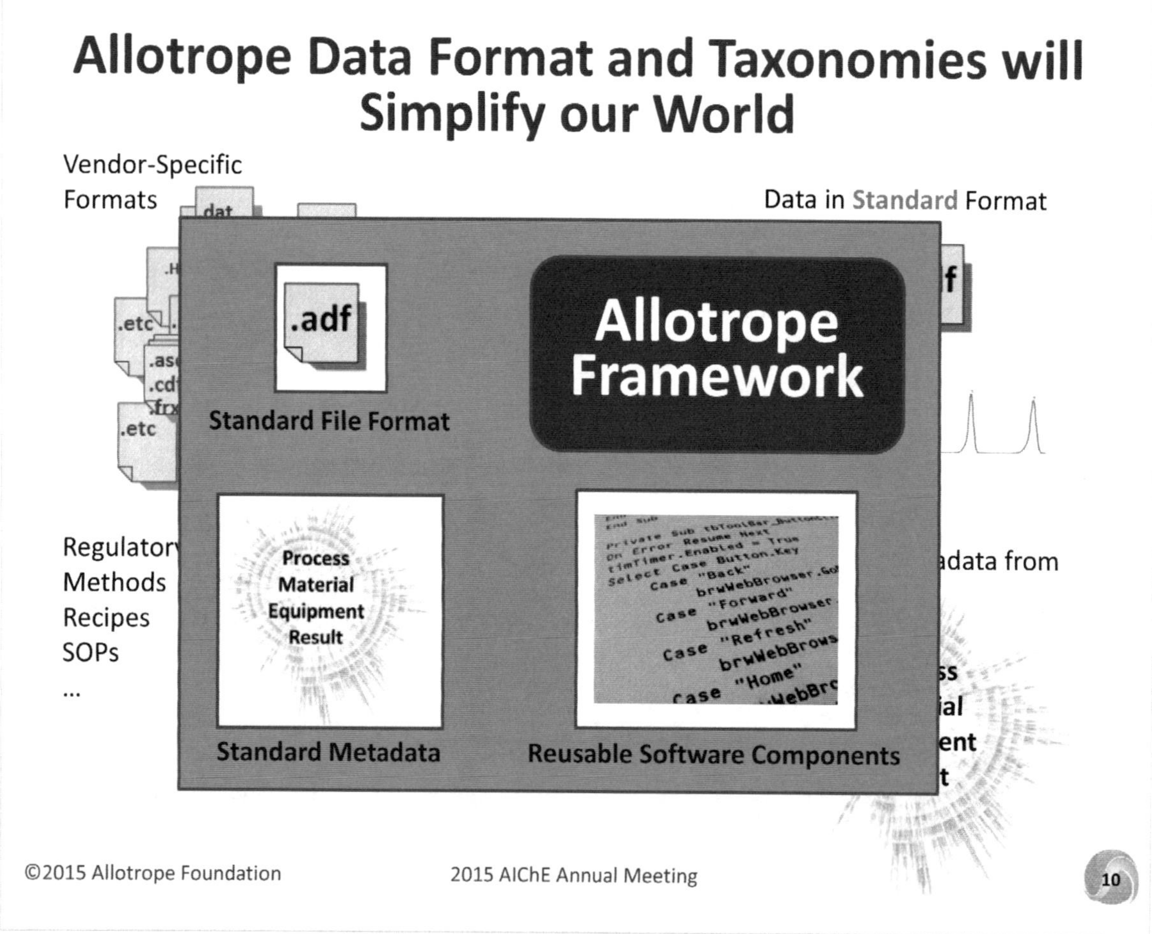
Allotrope Data Format and Taxonomies will Simplify our World
Vendor-Specific Formats
Data in Standard Format
.adf
Standard File Format
Allotrope Framework
Process
Material
Equipment
Result
Standard Metadata
Reusable Software Components
Regulator
Methods
Recipes
SOPs
...
©2015 Allotrope Foundation
2015 AIChE Annual Meeting
10

Scope: Not Just Analytical Data

Holistic solution across multiple data generation scenarios

- Measurement process
  offline, online, PAT
- Research through Manufacturing process
  chemistry, formulation, bioprocessing
- Records management
  record retention, regulatory submissions, reporting

Requirements from range of perspectives & roles

- Regulators, bench scientist, data analysts, modelers, manufacturing, archivists, IT

Allotrope Data Format (ADF):
Standardizes our Data

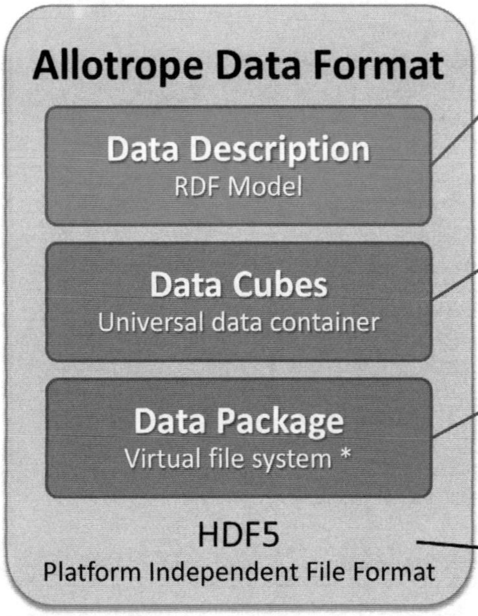

Contains:
- Method, instrument, sample, process, result, etc.
- Data cube metadata
- Data package metadata
- ...

Data represented by one- or multidimensional arrays.

Data represented by arbitrary formats, incl. native instrument formats, images, pdf, video, etc.

Specifically designed to store and organize large amounts of numerical data.

** Use is optional*

Allotrope Foundation Taxonomies (AFT):
Standardizes our Metadata

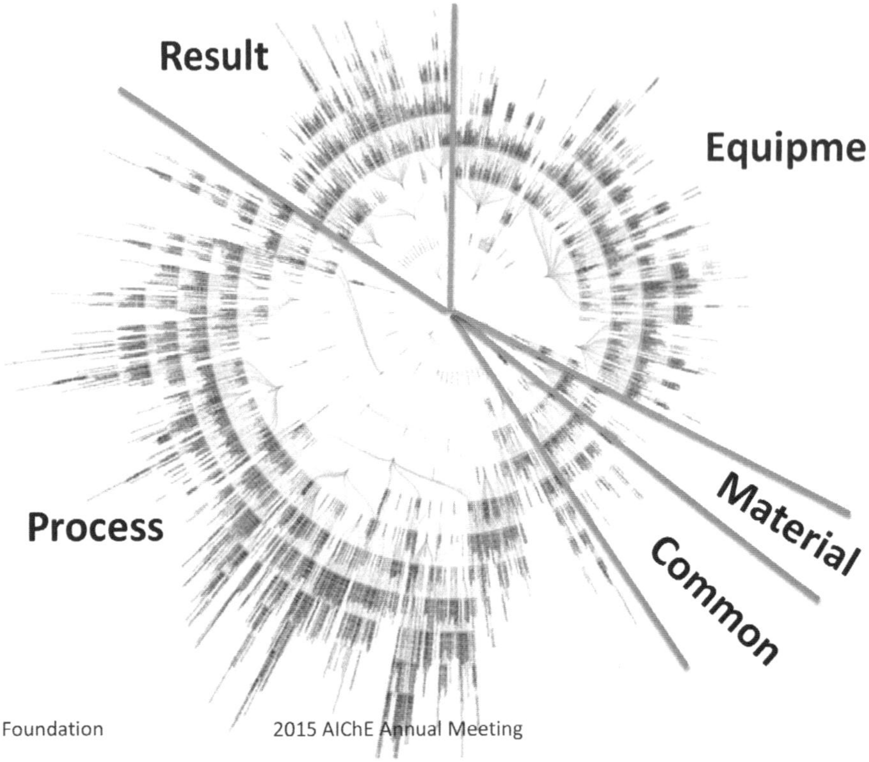

Allotrope Foundation Taxonomies (AFT):
Standardizes our Metadata

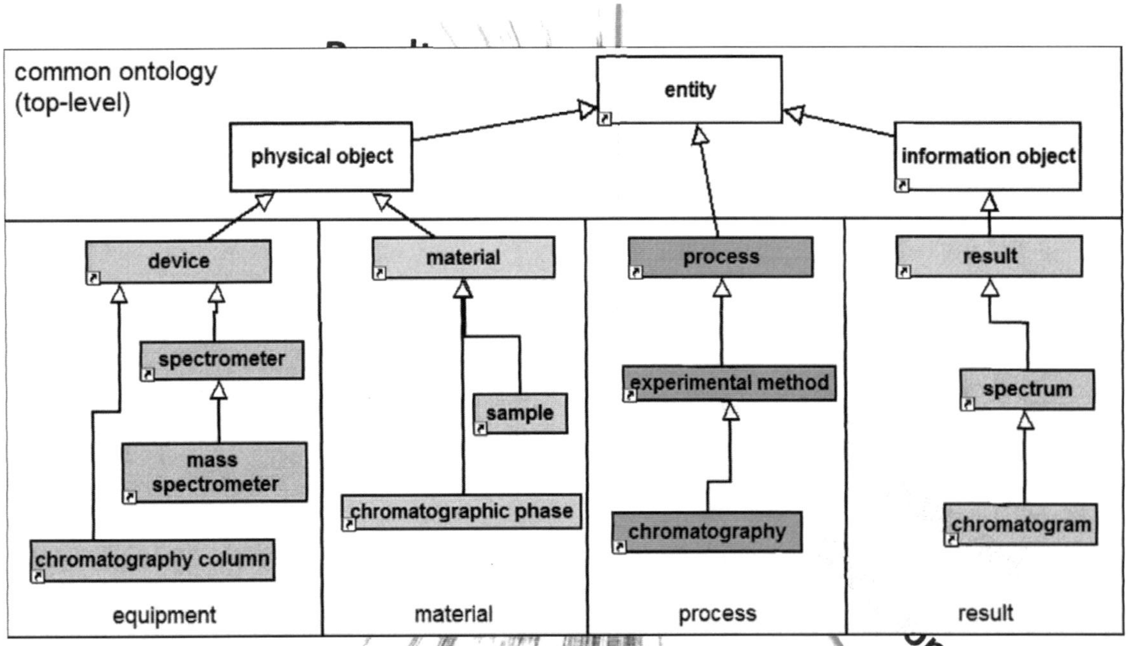

Example: ADF for NMR

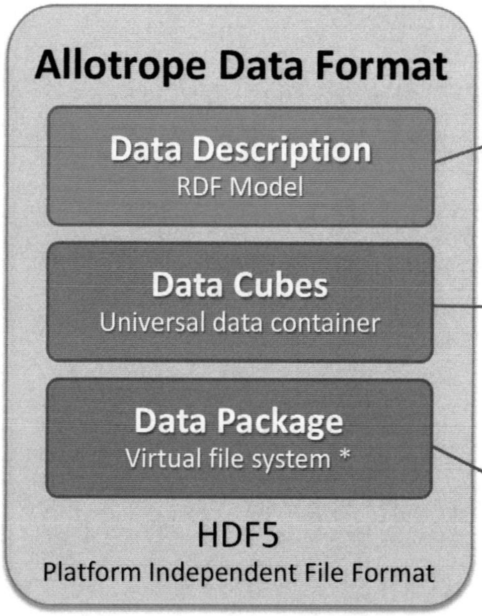

Contains semantic descriptions of:

- Context sensitive metadata – ex. sample ID
- Pulse programs, gradient programs
- Instrument parameters - power & pulse lengths, delays, other settings incl. selectable DF settings, LP settings, NUS parameters as needed for expt(s).
- Processing macros & parameters: apodization functions, ZF settings, etc.
- Peak lists, integrals, assignments, etc.

Contains binary representation of:

- FID(s) (1D/nD)
- Transformed data (maybe multiple complementary freq domain representations – ex. signal & resolution enhanced 1D's)

Contains intact files in virtual FS:

- Structure as MOL file
- PDF Reports of peak lists, integrals, assignments
- Report templates.
- PDF plots of region(s) of interest

Allotrope Capabilities Today

- **September 2015 internal release package**
  - Allotrope Data Format (ADF)
  - Allotrope Taxonomies (AFT)
  - Application Programming Interfaces and Class Libraries
  - Documentation

- **Supports the following techniques**
  - gas chromatography
  - Karl Fischer
  - liquid chromatography
  - mass spectrometry
  - NMR
  - thermogravimetric analysis
  - ultra violet spectroscopy
  - capillary electrophoresis
  - cell counter
  - cell culture analyzer
  - blood gas analysis
  - balance
  - pH

Integration Projects: Reduce it to Practice

- Divide and conquer: members share the costs and benefits
- Early adoption and acceleration of AF Framework development through delivery of Framework components
- Accelerate vendor adoption via collaboration on delivering specific capabilities through the integration projects
- Ensure validity of the Framework through real-world requirements and implementation
- Benefit the sponsoring companies in terms of shared development of new functionality & shared experience
- Drive integration of the Framework into the IT/IS Roadmap

 2015 AIChE Annual Meeting 17

Integration Projects: Reduce it to Practice

Research | Development | Commercial

Member	Instrument	Project
A	pH, Weighing, GC, Karl Fischer, TGA, NMR , Cell Density/Viability, Blood Gas Analyzer , Cell Culture Analyzer, Capillary Electrophoresis ... / Multiple types	Small and Large Molecule CMC
B	ICP-MS	Elemental Impurities
C	Bioanalyzer	Fermentation Process Control
D	HPLC-UV	CRO Integration
E	HPLC-MS	Structure ID
F	HPLC-UV/MS	Structure ID, Purification, Lead Profiling
G	HPLC-UV	Assay, Purity
H	HPLC-UV Balance	Drug Substance Release & Stability
I	HPLC-UV/MS	Method Screening
J	HPLC-UV	LC Data to ADF Converter/Adapter

Member | Instrument | non-GMP | GMP

 2015 AIChE Annual Meeting 18

Allotrope Foundation
Building an Open Framework for Laboratory Data

2016 AND BEYOND

GMP Compliance

- **Compliance (21 CFR Part 11)**
  - Digital and Electronic Signatures
  - Validation of the Framework
- **Security**
  - Encryption
- **Data Integrity**
  - Checksums
  - Audit Trail
  - Thread Safety
  - Veto Listener to Triple Store
  - Data Validation
  - Entailment Regime

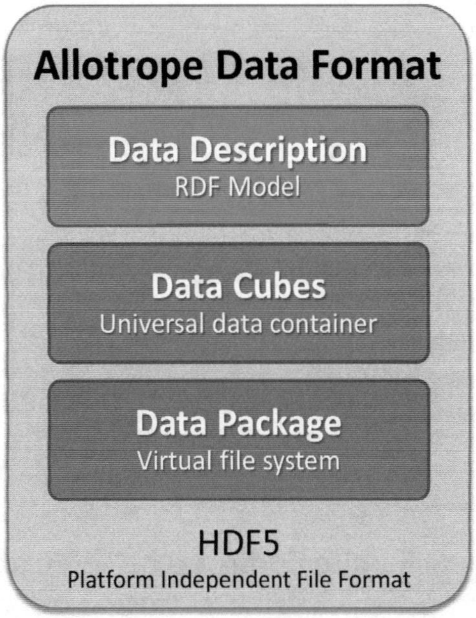

More Taxonomies Available in 1Q 2016

- **Instrument Taxonomies**
  - Protein Characterization System
  - Microplate Reader
  - Flow Cytometer
  - Protein Purification System
  - Millipore Filtration System
  - Differential Scanning Calorimeter (DSC)
  - X-Ray Powder Diffractometer (XRPD)
  - Particle Sizer
  - Roller Press
  - Tablet Press
  - Tablet Coater
  - Dynamic Vapor Sorption (DVS)
  - solid-state NMR (ssNMR)
  - Infrared Spectroscopy (IR)
- **Other Taxonomies**
  - IDMP-Identification of Medicinal Products
  - Extend Process (S88)
  - Extend Materials
  - Extend Equipment
  - Investigation/Study/Analysis
  - Phase of Development
  - Route of Administration
  - Product Quality Attributes
  - Container Categories
  - Chemical Reactions

Funded by Individual Allotrope Member

Shared with Allotrope Foundation

Growing an Ecosystem

- **Framework is applicable beyond pharmaceutical industry**
  - The data problems are universal, the instruments and software are the same, only the experiments and context are different
- **The Framework is as good as we make it**
  - New capabilities are realized only through new integration projects
- **Framework has been funded by Founding Members**
  - New members still welcome to join in 2016
- **Business Model will promote adoption and growth by vendor and end-user communities**
  - Licensing model will be simple
- **Public Commercial release in December 2016**
  - Integration of the Framework into your infrastructure will take time

Opportunities across R&D and Manufacturing

- **Allotrope Framework is designed to support *any* Process, *any* Material, *any* Equipment and *any* Result**

- **Members have an opportunity to address other big data challenges**
  - Offline analytical chemistry
  - PAT / Online
  - Process Chemistry Campaigns
  - Drug Product Manufacturing Campaigns
  - Safety Assessment
  - Bioanalysis – *e.g.*, DMPK, biologicals
  - ???

Next steps

- **Think about Allotrope & discuss it internally at your company**
  - What pain points do you want to fix?
  - Why would you want to join now?
  - Why not collaborate with your peers to create a sustainable solution for the industry?

- **Encourage your colleagues to attend one of the upcoming Allotrope Cross-Industry Workshops**
  - January 20, 2016: San Francisco, CA @ Genentech
  - April 2016: Ingelheim, Germany @ Boehringer Ingelheim
  - September 2016: Indianapolis, ID @ Eli Lilly and Co.

Thank you!

- Any questions, please contact the Secretariat at more.info@allotrope.org or james.vergis@dbr.com

- **2016 Workshops**
  - January 20, 2016: San Francisco, CA @ Genentech
  - April 2016: Ingelheim, Germany @ Boehringer Ingelheim
  - September 2016: Indianapolis, ID @ Eli Lilly and Co.

http://www.allotrope.org for more information and to register

What Will be Discussed Today

- ❑ Motivation for the Design Space Verification (DSV) Lab
- ❑ Scope of the DSV Pilot Project
- ❑ DSV Project implementation
- ❑ Data Standards and Data Flow
- ❑ Examples of how we plan to deploy the DSV Lab
- ❑ Key Accomplishments and Learnings

Motivation - Scale-up Verification

- ❑ ***Avoid surprises during scale-up that may lead to API supply issues, increased costs and longer timeline***
- ❑ ***Understand the capability of the commercial process and reduce failure rates***
- ❑ ***Implement a strategy that addresses concerns from regulators***

Would you trust data from this experiment?

Typical Regulatory Questions from recent GSK Filings

- ❑ *"If the large scale batches were manufactured within a narrow sub-range of the process parameters, then the full [ranges] cannot be approved on large scale and thus should be tightened accordingly. Alternatively, a [verification] strategy should be presented for future variation outside the set points already validated at commercial scale.*
- ❑ *"If the design space is not established with commercial scale batches, a discussion on the scale up aspects and a verification protocol for the scale dependant CQAs should be submitted. The design space verification protocol should allow to demonstrate that for any new operating ranges of the DS, product quality attributes are maintained."*

Motivation – The Informatics Challenge

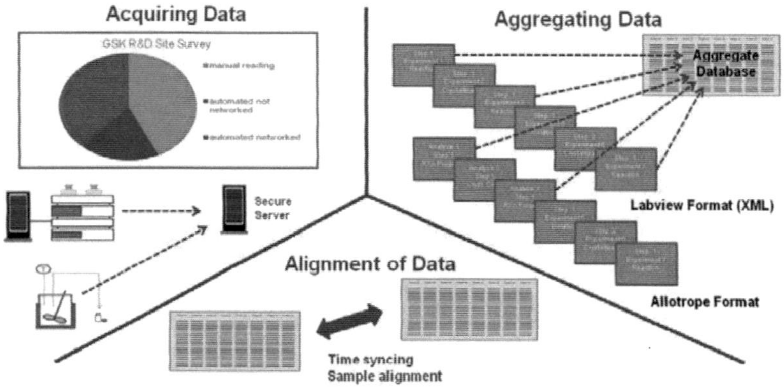

DSV Lab Project Scope

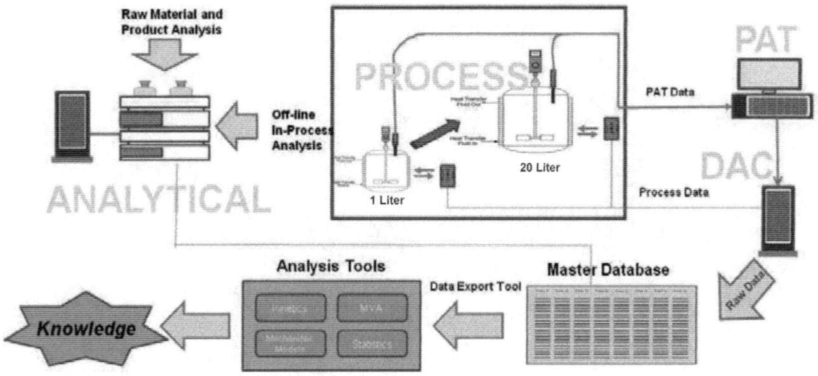

Equipment/Control System Interface

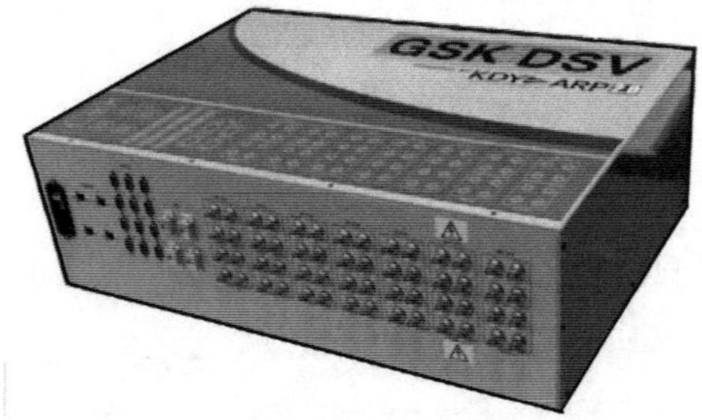

Connection Type	Number	Purpose
Ethernet	4	Connections to Labview control system and database
Serial Ports	12	Agitation control, dosing (e.g. Pumps, gravimetric, syringe, etc.)
USB Ports	4	Temperature control, PAT Systems
24V DC, 120V, RTD Power	Many	Turbidity, pH, conductivity, liquid flow, gas flow, level, etc.

Advanced Manufacturing Technology (AMT)

Project Support Model

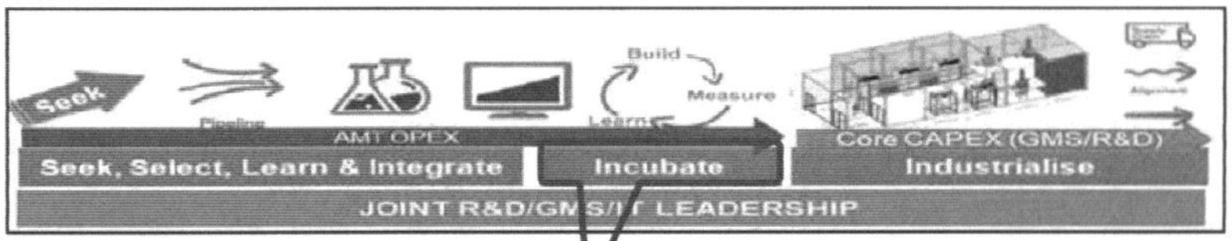

DSV Lab Project set-up in the ***Incubate*** framework:

- ❑ Pilot of ready-now technology in three phases
- ❑ Funded by AMT to demonstrate PoC
- ❑ Share resources AMT and other Development lines
- ❑ Read-out: of Phase 1 end of 4QR15
- ❑ Roll-out of Industrialization solution to be funded by R&D Product Development (PD)

Resource requirement:

- ❑ AMT provides Project Leader to drive activities
- ❑ Cost $190K
- ❑ PD provides access to Scale-up Lab and 0.8 FTE support until completion
- ❑ Product Quality & Risk Management provides single point of contact for regulator interactions - < 0.1FTE

DSV Lab Data Flowchart

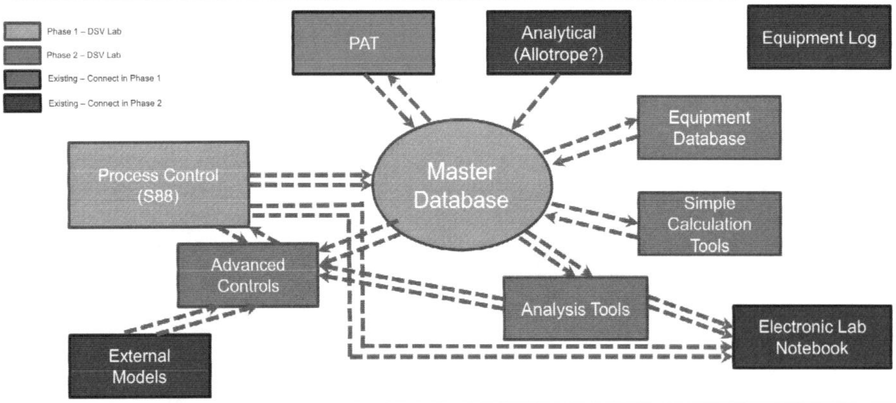

S88 Alignment – Process Data Standards

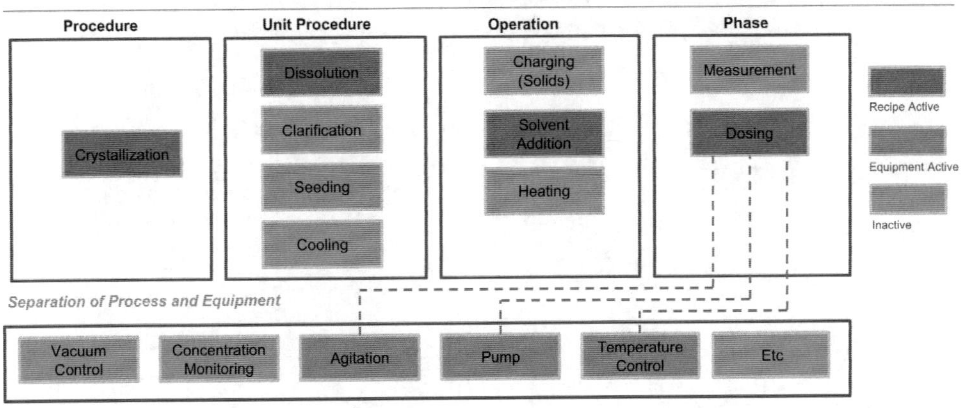

Note: Some Metadata entered upon experiment initiation including; Project Number, Step, User ID and Notebook/Experiment No.

DSV Lab Master Database

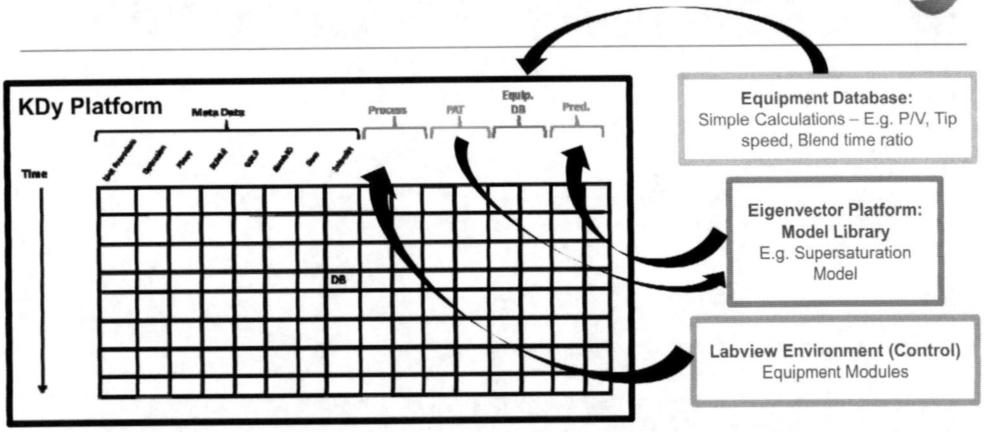

Scale-up Approach

Understanding and verifying scale-up

1. **Define design space at small scale**
   - Mechanistic model
   - DoE

2. **Define design space at intermediate scale**
   - Mechanistic model
   - DoE

3. **Scale (dis)similarity (MVA)**
   - Size and directions of factor effects
   - Reaction/ Cryst profile over time
   - Predictions at target scale

4. **Explore other sources of variability (MVA)**
   - Design factors sufficient?
   - Scale dependent factors
   - Natural variability of other process parameters/attributes

5. **Combine information at different scales (MVA of all historical experiments)**
   - Amount of CQA variability described
   - All experiments/scales described equally well
   - Size and directions of design effects
   - Natural variability effects
   - Factors changing with scale effects

6. **Target attribute value at large scale (MVA)**
   - Use interpretation of factor effects to obtain product with targeted attribute value
   - Simulations to identify process conditions with high likelihood to obtain product with targeted attribute value

7. **Root cause analysis (MVA) scale effects**
   - Link between process variation and reaction/crystallization profile
   - Model or control scale dependence

8. **Define design space based on multiple scales**
   - Bayesian approach incorporating between scale variability

3. Scale similarity

Relative size and direction of effects similar?

gsk

1L Crystallization

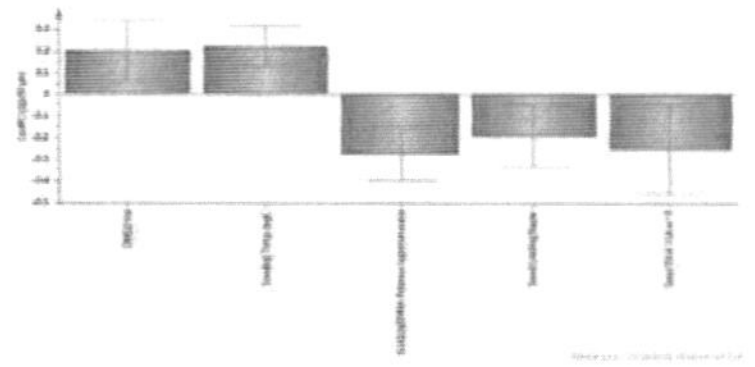

20L Crystallization

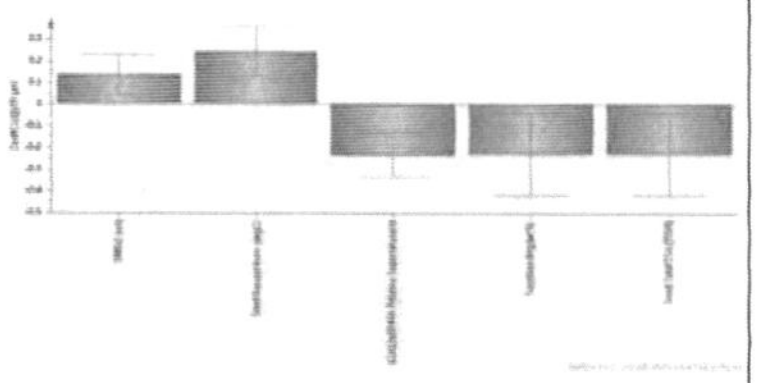

Example: effect of crystallization variation (5 factors) on x50 API

The direction of the effects is the same at both scales for all factors.

Only the relative size of the DMSO volume is different for the 20L scale caused by decreased stretch at that scale.

3. Scale similarity

Compare process profile over scales

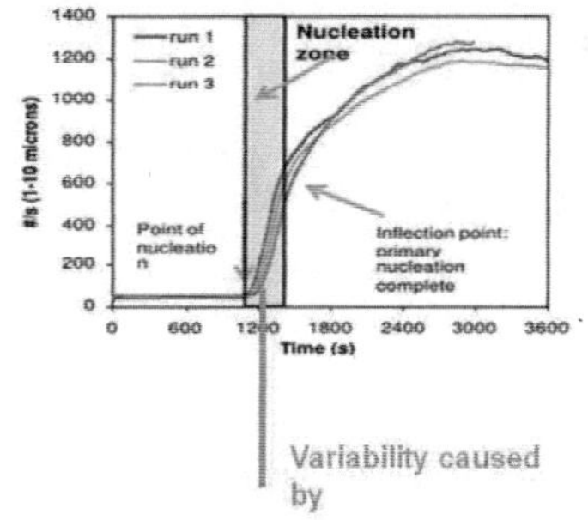

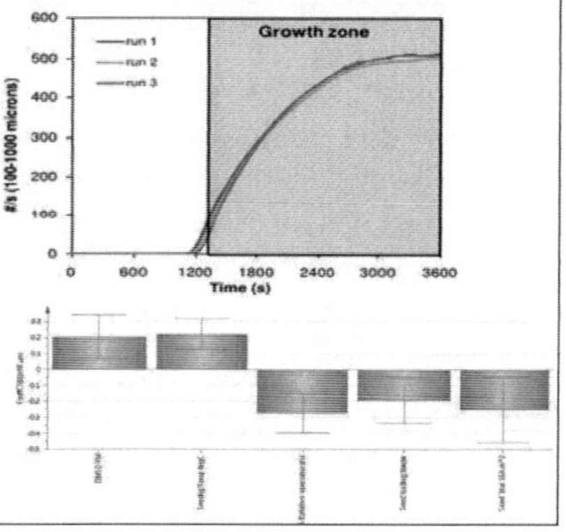

Illustrative example as no data available yet.

8. Define design space based on multiple scales

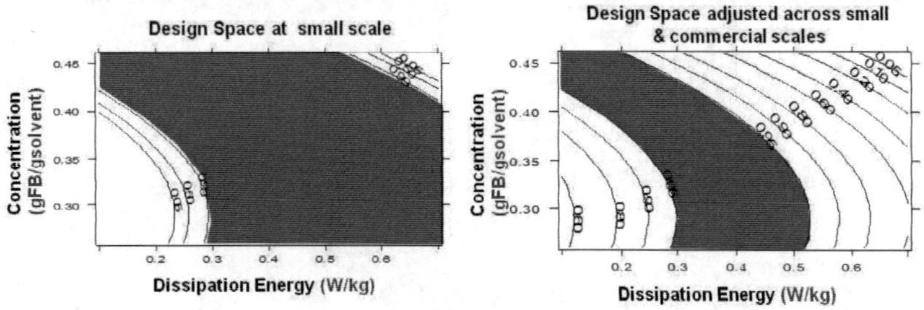

API crystallisation example demonstrated that design space was smaller (for the same level of assurance – 95%) when adjusting for commercial batches

Key Accomplishments and Learnings

- ❑ ***Informatics improvements*** – The pilot project will provide data needed to make intelligent investments in informatics improvements
- ❑ ***S88 implemented*** – We expect to be able to retrieve critical data more easily which will increase confidence in and understanding of scale-up
- ❑ ***Managing change*** – The project has triggered internal discussions about various informatics options (e.g. Process, eLNB or analytical centric systems)
- ❑ ***Implementation challenges*** – Improvements will require new ways of working by platform users. Good implementation and support plans are required to achieve success.
- ❑ ***Additional opportunities*** – We have identified many new improvement opportunities not apparent at project inception.

Acknowledgements

Sponsors;

- Vance Novack *(Global Head of API Chemistry)* and Phil Dell Orco (*Global Head of Particle and Process Engineering)*

Contributors

- Jeff Sterbenz and Rick Atwell - *Upper Merion Site Scale-up Lab*
- Allan Salant, Doug McVey, Justin Palpant - *KDY Automation*
- Melanie Dumarey and Mohammed Yahyah - *Manufacturing Simulation and Modeling*
- Jim Rydzak – *Analytical Sciences*
- Brian Lewis- *Research & Development Information Technology*
- API Project Team –*Huan Wang, Dan Patience et al*

416559 Challenges for Academic and Industrial Partners in Major Research Collaborations - a Case Study

Abstract:

A major challenge in the management and delivery of academic-industrial collaborations is the alignment of expectations of both the academic and industrial partners. As the number of partners increases, the scale and diversity of the collaboration can present hurdles to successful implementation of the collaboration.

This presentation will discuss some of these challenges with reference to the Synthesis and Solid State Pharmaceutical Center (SSPC) which has 9 funded academic research organisations and 22 industrial partners. Some strategies to successfully address these challenges will be discussed, from both an academic and an industrial perspective.

422671 Erc-Sops - a Pre-Competitive Model for Strategically-Driven Research and Technology Development and Commercialization

Abstract:

In this talk, we share the experience of the last 9 years of C-SOPS, an NSF-ERC focused on Pharmaceutical products and manufacturing processes.

Creation of the academic/industrial partnership for C-SOP started well ahead of NSF funding. Our first meetings with industry took place approximately one year before the pre-proposal was submitted, and about 2 years before our center was funded. In the initial discussion with pharmaceutical (and other) companies, we focused on primarily two questions - the scope of the research program, and the collaborative model. A strategy emerged from these consultations, indicating that we should focus on pre-competitive research topics of maximum interest to pharmaceutical manufacturers. Once these companies joined the center, we anticipated that many f their technology providers would also be motivated to participate.

This approach succeeded as intended. In nine years, C-SOPS has executed over 60 membership agreements, currently has 40 members, including not only pharmaceutical companies, but also suppliers of manufacturing equipment, control equipment, sensors and process analyzers, modeling software, and others. These members have contributed cash memberships in excess of $7 million, in kind equipment and software iciness in excess of $10 million, and have funded collaborative projects in excess of $20 million.

The pre-competitive model makes IP developed at the center available to all members on a royalty-free basis, provided that is used in-house for non-commercial purposes. Commercial use (defined narrowly as using center IP to define and enable a company's commercial product) requires a commercial license, which is offered simultaneously to all center members.

Overlaying the pre-competitive research program is a large and rapidly growing set of projects that are funded directly by member companies. These projects also benefit from center IP and know-how, but because they are funded directly by members, IP deriving from these projects is negotiated on a case by case basis. Whenever possible, these projects are funded via service agreements, and negotiations regarding commercial disposition of IP rights is delayed until after the IP is developed.

In our experience, this model has been an extremely useful component of our center engagement model and has helped attract substantial levels of support over almost a decade of uninterrupted growth.

The Role of NCATS, SBIR to Support Precompetitive Collaboration

Asaf Alimardanov, PhD

November 9, 2015

NCATS

NIH National Center for Advancing Translational Sciences

National Center for Advancing Translational Sciences (NCATS)

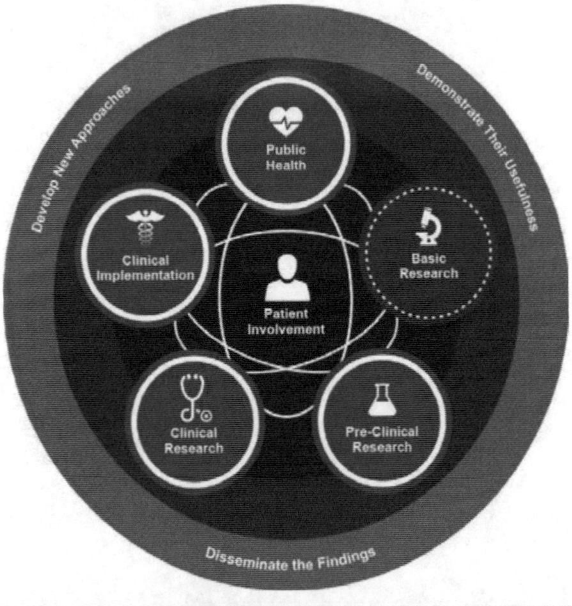

NIH National Center for Advancing Translational Sciences

NCATS

NCATS SBIR & STTR

FOCUS

To transform the translational science process so new treatments and cures for disease can be delivered to patients faster

GOAL

Seeks to increase small business participation in federally supported R&D and private-sector commercialization of technology

FUNDING

$18 million annually, allocated by NCATS for small business innovation research and technology transfer

What are the SBIR & STTR Programs?

Small Business Innovation Research (SBIR)

supports early-stage research and development projects at small businesses

SBIR & STTR Programs

help entrepreneurial researchers as they launch businesses, engage in research and development and seek to commercialize new products that will have public benefit

Small Business Technology Transfer (STTR)

helps small businesses formally collaborate with a research institution in Phase I and Phase II

Benefits of NIH SBIR & STTR Funding

- **One of the largest funding sources** of early stage life sciences in the country **(stable & predictable)**
- **IP rights** are **retained by the small business**
- **Not a loan, non-dilutive** capital
- Projects undergo **NIH's rigorous scientific peer review** process, which awardees leverage **to attract other funding and collaborations**

NIH National Center for Advancing Translational Sciences NCATS 5

NIH SBIR/STTR FY2016 Funding

Set aside requirements for Agencies will increase:

FY	Set-aside SBIR	Set-aside STTR
2013	2.7%	0.35%
2014	2.8%	0.40%
2015	2.9%	0.40%
2016 (current)	3.0%	0.45%
2017	3.2%	0.45%

NIH National Center for Advancing Translational Sciences NCATS 6

SBIR and STTR: One of the Largest Sources of Early- Stage Financing

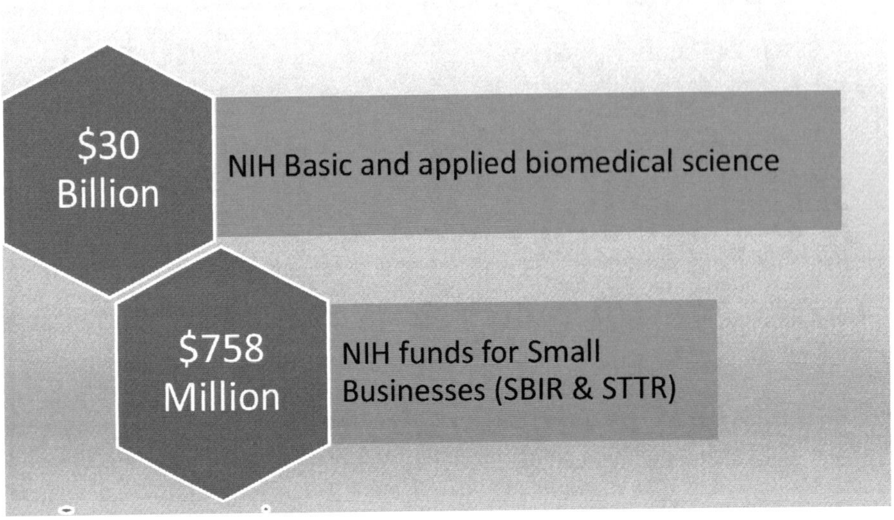

National Center for Advancing Translational Sciences

Budget Hard Caps

Award Guidelines

Program	Phase I	Phase II
SBIR	$150,000	$1,000,000
STTR	$150,000	$750,000

Award Hard Caps

Program	Phase I	Phase II
SBIR	$225,000	$1,500,000
STTR	$225,000	$1,500,000

See NOT-OD-15-101 for details

National Center for Advancing Translational Sciences

NCATS 8

Budget Hard Cap Waiver

- Agencies (not applicants) must apply for a waiver from the SBA for budgets over the hard cap
- SBA has approved a topics list for NIH for 2016
- NIH ICs may have their own budget guidelines
- Appendix A in "Program Description and Research Topics": https://sbir.nih.gov/funding#omni-sbir

Contact NIH Program Staff to discuss!

SBIR/STTR Program Structure Overview & Commercialization Process

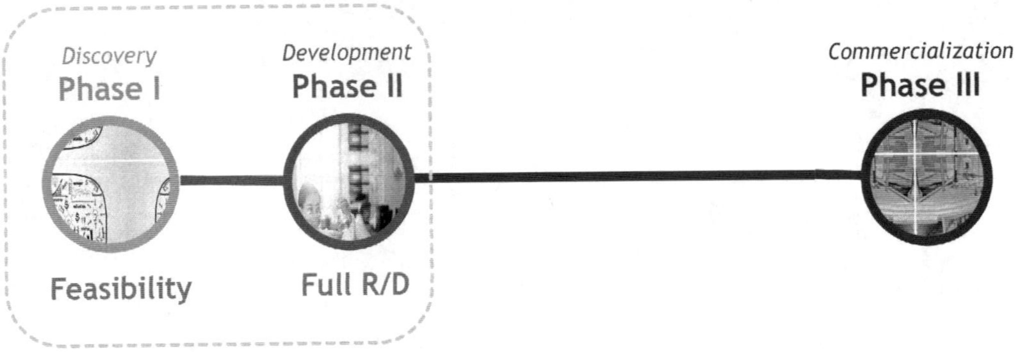

SBIR/STTR

Non-SBIR/STTR

Phase I: $150,000 total costs for 6 months (SBIR) or 1 year (STTR)

Phase II: Normally do not exceed $1,000,000 total costs for 2 years ($750,000 for STTR)

Fast Track SBIR Opportunity

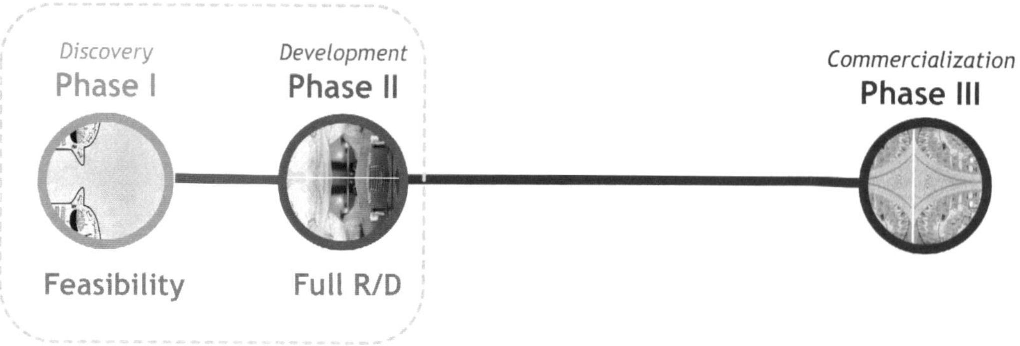

SBIR/STTR **Non-SBIR/STTR**

- Combines Phase I and II into one application
- NIH Administrative Review
- Phase II begins if milestones are met

***Note: Fast track designated for SBIR only, not accepting STTR applications*

SBIR Direct to Phase II

Non-SBIR/STTR **SBIR** **Non-SBIR/STTR**

- Company has completed Phase I R&D with non-SBIR/STTR funds
- Separate General Solicitation: PAR-14-088 | PAR-15-288
- Not all ICs participate: http://sbir.nih.gov/funding#phase2

NCATS Funding Overview

The NCATS SBIR and STTR programs support small businesses and research organizations advancing technologies that can be commercialized to improve translational science and health care

Omnibus Solicitation	• Investigator-initiated grant funding • Deadlines: April 5, September 5, January 5
Direct to Phase II Grant	• Designed to enhance the pace of technology development • Deadlines: April 5, September 5, January 5
Grant Solicitations in Targeted Areas	• Grant to advance a particular technology/ research area • Due dates may vary
Contract Solicitation	• Contract opportunity to advance areas of high research interest • Typically due in October or November

Eligible for SBIR Funding?

Criteria for applying to SBIR:

✓ U.S. businesses with **500 or fewer employees**

✓ **PI Primary employment with small business** at the time of the award and duration of the project

✓ **More than 50% U.S.-owned** by individuals and independently operated

OR

✓ More than 50% owned and controlled by other business concern/s that is/are > 50% owned and controlled by one or more individuals

OR

✓ More than 50% owned by multiple venture capital operating companies, hedge funds, private equity firms or any combination of these

Eligible for STTR Funding?

Criteria for applying to STTR:

✓ **An established cooperative research and development effort** delineated as:

- Minimum 40% by small business concern; minimum 30% by U.S. college or university, non-profit research organization or Federally-Funded R&D Center (FFRDC)

✓ **Formalized intellectual property agreement**

- Should provide the necessary IP rights in order to carry out follow-on R&D and commercialization

✓ **Primary employment of the principle investigator with either the small business or research institution**

Key Differences Between SBIR and STTR Funding

SBIR

- **Permits research institution partners (e.g., universities)**
- **Small business concern may outsource ~33% of Phase I activities and 50% of Phase II activities**

STTR

- **Requires research institution partners (e.g., universities)**
- **Minimum 40% of the work should be conducted by the small business concern (for profit), and minimum of 30% by a U.S. research institution (non-profit)**

Awards always made to small business

FEATURED SBIR EXAMPLE

NCATS

NIH National Center for Advancing Translational Sciences

17

Collaboration with IQ Consortium

- Emerging Technologies group: survey of IQ member companies
- Discussion at Emerging Technologies group meeting
- Two potential topics for SBIR identified
  - Online Real Time Metals Analysis at Low ppm Level
  - Simple and Robust Reaction Progress Analyzer

NIH National Center for Advancing Translational Sciences

NCATS

18

Online Real Time Metals Analysis at Low ppm Level

- Allow rapid measurement of trace metal levels (at least <5 ppm)
- Inexpensive, Small footprint, Co-localized
- Robust and have an ease of use advantage, i. e. not require any specialized training
- Ideally, able to be used for continuous monitoring of metal content in continuous processing operations
- Ideally, detect more than one transition metal typically used for catalysis (Pd, Rh, Cu, Zn, Fe, Ir, etc.)

Metals Analysis at Low ppm Level

- Strong response
- Unique approach in each proposal
- Thorough peer review
- Three applications funded for Phase I
- Feasibility work ongoing
- Potential pharma end users agreed to act as consultants

CURRENT NCATS FUNDING OPPORTUNITIES

NCATS

NIH National Center for Advancing Translational Sciences

21

Investigator Initiated Grants

Omnibus "Parent" SBIR/STTR Grant Solicitation

SBIR: PA-15-269 (R43/R44) **STTR:** PA-15-270 (R41/R42)

We encourage applications for topics within the NCATS mission

Read the "Program Descriptions and Research Topics" Section in the Solicitation for more details

Due January 5, April 5, September 5

NIH National Center for Advancing Translational Sciences

NCATS

22

Addressing Unmet Needs

NCATS SBIR/STTR Research Interests

Drug Discovery & Development → Diagnostics & Devices → Research & Bioinformatics → Clinical Research

National Center for Advancing Translational Sciences

NCATS 23

NCATS SBIR/STTR Topics of Interest

Drug Discovery & Development

- Small molecule and biologics analytical characterization
- Accelerate bioengineering approaches to the development and clinical application of biomedical materials, devices, therapeutics and/or diagnostics
- Technologies to determine alternative uses for existing therapeutic interventions
- Protein-protein interaction assays for high-throughput screening of rare disease related projects
- Tools and technologies to enable assaying of compound activity on currently "non-druggable" targets
- Innovative platforms for identification and prioritization of targets for therapeutic intervention with clear clinical impact
- Fluorescence probes to replace antibodies for determination of cellular protein translocation
- Co-crystallization high-throughput screening techniques
- Tools and technologies that increase the predictivity/efficiency of medicinal chemistry, biologic or other intervention optimization
- Use of continuous flow manufacturing technology to address therapeutics shortages
- Methodologies and technologies to substantially lower the cost of manufacturing biologics
- Development of novel technologies for enzyme replacement therapies to solve a major bottleneck in rare disease research
- Development of non-AB biologics, cell-based therapies and gene therapy discovery amid technology development

National Center for Advancing Translational Sciences

NCATS 24

NCATS SBIR/STTR Topics of Interest

Specialized Topics

- Platform delivery technologies for nucleic acid therapeutics
- Bioreactors for reparative medicine
- Development of appropriate pediatric formulations and drug delivery systems
- Onsite tools and technologies for heart, lung and blood clinical research point-of-care

NIH National Center for Advancing Translational Sciences NCATS 25

ADDITIONAL SMALL BUSINESS RESOURCES

NCATS

NIH National Center for Advancing Translational Sciences 26

SBIR/ STTR Technical Assistance Opportunities

NIH offers distinct technical assistance programs to SBIR Phase I and Phase II awardees

- Niche Assessment Program
- Commercialization Assistance Program
- I-Corps at NIH offers a nine-week boot camp to help awardees explore potential markets for their federally funded innovations

Applicants can utilize their own technical assistance vendor (must be included in the budget) in lieu of these programs

Dedicated NCATS Small Business Resources Site

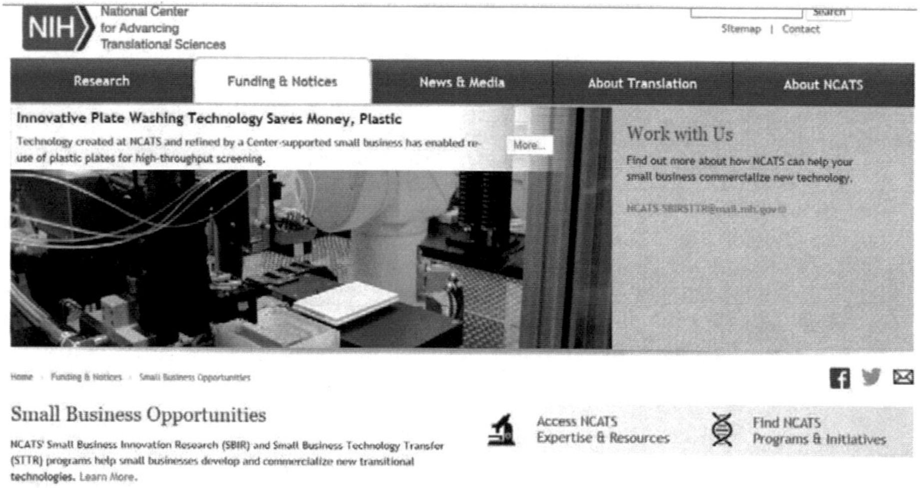

http://ncats.nih.gov/smallbusiness

Key NCATS Contacts and Resources

Email	**Lili Portilla, M.P.A.** NCATS-SBIRSTTR@mail.nih.gov
Online	http://ncats.nih.gov/smallbusiness
Stay Connected:	@ncats_nih_gov Join the conversation #NCATSsbir
Searching for funding opportunities?	http://grants.nih.gov/grants/guide/index.html **Use activity codes:** R41/R42 & R43/R44

NIH National Center for Advancing Translational Sciences NCATS 29

431932 Development and Scale up of a Continuous Flow Reaction for API Intermediate Manufacture

Abstract:

Continuous processing is a growing interest in pharmaceutical industry for the development and the manufacture of active pharmaceutical ingredient (API) and upstream intermediates. Enhanced heat and mass transfer rates inherent to continuous flow reactors are enabling chemical researchers to explore reactions that are traditionally difficult to investigate in common batch reactors. By selecting reactors with small internal volume and operating at the appropriate flow rates, it is possible to explore reactions at short reaction times. These features render flow reactors powerful tools for studying highly energetic reactions with unstable intermediates. Furthermore, it may be possible to gather critical reaction data more quickly with these reactors as the continuous flow operations enable rapid, sequential experimentation.

We present a lab-based flow reactor system used to investigate the Aldol reaction used in the production of an API intermediate. This system was utilized to understand the impact of various reaction parameters on conversion and selectivity. To better understand and predict the reaction performance during scale up, the reaction was studied in the lab under various mixing conditions. The lab-scale results are in agreement with the process performance at pilot scale.

433910 Development and Implementation of Flow Chemistry for a Curtius Rearrangement

Abstract:

The Curtius rearrangement is the conversion of a carboxylic acid moiety into an amine with retention of stereochemistry via a high energy acyl azide intermediate. An investigation of the Curtius rearrangement to generate an asymmetric isocyanate (Curtius product) via the acyl azide and reaction of the isocyanate with BnOH will be described. A flow procedure to produce the Curtius product was developed. Attempts to conduct the entire sequence from trans acid to amino ester using flow chemistry will be described. Multikilograms of the amino ester have been produced safely through the Curtius reaction using batch or flow processing in high overall yield.

410930 Development and Scale-up of Acrylamide Formation Using an Impinging Jet

Abstract:

In a synthetic step of a pre-clinical drug candidate, a problematic dimer impurity was observed. Early work identified the origin of this dimer impurity as reaction between the starting material and product, indicating that efficient mixing was critical to achieving low levels of this impurity. A fit-for-purpose flow chemistry solution using an impinging jet was developed and implemented for this process. Kinetics work performed on this reaction also led to mechanistic understanding for the formation of additional impurities, including a chloro impurity that was identified as a GTI (genotoxic impurity). Reaction kinetics and response surface simulations were used to reduce the chloro impurity to acceptable levels. Using these insights, this process was successfully scaled to 12 kg in a cGMP manufacturing environment.

The work presented here was sponsored by AbbVie. AbbVie contributed to the design, research, and interpretation of data, writing, reviewing, and approving the presentation. Daniel Caspi, Benoit Cardinal-David and Alex Huters are employees of AbbVie.

416213 Robust Parallelization of Triphasic Micro/Millireactors for Sustainable Pharmaceutical Manufacturing

Abstract:

Robust Parallelization of Triphasic Micro/Millireactors for Sustainable Pharmaceutical Manufacturing

Swee Kun Yap, Wai Kuan Wong, Nicholas Xiang Yang Ng, Ning Yan, and Saif A. Khan

Department of Chemical and Biomolecular Engineering

National University of Singapore, SINGAPORE

Metal-catalyzed gas-liquid reactions, such as hydrogenation, carbonylation and halogenations are ubiquitous in pharmaceutical and fine chemical manufacturing. Such reactions are typically carried out in stirred batch reactors, and face several challenges in the form of mass and heat transport limitations, which often require energy intensive solutions such as high stirring rates and pressurization. There is considerable interest in continuous-flow micro/millireactor technology for gas-liquid reactions due to the tremendous acceleration in heat and mass transport possible in such systems, in addition to the prospect of facile phase separation.[1] We have previously demonstrated the design and application of a triphasic segmented flow millireactor for rapid nanoparticle-catalyzed reactions with facile catalyst recovery.[2,3] Here, we present the first demonstration of robust eight-fold parallelization of triphasic millireactors for sustainable pharmaceutical manufacturing. Specifically we demonstrate a novel parallelized reactor system for rapid hydrogenation reactions with facile catalyst recovery and recycle. Parallel operation of multiphase microfluidic circuits presents several unique challenges over their single-phase counterparts; in particular, nonlinear pressure-flow characteristics lead to severe problems with robust and equal routing of the same multiphase flow in the various arms of a parallelized flow network. Biphasic circuits have been studied extensively in recent years, and parallelized biphasic circuits have been demonstrated recently.[4] The dynamics of triphasic circuits, comprising a gas and two liquid phases, are considerably more complex. In our demonstration of numbering-up to eight parallel *triphasic* flow millireactors, the segmented flow in each millireactor consists of aqueous NPs, organic substrate and hydrogen gas at near atmospheric pressure. Feed solutions (aqueous NPs and organic substrate) are drawn from their respective reservoirs by means of peristaltic pumps, the latter being preferred over syringe pumps due to limited syringe capacities. However, flow rate fluctuations are typical of peristaltic pumps, and these fluctuations dramatically influence the stability of the three-phase flows in the parallelized reactor system. We also demonstrate the design of inline hydraulic damper systems for fluid delivery, which are able to damp out pressure fluctuations from the pumps, thereby allowing for the formation of a smooth triphasic flow in all the eight millireactors. We model, from the principles of fluid mechanics, the flow profile of fluid in the hydraulic dampers to better understand the influence of tube material in regulating pressure fluctuations in a parallelized multiphase flow system.[5] Our system is then applied in the hydrogenation of nitrobenzene, a model pharmaceutical substrate. The parallelized system yields a consistent conversion of $(80 \pm 4)\%$ across all eight reactors over the course of several hours, with a throughput of ~10 mL/h. Finally, we demonstrate continuous catalyst recovery and recycle, where the aqueous PtNPs are subsequently continuously recycled. In summary, we present a robust and general scheme for parallelization of triphasic flows, and demonstrate an industrially relevant application in sustainable pharmaceutical manufacturing.

REFERENCES:

[1] E.R. Murphy, J.R. Martinelli, N. Zaborenko, S.L. Buchwald, K.F. Jensen, Angew. Chem. Int. Ed. 46, pp. 1734-1737, 2007.

[2] S.K. Yap, Y. Yuan, L. Zheng, W.K. Wong, J. Zhang, N. Yan, S.A. Khan, Green Chem. 16, pp. 4654-4658, 2014.

[3] S.K. Yap, Y. Yuan, L. Zheng, W.K. Wong, N. Yan, S.A. Khan, J. Flow Chem. 4, pp. 200-205, 2014.

[4] M. Al-Rawashdeh, L.J.M. Fluitsma, T.A. Nijhuis, E.V. Rebrov, V. Hessel and J.C. Schouten, Chem. Eng J., 181-182, 549 (2012).

[5] "The Physics of Pulsatile Flow," M. Zamir, Biological physics series, Springer, 2000

416384 Multi-Step Continuous GMP Production of a Highly Potent Compound

Abstract:

The development of a GMP continuous process to synthesize a highly potent compound is disclosed. A dedicated, inexpensive equipment set run in laboratory fume hoods was utilized on manufacturing scale with in-process monitoring. This process included four unit operations including a continuous crystallization and filtration. Two chemical reactions were run plug flow tube reactors. One of these PFRs was designed to strip off a gas byproduct continuously during reaction and decouple liquid residence time from gas residence time. Continuous anti-solvent driven crystallization was done using two MSMPRs in series. Continuous filtration, washing, drying, and re-dissolving eliminated the isolation of a potent intermediate. Throughput was 4 kg per day in-situ penultimate from the GMP manufacturing plant. Finally, the robustness of the process and a comparison of lab and commercial scales highlighting the challenges that were encountered will be discussed.

416697 Continuous Processing: Application to Process Design and Material Supply

Abstract:

Continuous processing represents a powerful methodology that can be applied both to process design and material supply from reactions to isolations. Within process design, augmenting typical approaches with continuous operation enhances the mapping of the design space. Specific points within the design space, represented by each steady state, can be probed and characterized generating deeper understanding and allowing for the extraction of kinetic information not otherwise readily accessible. Continuous operation also increases the options with regard to material supply offering advantages with regard to impurity profile, control of solid state attributes such as PSD and form, capacity and safety. However, given that continuous processing is still within its infancy with regard to adaption by the pharma industry, there are significant technical and regulatory challenges to be overcome to successfully supply material. A number of case studies applying continuous processing to process design and material supply will be presented.

416924 Preparation of Enteric Capsules Using a Fugitive Salt

Abstract:

Preparation of Enteric Capsules Using a Fugitive Salt

Jin Zhao[1], Roland Adden[2], Robert Schmitt[1], Wes Spaulding[1], Uma Shrestha[1], Debbie Holbrook[1], Josh Imperial[1], Shrikant Khot[1]

(1) Pharmaceutical Excipients R&D, Dow Pharma & Food Solutions, Midland, MI 48674

(2) Food & Nutrition R&D, Dow Pharma & Food Solutions, Bomlitz, ND 29699, Germany

To protect acid sensitive active pharmaceutical ingredients (API), current technology typically involves coating multi-particulates, capsules or tablets with a known enteric polymer. Such coating processes usually involve organic solvents that are flammable, toxic, and explosive, or involve aqueous dispersions that are hard to handle, tend to block coating nozzles and have poor stability(*1-3*). To overcome these drawbacks, films and capsules were directly prepared using aqueous enteric solutions based on a fugitive salt.

Films and capsule shells were prepared using aqueous hypromellose acetate succinate (HPMCAS) solutions based on a model fugitive salt, ammonium bicarbonate. The rheology of such aqueous solutions was adjusted to fit manufacturing requirements by varying the polymer concentration and the addition of triethyl citrate. The resulting films demonstrated comparable mechanical strength to films made from organic solutions. The API diffusion across the film also demonstrated acid resistance that satisfied USP enteric requirement.

HPMCAS capsules were produced by a dipping process where the aqueous solution with model fugitive salt enhanced the dipping and drying process (Figure 1a). The residual by-product of the neutralization reaction such as H_2CO_3 can act as plasticizer via association with the succinic groups – C(O) – R – COOH. When the aqueous composition is dried on the molding pin, the residual ammonium bicarbonate, decomposes at the drying temperature. Accordingly, the evaporation of NH_3, CO_2 and H_2O facilitates gelation of the polymer on the mold pin resulting in a homogeneous film. The resulting dried capsule does not comprise substantial amounts of residual alkaline material. Even though the gel strength of the aqueous HPMCAS solution is significantly lower than that of HPMC solutions, the low gelation temperature and quick formation of the HPMCAS film leads to a straightforward capsule manufacture technique advantaged over the current state of the art (*4*). The dissolution of the enteric capsules and subsequent API release also demonstrated the necessary acid resistance to satisfy USP enteric requirements (Figure 1b).

The findings from this research can help pharmaceutical scientists utilize truly enteric capsules as opposed to enteric coated conventional capsules. This can protect APIs from degradation in the acidic gastric environment, prevent irritation of the gastric mucosa that often leads to ulceration, and achieve delayed and targeted release in the small intestine.

(a)

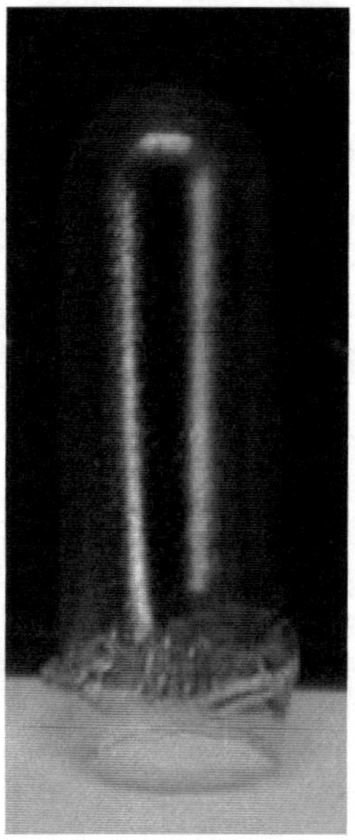

(b)

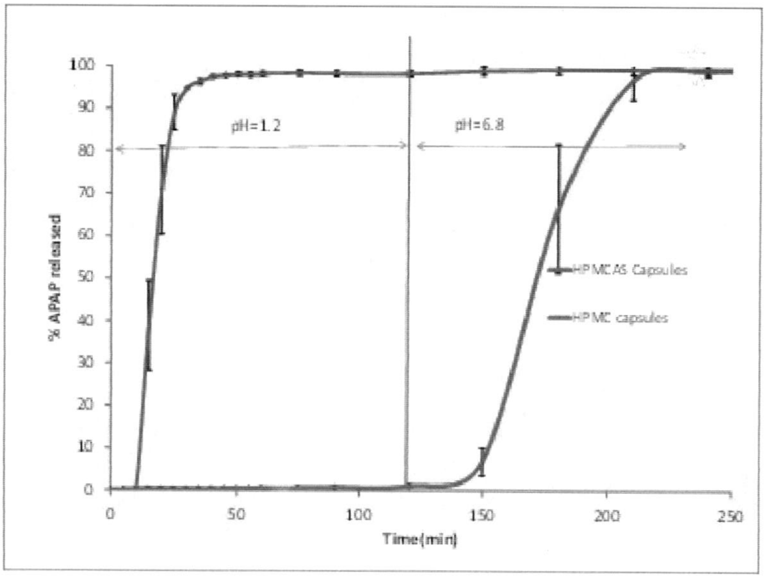

Figure 1. Enteric capsules made from HPMCAS aqueous solutions (a) and demonstrated enteric performance acetaminophen loaded capsules (b).

Acknowledgements

The authors wish to thank Shrikant Khot for his insights and suggestions throughout this research.

References

1. M. A. Osman, R. B. Patel, D. S. Irwin, P. G. Welling, Absorption of theophylline from enteric coated and sustained release formulations in fasted and non-fasted subjects. *Biopharmaceutics & drug disposition* **4**, 63-72 (1983).

2. M. A. Osman, R. B. Patel, D. S. Irwin, P. G. Welling, Absorption of theophylline from enteric coated and sustained release formulations in fasted and non-fasted subjects. *Biopharmaceutics & drug disposition* **4**, 63-72 (1983)

3. P. Mura, F. Maestrelli, M. Cirri, M. L. Gonzalez Rodriguez, A. M. Rabasco Alvarez, Development of enteric-coated pectin-based matrix tablets for colonic delivery of theophylline. *Journal of drug targeting* **11**, 365-371 (2003).

4. D. Cade, H. straub. (Capsugel France SAS, WO2013/164121).

419844 High-Throughput Polymer Screening for Oral Drug Delivery: From Discovery to Design

Abstract:

Spray-dried dispersions are fascinating polymer–drug mixtures that exploit the amorphous state of a hydrophobic drug to dramatically elevate its aqueous solubility above equilibrium for oral administration. However, these materials have limited fundamental understanding of what polymeric attributes prolong drug supersaturation and combat crystallization in solution. We present an automated protocol to combinatorially explore a vast architectural parameter space for five polymer platforms and elucidate how specific microstructures dictate drug dissolution. Over 60 reversible addition-fragmentation chain transfer (RAFT) polymerizations were conducted in three high-throughput experiments, varying constituent properties such as polymer amphilicity, thermoresponsive phase behavior, and hydrogen bonding capability. The intermolecular interactions of well–defined lead compounds with model drug phenytoin were screened and studied in vitro. This study reports the first results of a polymer–focused screening assay that successfully solubilized an otherwise-intractable drug for oral delivery.

429258 Improving Drug Delivery By Employing Terahertz Technologies to Monitor Manufacturing Processes

Abstract:

Terahertz spectroscopy and imaging are powerful new techniques for monitoring of pharmaceutical manufacturing processes. A number of parameters that critically affect drug delivery are able to be measured and continuously evaluated by these methods. These include bulk physical properties such as density, porosity, solid fraction and hardness as well as chemical properties such as crystalline/amorphous state of an API or blend uniformity. Furthermore, the unique ability of terahertz light to penetrate deeply into opaque materials allows one to determine the thickness and uniformity of functional coatings on tablets.

Such methods are easily integrated into the pharmaceutical manufacturing process. Their rapid and non-destructive nature enables them to be employed in-line to assess and monitor key Critical Quality Attributes (CQAs) and maintain product quality.

This talk gives an overview of a number of applications where CQAs critical to effective drug delivery have been measured using these techniques. These include:

- Continuous, in-line monitoring of roller compaction ribbon density
- Measurement of tablet porosity and hardness and their correlation to dissolution properties
- Rapid and precise quantification of percent crystallinity of API in a mixture
- Determination of coating thickness and uniformity in functionally coated tablets

Due to these diverse applications and simplicity of use, Terahertz Process Analytical Technology (PAT) is rapidly becoming a powerful and effective tool for obtaining optimal and consistent drug delivery characteristics.

423774 Hot Melt Coating with Lipid Excipients for User-Friendly Solid Oral Dosage Forms

Abstract:

Hot melt coating with lipid excipients for user-friendly solid oral dosage forms

[1]Diogo Gomes Lopes, [2]Karin becker, [3]Detlev Haack, [4]Michael Stehr, [2]Andreas Zimmer, [1]Sharareh Salar-Behzadi*

1- Research Center Pharmaceutical Engineering (RCPE) GmbH, Graz, Austria

2- Institute of Pharmaceutical Science, Department of Pharmaceutical Technology, University of Graz, 8010 Graz, Austria

3- HERMES ARZNEIMITTEL GmbH, Großhesselohe, Germany

4- CREMER OLEO GmbH & Co. KG, Witten, Germany

The simple and effective orally delivery of an active pharmaceutical ingredient (API) is an important factor influencing treatment efficacy, tolerability and patient compliance. To further improve the user-friendliness and acceptance of solid oral dosage forms, multiparticulate systems are becoming more common. As they have proven especially popular with patients requiring regular doses or those that find it difficult to swallow solid tablets, such as children, the elderly or individuals suffering from esophagitis. However, for this method to prove effective, most APIs must be effectively coated. Traditional methods to coat solid oral dosage forms mostly use solvents, an approach that suffers from a range of drawbacks including environmental concerns, lengthy processing times and the risk of dissolving drug. Hot melt coating (HMC) is an innovative and effective alternative that is reliable, cheap and fast, while simultaneously offering an opportunity to better manipulate and control characteristics such as taste, stability and release rate.

In this work, multiparticulate systems of different APIs are coated with lipid formulations via HMC. Drug release profiles and the solid state of triacylglycerides (TAGs) based excipients are monitored over storage. Small angle X-ray scattering, scanning electron microscopy and differential scanning calorimetry are used to characterize polymorphism, surface disruption (blooming) and complex networks of crystallites.

Our results show that surface blooming does not play a critical role in the drug release from TAGs coatings, but that the apparent diffusion coefficient of drugs can be dramatically reduced by increasing TAGs crystallite size and resulting tortuosity. This work brings new insights on the drug release from solid lipid dosage forms being an important step towards avoiding conventional complex formulations and preventing the overuse of excipients with unknown toxicity.

428444 Evaluation of Static or Dynamic Curing Process for Improved Controlled Release, Dry Powder Coatings of Ethylcellulose Using a Rotor Coater

Tuesday, November 10, 2015: 10:10 AM
Ballroom D (Salt Palace Convention Center)

Shawn Engels[1], Cody A. Schoener[2], Uma Shrestha[3], Brian Jensen[1] and Kody Bellach[4], (1)Freund-Vector Corporation, Marion, IA, (2)Dow Pharma & Food Solutions R&D, The Dow Chemical Company, Midland, MI, (3)Dow Pharma and Food Solutions, The Dow Chemical Co., Midland, MI, (4)Freund-Vector Corporation, marion, IA

Evaluation of Static or Dynamic Curing Process for Improved Controlled Release, Dry Powder Coatings of Ethylcellulose using a Rotor Coater

Shawn Engels[‡]*, Cody Schoener*[†]*, Uma Shrestha*[†]*, Kody Bellach*[‡]*, and Brian Jensen*[‡]

[‡]*Freund-Vector Corporation, Marion, IA 52302, USA*

[†]*The Dow Chemical Company, Midland, MI 48674 USA*

Purpose: To assess how different curing conditions influence controlled drug release from ethylcellulose, high productivity coatings using a rotor process technology. The rotor process technology enables the user to achieve high levels of weight gain in a short period of time while still demonstrating sustained drug release with micronized ethylcellulose. For this work, static and dynamic curing was assessed at different temperatures and lengths of time to improve film coalescence to provide more stable dosage forms. Static curing was completed in an oven while dynamic curing was completed in the rotor coater. Optimized curing conditions and times will enable formulators to achieve the most robust product.

Methods: Acetominophen (APAP) was dry powder coated onto sugar spheres (#20 – 25; Colorcon Inc., USA) using polyvinyl pyrrolidone (K2932, ISP Inc., USA) as binder on a Granurex GXR-35 rotor process (Freund-Vector Corp., USA). A drug to binder ratio of 97:3 w/w was maintained. Experimental ethylcellulose samples (The Dow Chemical Company, USA) were dry powder coated onto the APAP beads with triethyl citrate (TEC) as a plasticizer using the same rotor process as above. An ethylcellulose to TEC ratio of 3:1 w/w was maintained throughout the coating. Once the APAP sugar spheres were coated to 20% weight gain of ethylcellulose, samples were either removed and static cured or kept in the rotor coater and cured. For static curing, samples were placed on trays and placed inside an oven for 2 hours at 60°C. For dynamic curing, samples continued to rotate at 250 rpm while forced, hot air was applied from above at 40 or 60°C between 15 and 60 minutes. Dissolution and SEM were evaluated for curing influence on drug release. USP Apparatus I (baskets) was utilized for in-vitro dissolution studies and drug release data was compared using similarity factor (f_2) analysis.

Results: Initial experiments demonstrate that dynamic curing provided more controlled drug release compared to static curing (Figure 1A). Dynamic curing for at 40°C for 15 minutes was significantly slower (by f_2 analysis) in drug release compared to static curing at 60°C for 2 hours. This indicated the combination of the force hot air plus the frictional movement caused by the rotor bed movement enables improved film coalescence versus a static tray of sample placed in the oven. Dynamic curing at 40°C for 15 minutes versus dynamic curing at 60°C for 60 minutes was similar based on f_2 analysis indicating a short cure time of at least 15 minutes might provide full film coalescence and a stable coating. Further curing assessment on drug release will be presented on samples which were placed on stability for 1, 3, and 6 months in storage conditions of either 30°C and 65% relative humidity or 40°C and 75% relative humidity.

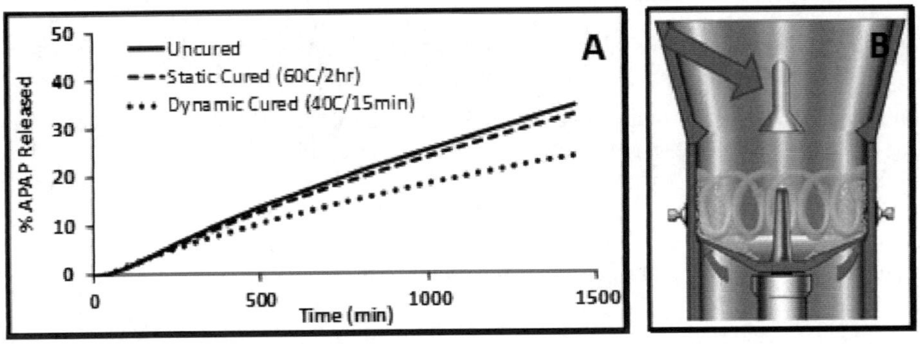

Figure 1. (A) Influence of static vs dynamic curing on APAP drug release and (B) cross-sectional view of rotor coater. Hot air is forced upon the product bed from above (large arrow at top).

Conclusion: Using the rotor process and force hot air, formulators can achieve sustained release and fully coalesced films by dynamically curing. Dynamic curing as short as 15 minutes at 40°C can significantly reduce curing times compared to static curing. The combination of dynamic curing and dry powder coating using the rotor process and micronized ethylcellulose also provides high productivity opportunities for pharmaceutical companies when compared to traditional organic or aqueous based ethylcellulose wurster spray coatings.

423668 Gluing Pills: A Novel Approach for Processing of Multi-Layer Tablets

Abstract:

Gluing Pills: A Novel Approach for Processing of Multi-Layer Tablets

[1]Sharareh Salar-Behzadi, [1]Sandra Stranzinger, [2]Josef Lingitz, [1]Stephan Sacher, [1,3]Johannes Khinast

1- Research Center Pharmaceutical Engineering (RCPE) GmbH, Graz, Austria

2- M & R Automation GmbH, Graz, Austria

3- Institute for Process and Particle Engineering, Graz University of Technology, Inffeldgasse 13/3, 8010 Graz Austria

Multi-layer tablet have several advantages over conventional tablets. For example, multi-layer tablet dosage forms can be designed as osmotically-controlled tablets, fixed-dose combinations for delivering incompatible APIs, controlling the releases of multiple APIs at different rates and releasing actives at different sites.

In the production of multi-layer tablets it is critical to control the weight of each layer and to assure a good adhesion between layers. It is also desirable to achieve a clear demarcation plane between the layers, since this is not only appealing but also assures the absence of cross-contamination.

The conventional technology to produce multi-layer tablets is a special rotary press, where each layer has a special compression step. The weight control of this technology is rather complex, making use of either a force control system or tablet thickness. The adhesion between layers is dependent on the material properties and requires a certain surface roughness of the initial layer for particle interlocking and adhesion with the second layer. Furthermore, fines below an average size of 150 microns should be avoided to provide a sharp resolution between the two layers. Problems are often encountered. Inadequate individual layer weight control, disproportionate layers-weight ratio coupled with low drug loads, cross contamination between the layers, insufficient multi-layer tablet hardness and a tendency of delamination of the layers are some of the disadvantages of rotary tablet multi-layer compaction.

In this study we present an innovative approach for the processing of multi-layer tablets. Each layer will be compressed separately, in the form of thin tablet bodies. Then, these tablet bodies will be glued to each other in an automated process. The adhesion between layers depends mainly on the properties of the adhesion gel, physical properties of tablets and the homogeneity of the adhesive layer.

To investigate the impact of these properties, tablets were produced with different excipients and different compaction forces. The impact of potentially critical material attributes (CMA), such as the degree of the plasticity of excipients, particle size distribution, compaction force and punch velocity on critical quality attributes (CQA) of tablets (rawness, specific surface area, hardness and friability), were investigated. Afterwards, these CQAs together with the type of gel binder, its concentration and viscosity were considered as CMAs for glued tablets and their impact on the quality of the adhesion between layers was investigated. Moreover, the penetration and cross-contamination between layers have been studied using a worst case scenario, i.e., tablets with high concentrations of water-soluble API, glued with water soluble gel binder.

This innovative and fast gluing method offers significant advantages, e.g., the ability to control the weight and compaction force in each tableting process, thus avoiding the variability in tablet hardness, disintegration time and dissolution profile of each layer. Moreover, this approach enabled us to design multilayer-tablets with advanced performance and low risk of cross contamination between layers.

434371 Continuous Plug-Flow Crystallization with Anti-Fouling Control

Abstract:

While plug-flow crystallization (PFC) has been touted as a promising continuous pharmaceutical unit operation, it is prone to fouling or "encrustation", a phenomena by which uncontrolled crystallization occurs on the crystallizer's surface. This results in operational issues such as flow blockage, increased thermal resistance, and reduced supersaturation, which in turn lead to limited continuous operation and reduced crystal quality and yield. Here, we introduce a model-based and model-free anti-fouling control (AFC) via spatial and temporal heating and cooling cycle. The two methods are compared with optimized PFC operation without AFC under two conditions: (1) maximization of crystal growth in the presence of encrustation and (2) minimization of encrustation while maintaining sufficient crystal growth. Specifically, the study contrasts an open-loop control approach, analogous to quality-by-design (QbD), with a feedback control approach, referred to herein as quality-by-control (QbC) concept. The computational and experimental results highlight the pros and cons of the different AFC designs as well as demonstrates the effect of PFC designs in terms of number of temperature segmentations, the segment length and the cycling frequency between the heating and cooling cycles, to the control performance.

416773 Mechanistic Details and Process Implications of Azide Mediated Conversion Route of Carboxylic Acid to Isocyanate

Abstract:

Mechanistic Details and Process Implications of Azide Mediated Conversion Route of Carboxylic Acid to Isocyanate

Anuj A. Verma, Herbert Ogutu, Yuanyuan Li, Ann M. Czyzewski, Su Yu, & Xiu C. Wang

Process Research & Development, AbbVie Inc., North Chicago, IL

The conversion of a carboxylic acid to an isocyanate via the Curtius rearrangement is a commercially attractive route in the pharmaceutical and chemical industries. Traditionally, the carboxylic acid is converted to acyl azide at 20°C by using diphenyl phosphoryl azide (DPPA) as the azide source. The acyl azide is subsequently heated to enable Curtius rearrangement to form isocyanate. A safety concern is identified in handling a pot of acyl azide across scales of operation and it is desired to minimize the inventory of that azide. Herein, we describe a process which minimizes that inventory. The process synthesis was accomplished by developing a generalized azide chemistry/reaction engineering platform (attached to Curtius reactions) to critically evaluate and identify optimal processes which are inherently safe at different scales of operation. The platform involves a detailed mechanistic understanding of product formation along-with impurity generation, and represents a step above traditional kinetic modeling by *elucidating optimal reagent selection rules* in addition to traditional reaction engineering solutions to reduce impurity generation and increase the product formation rate.

AbbVie contributed to the design, research, and interpretation of data, writing, reviewing, and approving the publication. Anuj A. Verma, Herbert Ogutu, Yuanyuan Li, Ann M. Czyzewski, Su Yu, and Xiu C. Wang are AbbVie employees who worked on this project.

Validation of a New Product Introduction Involving Continuous Unit Operations in a Batch/Continuous Hybrid Process

Ashley Humenik, Process Engineer
Eli Lilly S.A. – Irish Branch

AIChE Annual Meeting, Salt Lake City
10th November 2015

Lilly

New Product Introduction

- Technical Transfer from Development to Manufacturing
- Validated at Kinsale in 2015
- 2 Continuous Unit Operations:
  - Flow Palladium Adsorption Removal in Packed Columns
  - Thermal Cleavage of the BOC protection group in Plug Flow Reactor

Continuous Process Flow Drawing

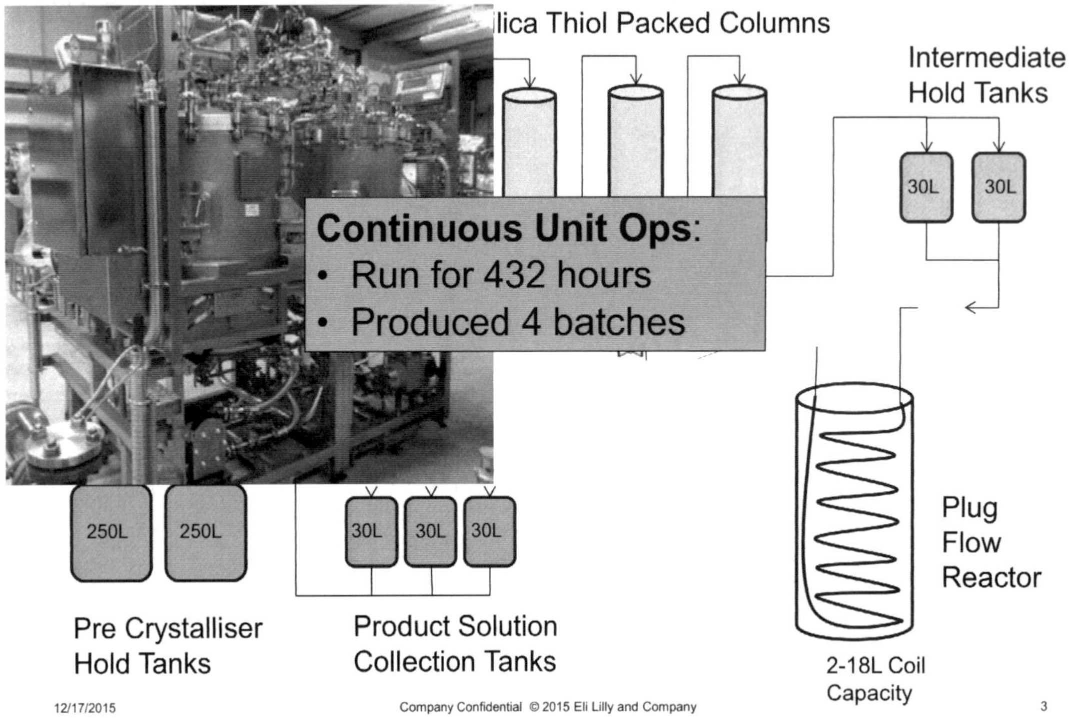

Palladium Removal

- Mechanistic model determined proven acceptable ranges (PAR)
- Two process parameters have multivariate interaction: Adsorption temperature & Superficial flow velocity (v)
- Assumptions: Campaign size & Equipment setup

$$v = \frac{L_C}{\tau_C}$$

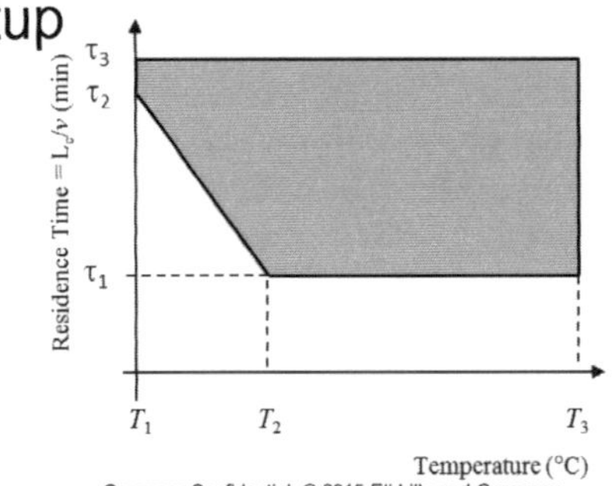

≥12.13kg silica thiol

Column Setup & Considerations

- Silica Thiol Particle Size Distribution (PSD)
- Silica Thiol Resin Stability
- Height of packing
- Flowrate
- Isothermal operation – Heat tracing installed
- Column flushing & Gas displacement
- Negligible axial dispersion (approaching plug flow, no bypassing)

Thermal BOC Cleavage

- Mechanistic model determined proven acceptable ranges (PAR)
- Two process parameters have multivariate interaction: Reaction temperature & Residence time
- Assumptions: Single liquid phase, Plug flow, Steady State

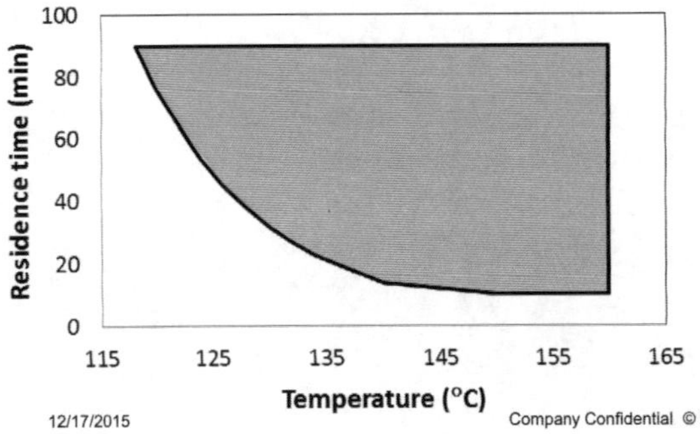

6L volume coil
3/8” Hastelloy tube

Characterisation of System Residence Time Distribution (RTD)

- Flushed system with pure THF until baseline was achieved
- Solvent Switch to 90:10 THF:H_2O
- FTIR used to determine residence time (τ) at 50% of the water response and system volume ($V = \tau \times q$)

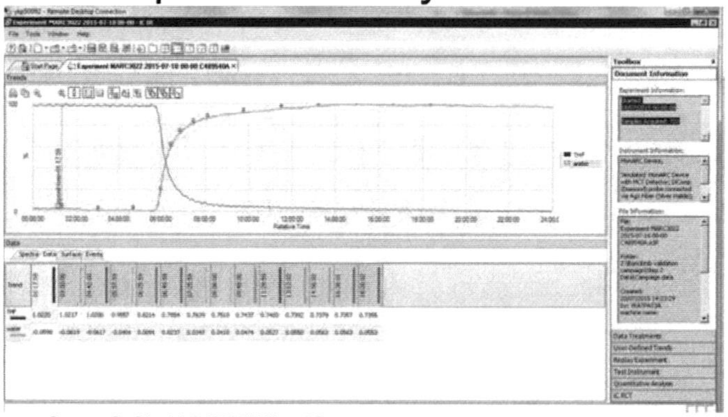

12/17/2015 Company Confidential © 2015 Eli Lilly and Company 7

Batch Definition

- Offline samples analysed using Karl Fischer (KF) Method
- Time tc ...)% of the origina
- Calcula ... ieve above
  - $N =$
- Calcula
  - $\frac{Sys}{Process\ Flowrate\ (\frac{ml}{min})}$ X N

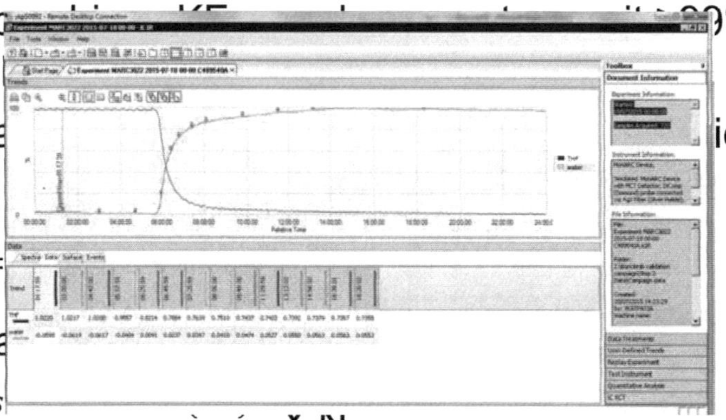

12/17/2015 Company Confidential © 2015 Eli Lilly and Company 8

Safety Considerations

- Solution transfer at high pressure (300psig) and high temperature (140°C)
  - Potential spill leading to fire or personnel injury
  - Potential overpressurisation of tubular reactor
  - Potential leak leading to vapour cloud explosion (VCE)

Unit Operations Setup & Considerations

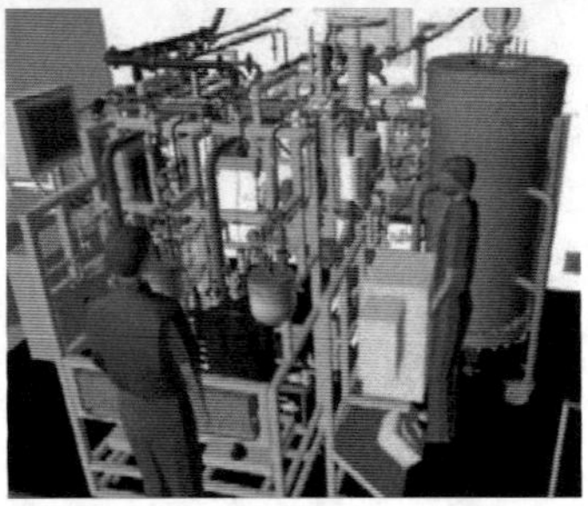

- Same mass flowrate
- Intermediate hold tanks
  - Not plug flow
  - Increased dispersion
  - Cascade control
  - Reactor flow range (not setpoint)
  - Minimum weight setpoint (2kg)
    - Reduced Batch to Batch Carryover
- 300 psig back pressure
  - Maintain evolved CO_2 in solution

Plant Operations & Process Monitoring

- Continuous = 'Intentional' Bottleneck
- Simultaneous activities:
  - Front end batch reactions
  - Continuous
  - Back end batch isolation

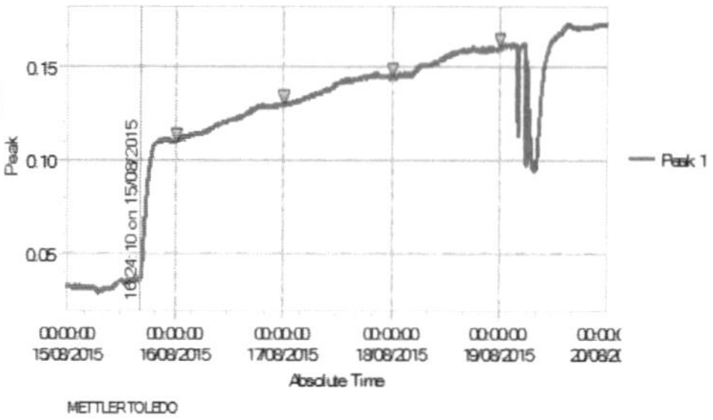

- 'DECISION' phase
  - Process monitoring of continuous unit ops

Unplanned Stoppage Considerations

- Stoppages will occur
  - Little time to decide best response
- Option to restart or push-out
  - Possible segregation of material
- Agree a decision tree to handle various cases
- 'Short stop' of <u>less than 30 minutes</u> duration
  - Testing completed during Solvent Run showed all process parameters remained within acceptable range.

Conclusion

- Successfully validated 2 manufacturing steps
  - 2 continuous unit operations
  - 4 batches
  - Quality material meeting specification
  - No safety issues

Acknowledgements

- E. Crick
- K. Seibert
- C. Luciana
- T. Wilson
- G. Flanagan
- C. O'Mahony
- S. Moursy
- R. Spencer
- R. Moylan
- A. Corrigan
- R. O'Sullivan
- M. Collins
- E. Cotter
- J. Cunningham
- C. Duffy
- B. Healy
- C. Healy
- B. Murphy
- P. Desmond
- S. Jeffery
- I. Mossop
- L. Maguire
- B. Walsh
- J. Cashman
- H. Moynihan
- V. Stallard
- K. Lydon
- P. Harrington
- P. Harrigan
- L. McCarthy
- J. Grace
- D. Smith

Questions?

gsk do more feel better live longer

The Development of an Integrated Control Strategy for a Continuous Primary Pharmaceutical Process

Hannah Robinson
GlaxoSmithKine

Applications of Continuous Processing in the manufacture of Pharmaceuticals: Drug Substance Part II
AIChE Annual Meeting Nov 2015

Agenda

gsk

- Introduction
- Case Study
  - Development to Industrialisation of a Flow Process
- Summary

Introduction

Introducing Continuous to Pharma

Batch

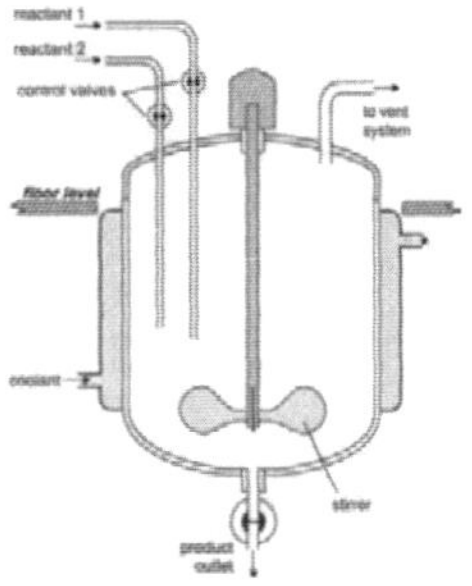

- Historical workhorse of Pharma
- Control is scale dependent
- Quality mindset batch dominated

Continuous

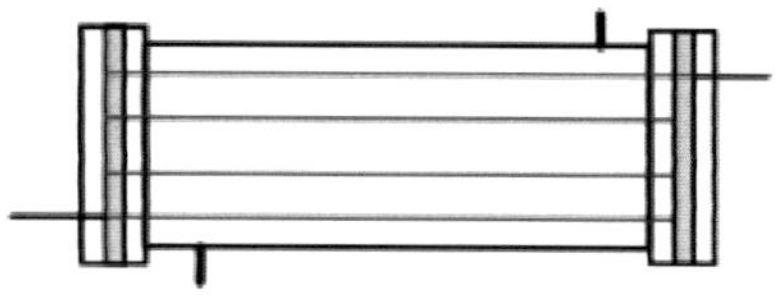

- Workhorse of other industries
- Uniform control easier to achieve
- Constant production provides challenge to traditional quality approaches

Case study – Development to Industrialisation of a Flow Process

Process Overview

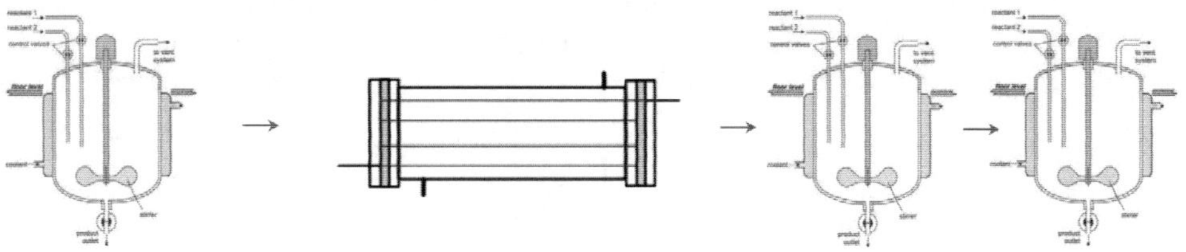

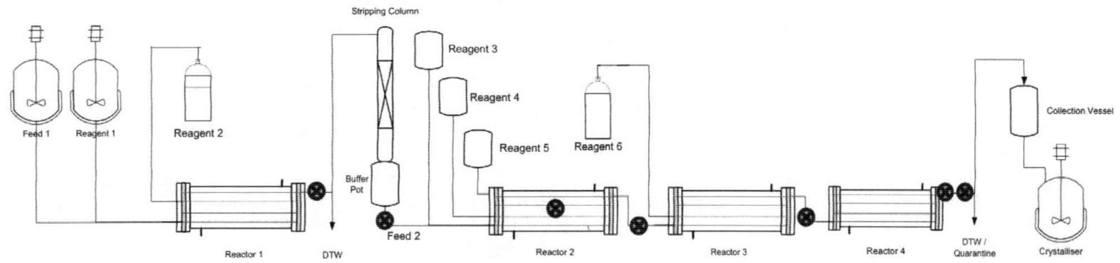

Process Understanding

- Kinetics
- Calorimetry
- Physical property behaviour of streams
- Mixing sensitivity

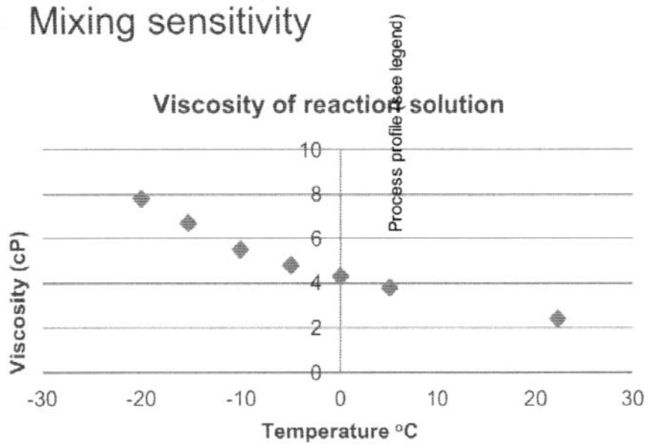

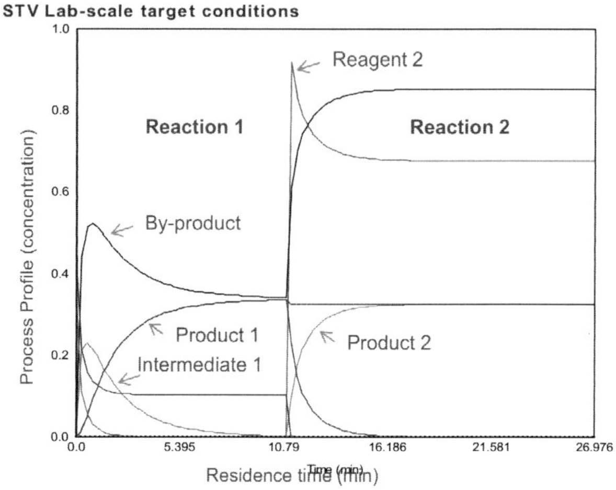

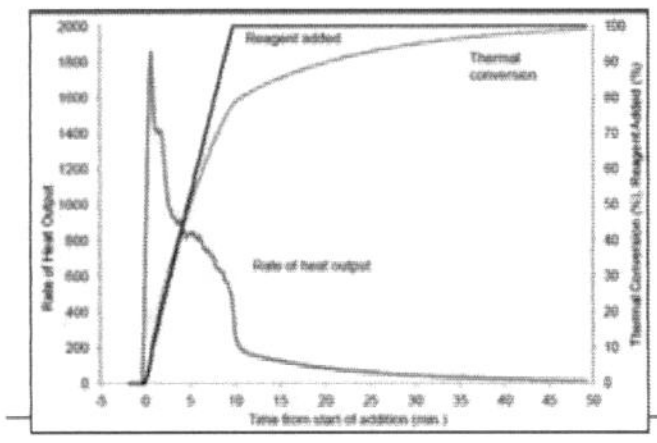

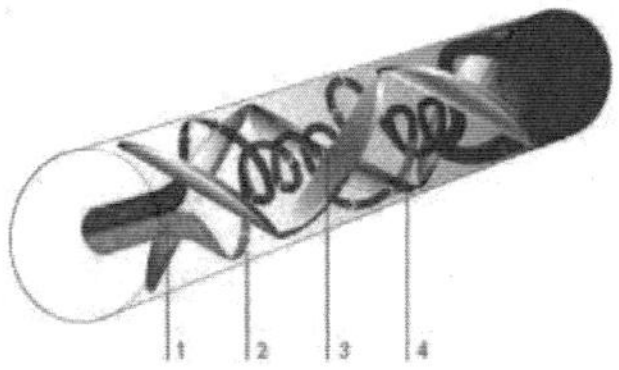

Equipment Characterisation

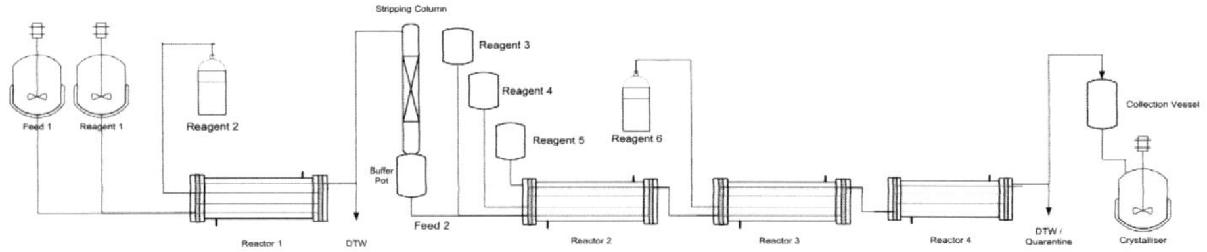

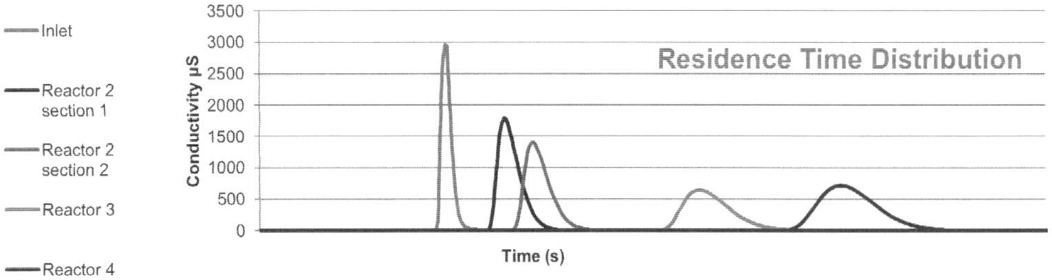

- Mixing intensity and Re
- Reactor volumes and surface area
- Heat transfer capability
- Residence time distribution

Lab processing data

Online UPLC

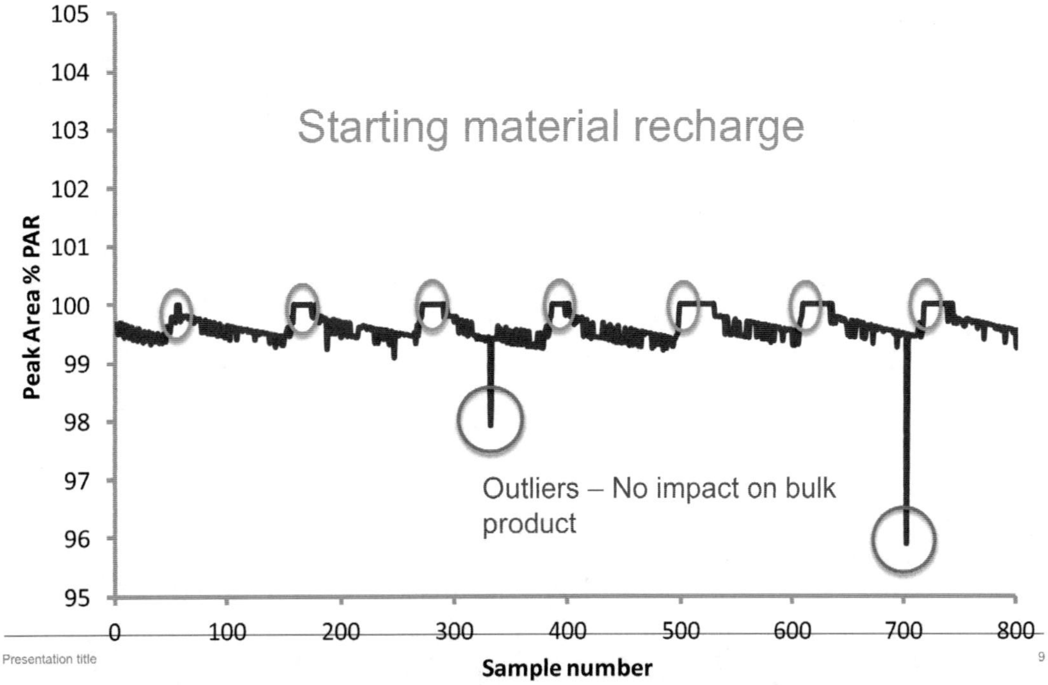

Presentation title 9

Transfer to Manufacturing

Operational controls

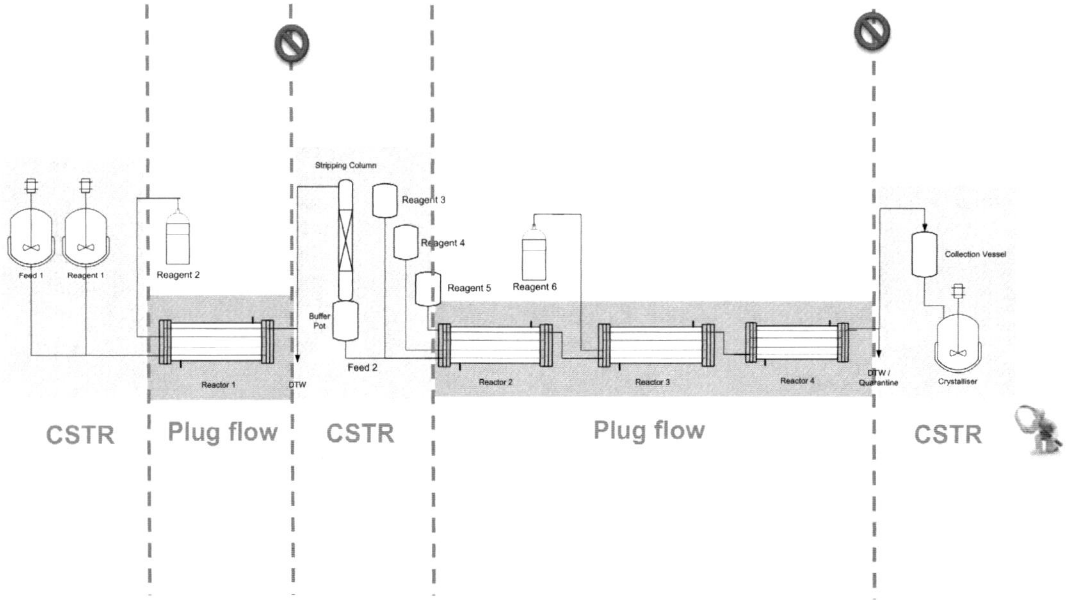

Operational controls

Axial Dispersion model

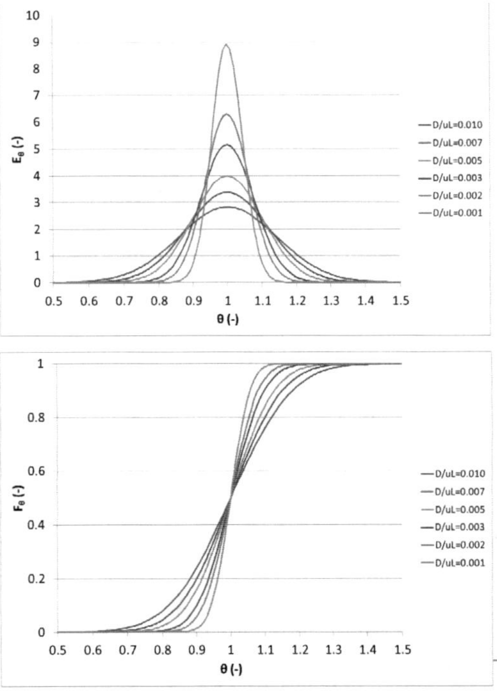

- Quality control based on combination of <u>dispersion and time</u> to meet a pre-defined acceptable quality
- Acceptable quality defined by process requirements of design space
- Characterisation of equipment system → D/uL < 0.01 for equations to hold

Totalised residual flow post residence time θ

D/uL	0.01	0.004	0.001
θ=1.0	135g	81g	47g
θ=1.1	48g	12g	0.9g
θ=1.2	12g	0.6g	<0.001g
θ=1.3	2g	0.01g	Trace
θ=1.4	0.2g	<0.001g	Trace

Control phases

Disturbances

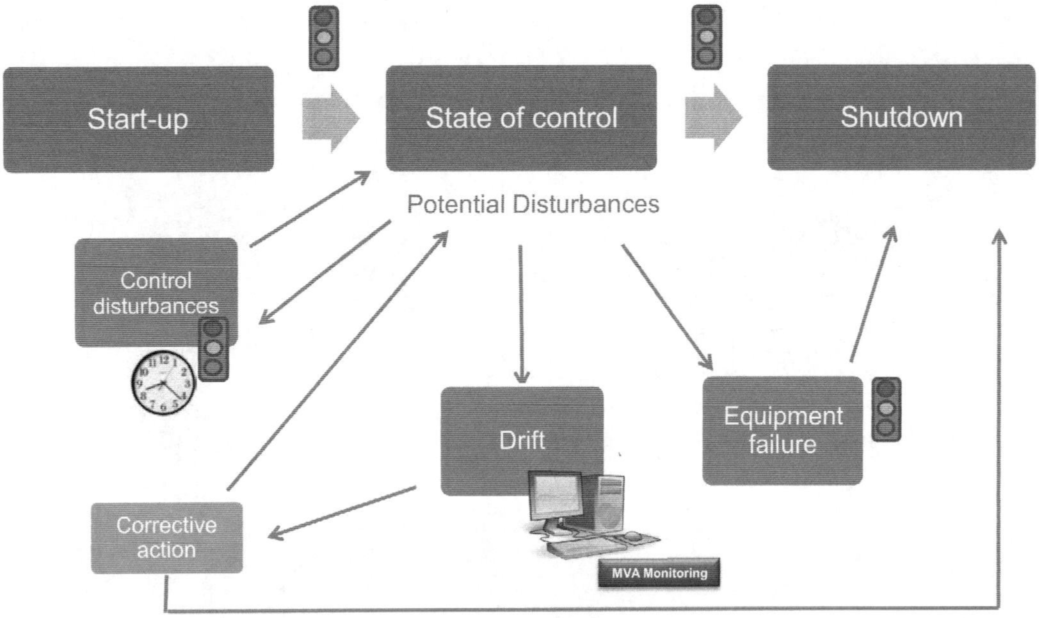

MVA monitoring

3 hours post model

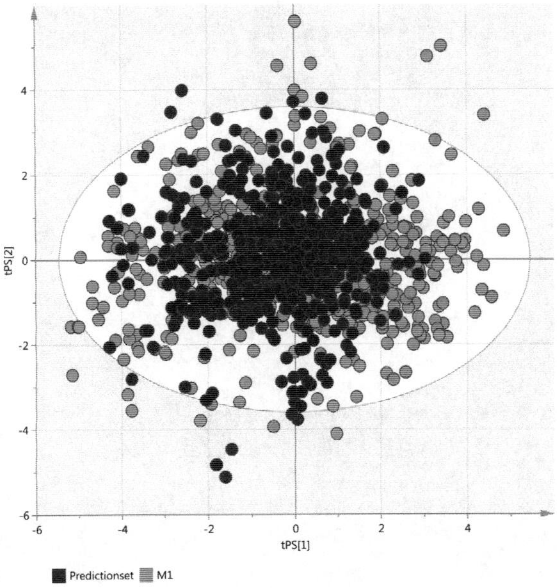

MVA monitoring

3 hours post model

14 hours post model

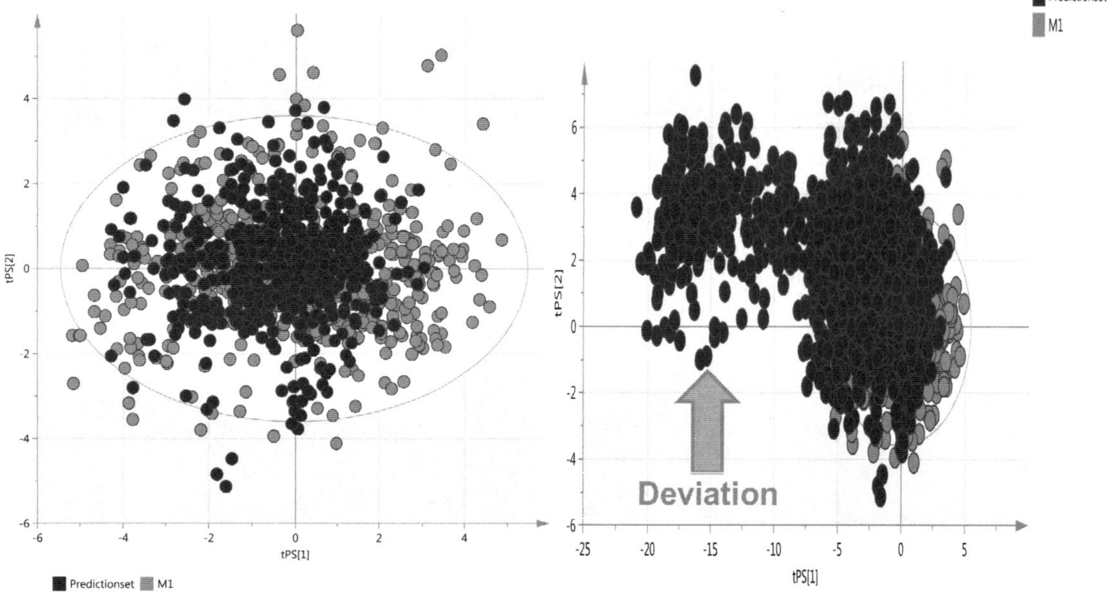

MVA monitoring

14 hours post model

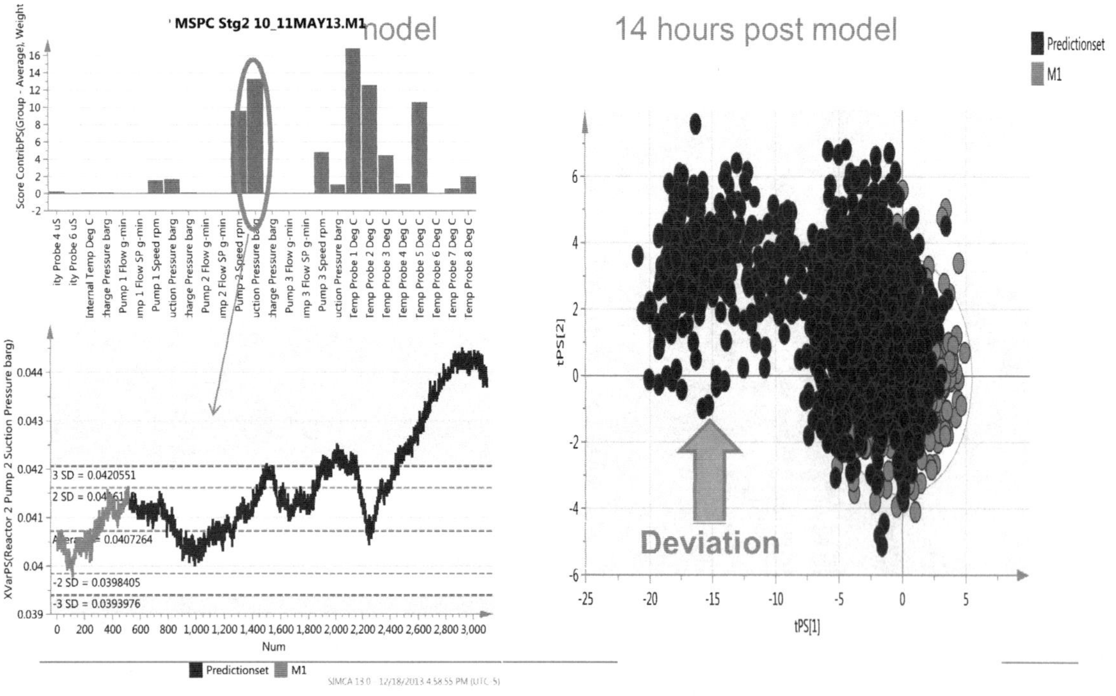

Control system infrastructure

Quality Traffic Light	Quality alarms activated depending on sequence phase Quality met? Yes / No → Collect / DTW
Ratio control	Active during controlled state, To maintain stoichiometry, To allow co-ordinated step up step down of each sectionof the process
Alarms, trips and interlocks / Sequence Phases	Safety system active throughout Sequence phases drive allowable activities
CM / EM	Individual control loops – maintain SP
MVA process monitoring	Trending of tags, to detect system drift → operational control Prompt for operations to look at control system to establish need for manual action
System Characterisation RTD	Providing timings for start-up, shutdown, alarm settings and DTW actions

Control trends

Start-up Sequence

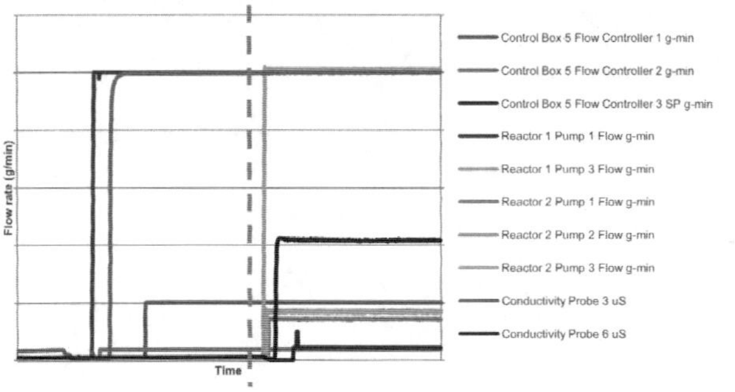

Controlled state

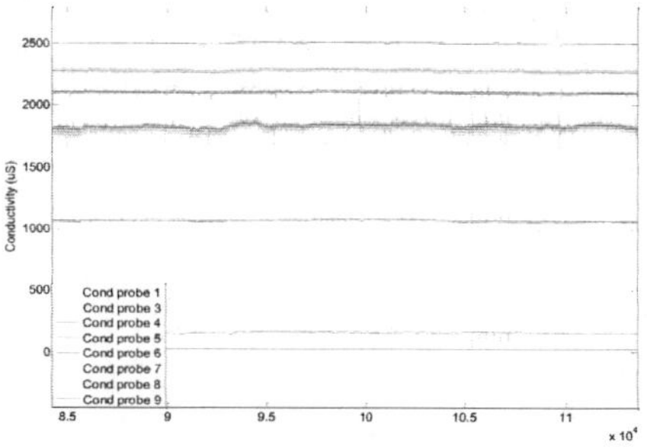

Control Failure

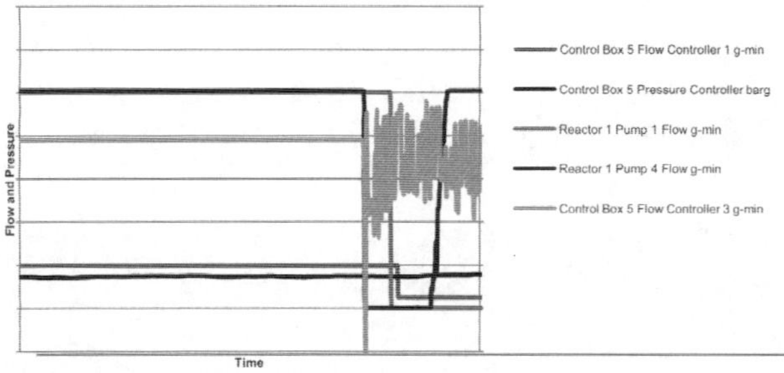

Control Blip

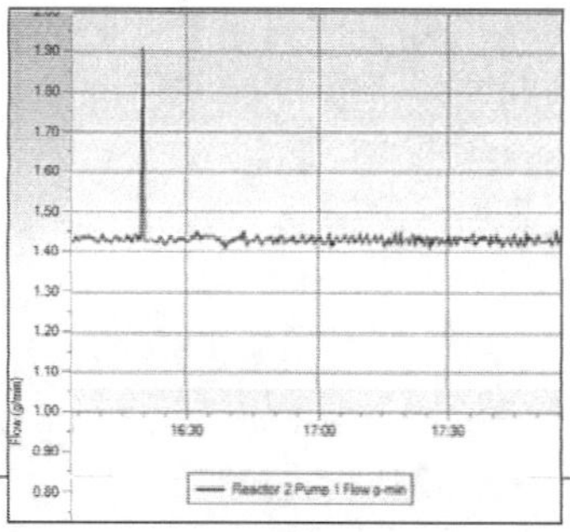

Summary

Summary

- Use of continuous technology processes within the Pharma industry requires a change in batch operation mindset
  - Some additional quality levers
  - Some additional challenges

- Continuous processes offer many opportunities for utilisation of combined process and equipment understanding
  - Predictive modelling
  - Scaling capability
  - Operational controls

- Continuous processes offer a wealth of data which can be used pro-actively to manage the process and equipment
  - Data display
  - Use of MVA

Manufacturing Installation

Acknowledgements

- Malcolm Berry
- Peter Shapland
- Mark Hughes
- Matt John
- Gill Turner
- Irene Areri
- Peter Hamilton
- Melanie Dumarey
- Yahyah Yahyah
- Neil Hodnett
- Andrew Rutter
- Gareth Alford
- Dave Johns
- Rob Wadley
- Heather Todd
- Peter Clements
- Ben Hagger
- Mathew Callaghan-Till
- Lesley Wong
- Wei-Feng Tan

427812 Development & Application of Continuous Crystallization in an Integrated Continuous Pharmaceutical Pilot Plant

Abstract:

Development & Application of Continuous Crystallization in an Integrated Continuous Pharmaceutical Pilot Plant

Haitao Zhang, Richard Lakerveld, Patrick Heider, Justin Quon, James Evans, Allan S. Myerson, Bernhardt Trout

Novartis-MIT Center for Continuous Manufacturing

Department of Chemical Engineering, Massachusetts Institute of Technology, E19-502b, 77 Massachusetts Avenue, Cambridge, Massachusetts 02139, USA.

Abstract

The concept of continuous manufacturing to address key challenges faced by the pharmaceutical industry has gained significant interest. A number of research programs related to this topic have been sponsored by pharmaceutical companies, including Pfizer, Novartis, Merck, and Eli Lilly, etc. The Novartis-MIT Center for Continuous Manufacturing has developed a seamless, end-to-end pilot plant for continuous manufacturing of pharmaceuticals including new pieces of technologies to support integrated continuous manufacturing (ICM). Continuous crystallization plays a critical role within ICM processes in terms of final product quality control. Crystallization processes are designed to obtain active pharmaceutical ingredients (API) crystals with controlled purity, size, shape, and polymorphic form in high yield. Knowledge of the continuous crystallization process conditions required to fabricate crystals with controlled characteristics is critical during process development. Real-time control using process analytical technology (PAT) tools is required for the implementation of continuous crystallization within integrated continuous manufacturing (ICM) of pharmaceuticals. However, appropriate selection of PAT tools is challenging, and the design and operation of automated control loops for continuous crystallization within a continuous pharmaceutical process brings forward important questions.

The aim of this contribution is to investigate the performance of a continuous crystallization step operated for a prolonged period of time within an ICM process. The integration of different unit operations and the steady-state operation of those unit operations are of primary concern. The aim is to keep the critical material attributes (CMAs) for each unit operation including the continuous crystallization units within an accepted range at all times through the control of critical process parameters (CPPs). Consequently, the control strategy should possess the capability to tolerate variations from upstream units and should minimize disturbances that propagate to downstream units. Therefore, the selection and implementation of PAT tools and the design of a control strategy are of vital importance for successful ICM of pharmaceuticals. Such a control strategy would rely on automated feedforward and feedback control loops.

Herein, we discusses the process design and operation of a continuous reactive crystallization of aliskiren hemifumarate as part of an ICM pilot plant. Several PAT tools were used within automated control loops to satisfy specifications on the critical materials attributes of the crystallization process. The operational performance of the process was maintained for periods of time over 100 h. The purity of the targeted product exceeded 99%, and the process yield reached 91.4% with the physical properties in specification.

Continuous and Discontinuous Anti-Solvent Crystallization:

An Integrated Real-time Process Monitoring and Modeling Approach

Huiquan Wu, Ph.D.

Office of Process and Facilities/Office of Testing and Research
OPQ/CDER/FDA
AIChE Annual Meeting, Salt Lake City, UT, Nov. 10, 2015

1

U.S. Food and Drug Administration
Protecting and Promoting Public Health
www.fda.gov

Disclaimer

The views and opinions expressed in this presentation are only of author, and should not be construed to represent FDA views or policies

2

Outline

- Why crystallization matters?
- Continuous and Discontinuous Anti-solvent Crystallization
- Modified classical nucleation theory model
- Conclusions
- Acknowledgement

3

Why Crystallization Matters?

- The drug substance's quality attributes are directly impacted by drug substance crystallization process
  - Crystallinity
  - Solubility
  - Purity
  - Size and Shape
  - Solid state and polymorphic forms
  - Hygroscopicity
  - Residual solvent level
  - Impurity level
- Link to the critical quality attributes (CQAs) of the final drug product

4

U.S. Food and Drug Administration
Protecting and Promoting Public Health
www.fda.gov

FDA Research on Crystallization Process

- Focus
  - Provide scientific basis and case studies to support FDA's Pharmaceutical Quality Initiatives
  - Gain Process Understanding and Product Understanding
- Work done so far
  - Model system: Naproxen-Eudragit L100-Alcohol-Water
  - Evaluate discontinuous and continuous crystallization
    - Discontinuous: Stepwise addition in multiple aliquots of a fixed amount of antisolvent
    - Continuous: Addition of the same amount of antisolvent at a fixed flow rate
    - All other conditions the same
  - Establish design space and nucleation mechanisms
  - Model construction and verification
- Work to be done in the future
  - Continuous crystallization with other formulation systems

5

U.S. Food and Drug Administration
Protecting and Promoting Public Health
www.fda.gov

An Integrated Approach to Pharmaceutical Crystallization

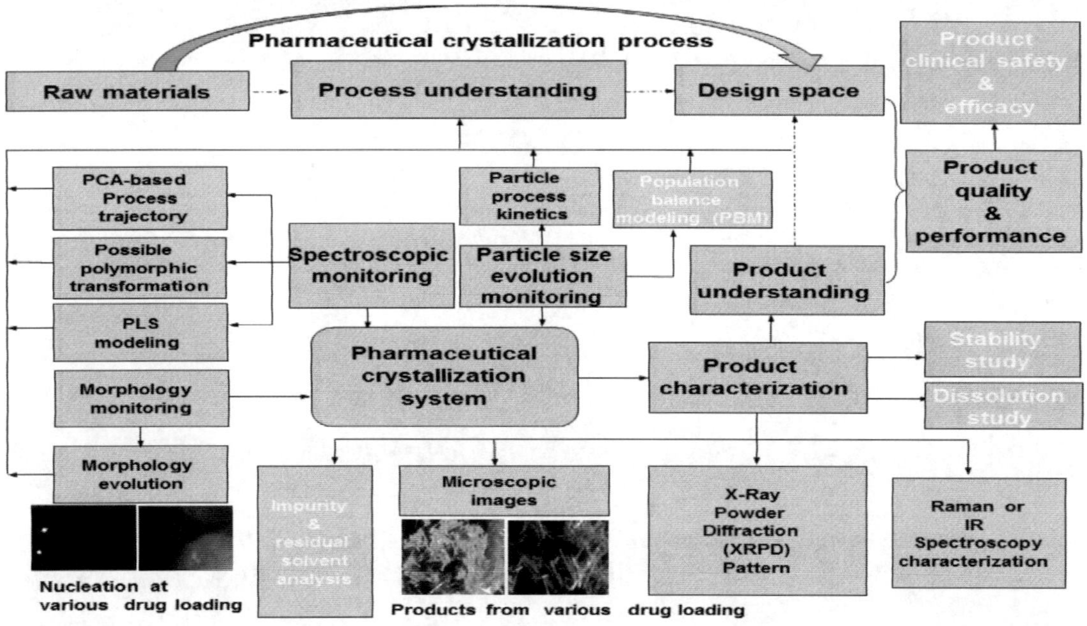

6

Schematic of Experimental Setup

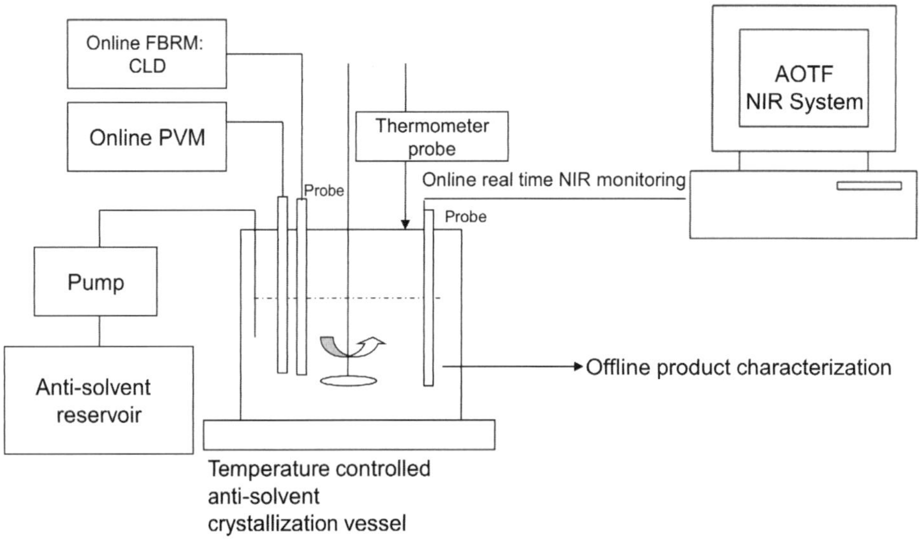

7

Real Time FBRM Chord-Length-Distribution

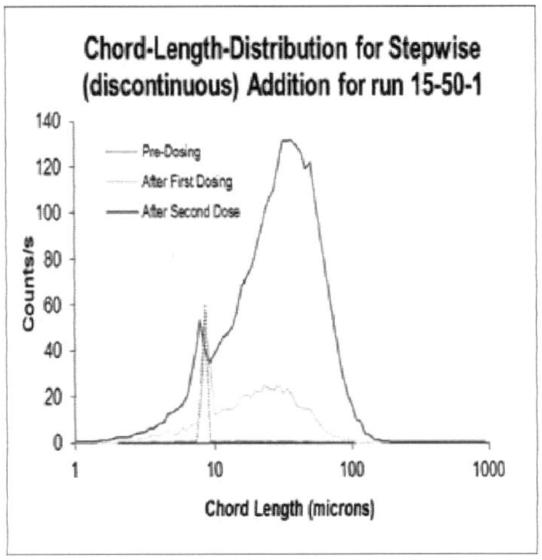

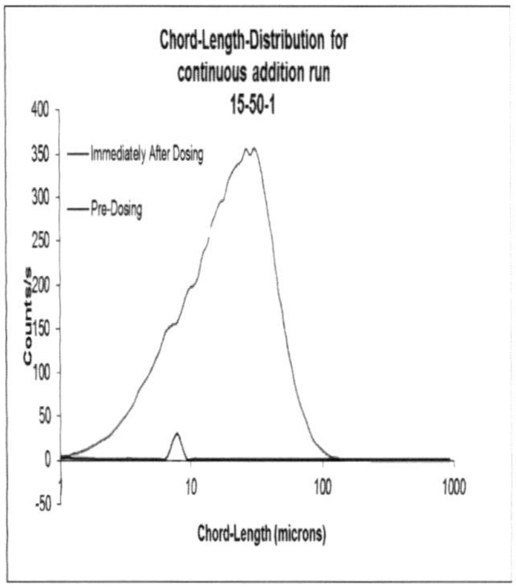

- Continuous addition mode demonstrated a faster generation rate for fine chords (i.e., fine crystals) than discontinuous addition mode

8

Real Time PVM Images

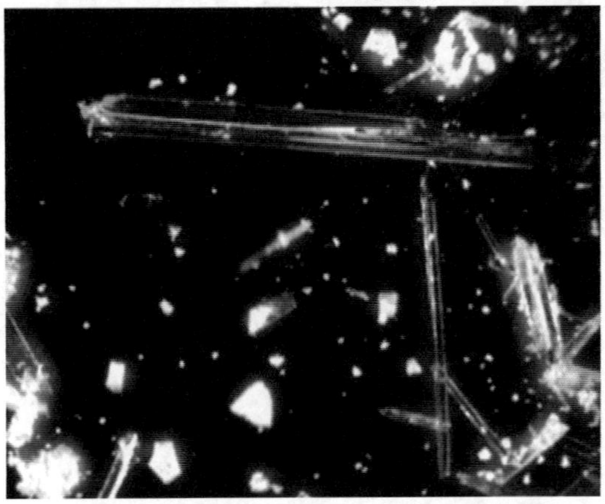

After discontinuous addition of 150 ml anti-solvent at 15^{0}C and 50 rpm

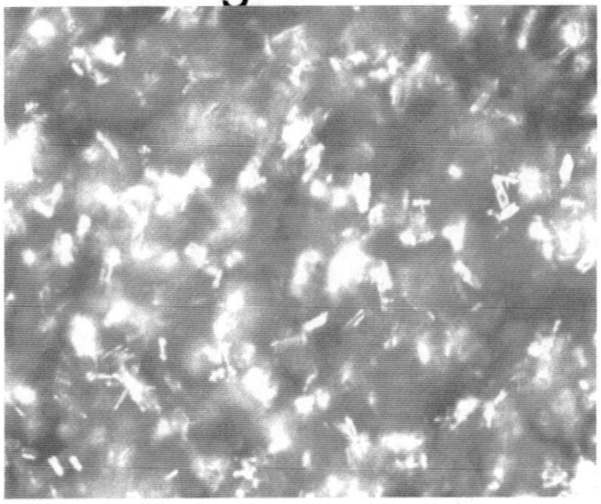

After continuous addition of 150 ml anti-solvent at 15^{0}C and 50 rpm

- Continuous addition mode can generate more uniform CSD than discontinuous addition mode =>a better controllability of the physicochemical properties of the crystalline product

9

Induction Time vs. Drug Loading (Discontinuous)

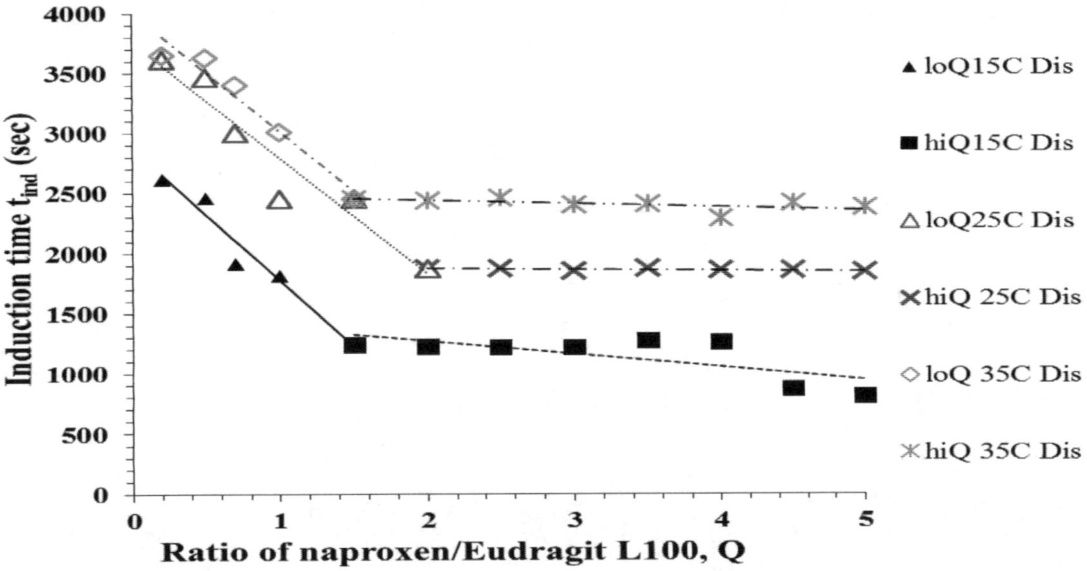

10

U.S. Food and Drug Administration
Protecting and Promoting Public Health
www.fda.gov

t_{ind} Measurement Verification (Discontinuous)

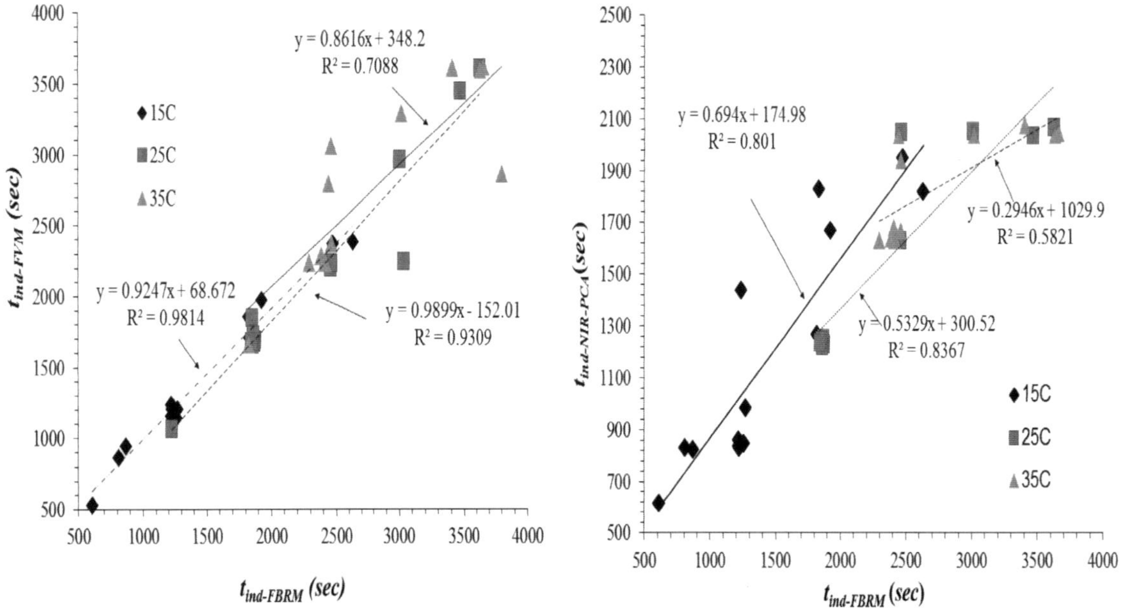

11

U.S. Food and Drug Administration
Protecting and Promoting Public Health
www.fda.gov

Classical Nucleation Theory (CNT) and Modified CNT (MCNT)

- Solubility calculated by Jouyban-Soltanpour-Acree Equation (2010)
- Super-saturation calculated by mass balance
- For homogeneous nucleation mechanism

$$\ln(t_{ind}V) \approx ln(\frac{N_c}{A_{hom,J}}) + \frac{16\pi\gamma^3 v^2}{3(kT)^3} \frac{1}{(lnS)^2}$$

- For heterogeneous nucleation mechanism

$$\ln(t_{ind}V) \approx ln(\frac{N_c}{A_{het,J}}) + \frac{16\pi\gamma^3 v^2}{3(kT)^3} f(\theta) \frac{1}{(lnS)^2}$$

12

U.S. Food and Drug Administration
Protecting and Promoting Public Health
www.fda.gov

$\ln(t_{ind\text{-}FBRM}V)$ vs. $(\ln S)^{-2}$ for Discontinuous

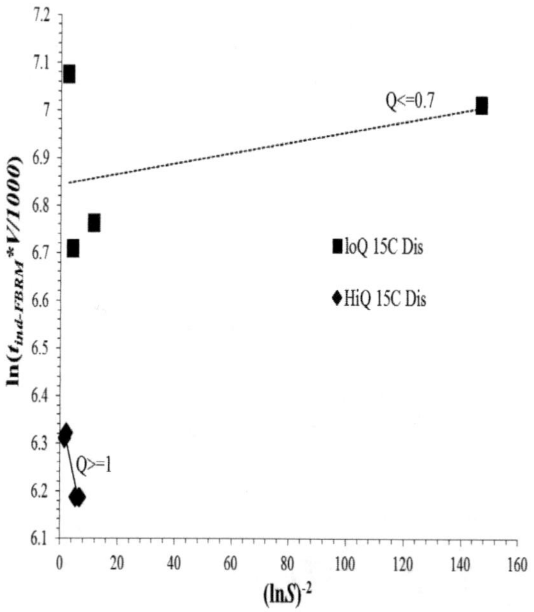

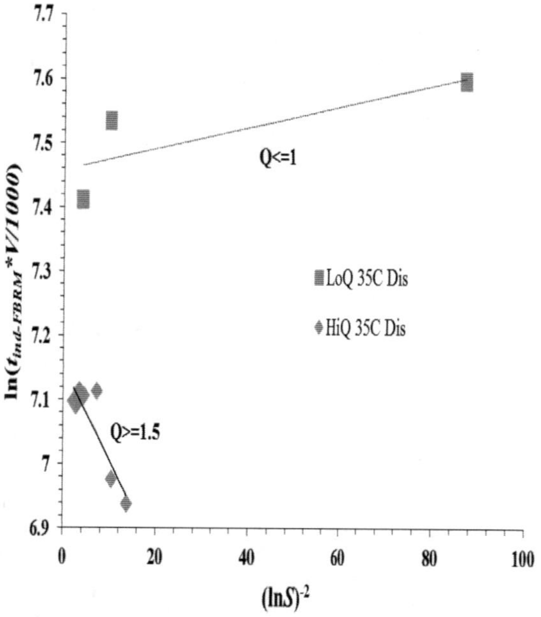

13

U.S. Food and Drug Administration
Protecting and Promoting Public Health
www.fda.gov

$\ln(t_{ind\text{-}NIR\text{-}PCA}V)$ vs. $(\ln S)^{-2}$ for Discontinuous

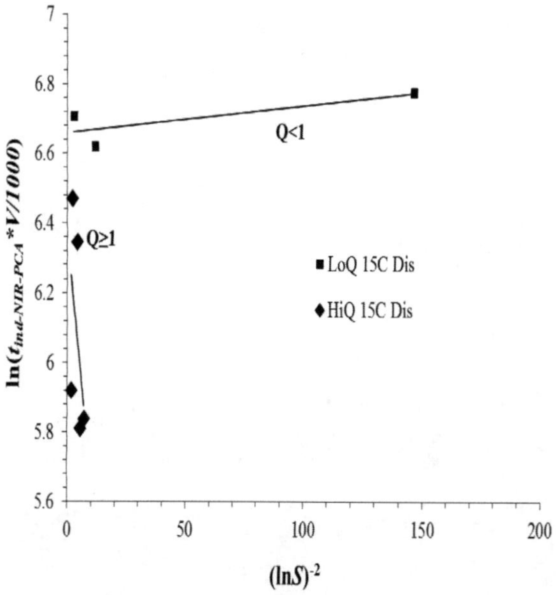

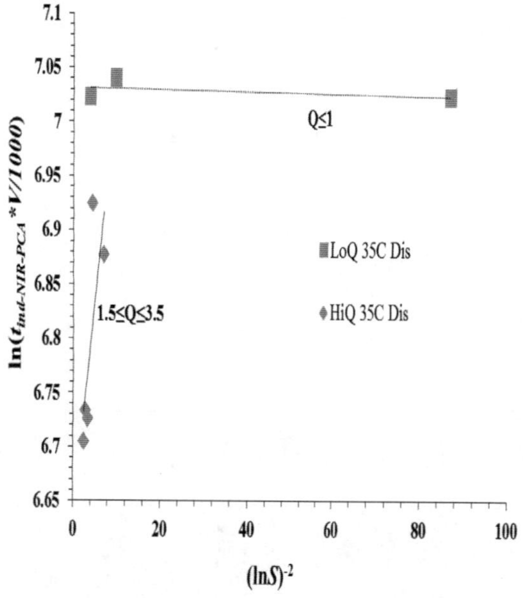

14

U.S. Food and Drug Administration
Protecting and Promoting Public Health
www.fda.gov

Continuous Addition of Anti-Solvent

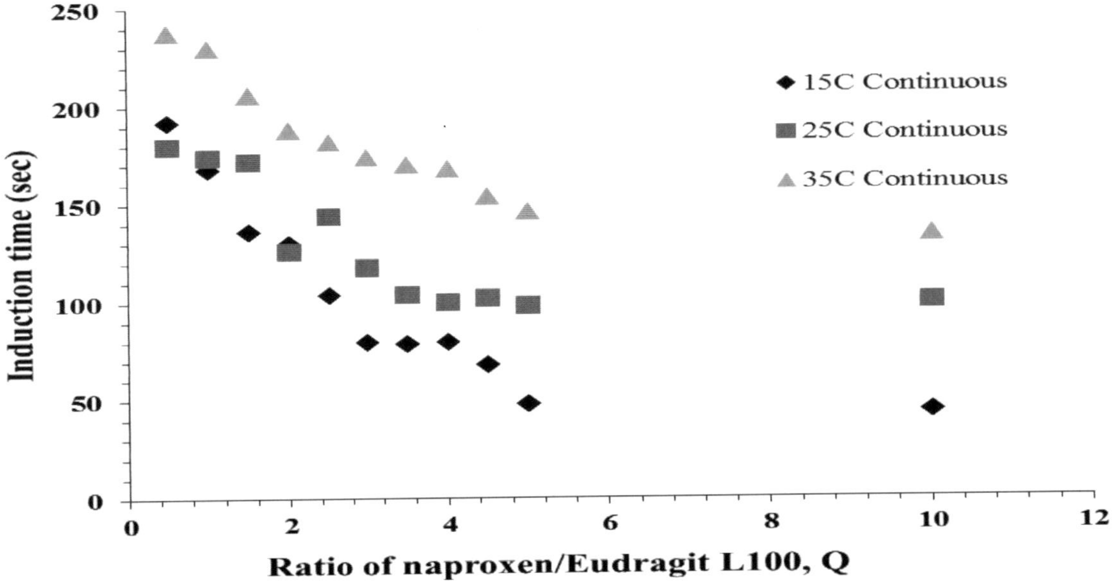

15

U.S. Food and Drug Administration
Protecting and Promoting Public Health
www.fda.gov

Continuous Addition: CNT

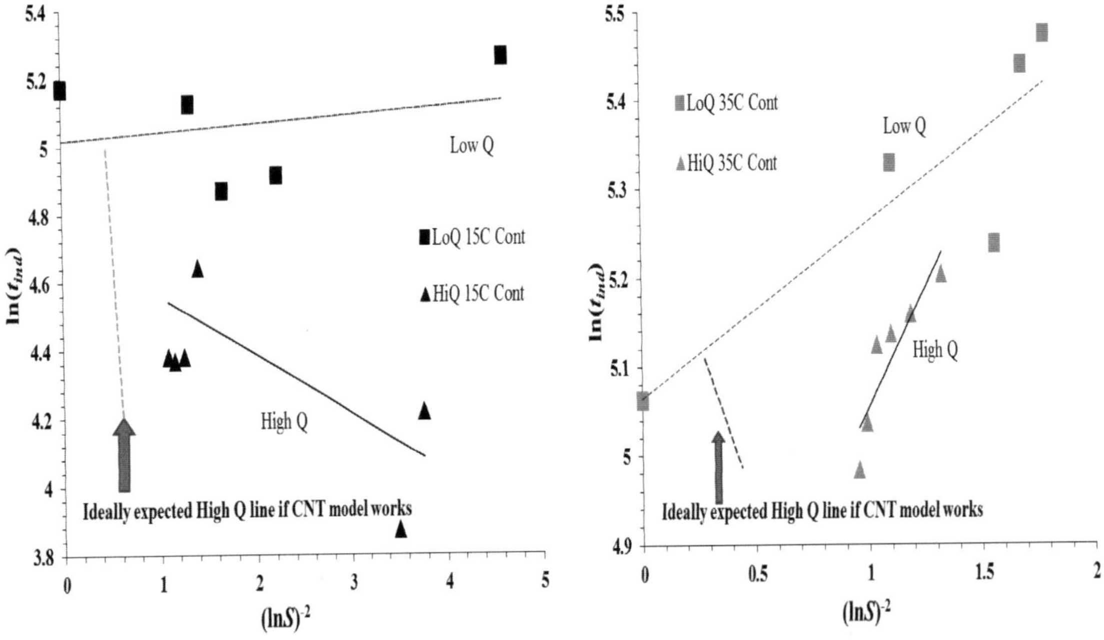

16

U.S. Food and Drug Administration
Protecting and Promoting Public Health
www.fda.gov

Continuous Addition: MCNT

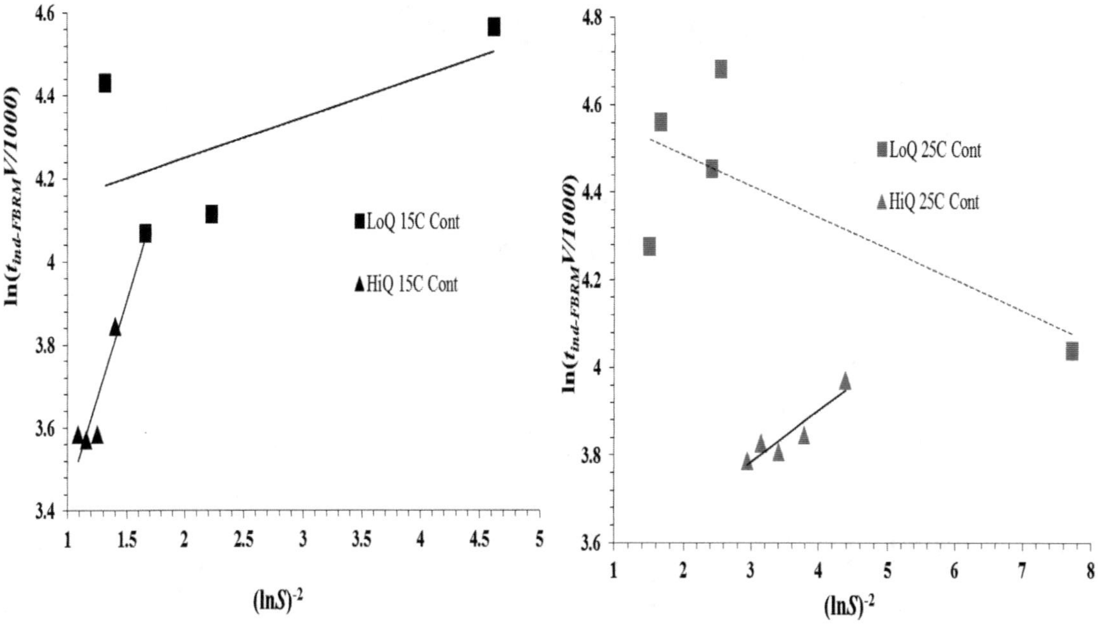

17

U.S. Food and Drug Administration
Protecting and Promoting Public Health
www.fda.gov

Conclusions

- Integrated PAT measurements (FBRM, PVM and NIR-PCA) together with modified classic nucleation theory can be used to model both continuous and discontinuous antisolvent crystallization process
- Modified classic nucleation theory can capture the dynamic features of continuous and discontinuous anti-solvent addition
- Continuous crystallization can offer advantages over discontinuous crystallization
  - Achieve a desired and more uniform particle size distribution
  - Generate fine crystals faster
- Above approaches offer the opportunity to develop continuous drug substance manufacturing processes

18

U.S. Food and Drug Administration
Protecting and Promoting Public Health
www.fda.gov

Acknowledgements

- FDA CDER OPQ Office of Testing and Research
  - ORISE Fellows: Maury White, Ryan Foringer
  - Cindy Buhse, Vincent Vilker (Retired), M.A. Khan(moved)
- FDA CDER OPQ Office of Process and Facilities
  - Rapti Madurawe, Jennifer Maguire, Robert Iser
- FDA CDER RSR 12-42; RSR-13-32; FDA CDER OTR PAT-12

19

U.S. Food and Drug Administration
Protecting and Promoting Public Health
www.fda.gov

References

- H. Wu and M. Khan, 2010. *J. Pharm. Sci.*, 99(3): 1516-1534.
- H. Wu, M. White, M.A. Khan, 2011. *Int. J. Pharm.*, 405:63-78.
- H. Wu and M.A. Khan, 2011. *J. Pharm. Sci.*, 100 (5): 1969-1986.
- H.Wu, M. White, R. Berendt, R. Foringer, M.A. Khan, 2014. *Ind. Eng. Chem. Res.*, 53 (4) :1688–1701.
- H. Wu, Z. Dong, H. Li, M.A. Khan, 2015. *Org. Process Res. Dev.*, 19(1) : 89-101.
- H. Wu, B. Kurtyka, G. Wu, J. Maguire, R. Madurawe, C. M.V.Moore, R. Iser, M.A. Khan, R.C. Lyon, P. Faustino, L. Buhse, S. Lee, L. Yu. *European Pharmaceutical Review*, 2015, Nov. Issue.

20

431784 Process Modeling and Simulation for Continuous Pharmaceutical Manufacturing of Diphenhydramine

Abstract:

Continuous Pharmaceutical Manufacturing (CPM) has been recognized as an increasingly promising multidisciplinary research domain, addressing the significant obstacles hampering sustainable profitability (increasing R&D costs, globalized competition, licensing protocols, legal and regulatory landscape changes). Introducing a viable alternative to traditional batch production processes, CPM has an expanding potential capable of achieving lower costs (Schaber et al., 2011), higher yields, more economical heat and solvent use, high mass transfer rates, rapid scalability and elimination of intermediate storage (Calabrese and Pissavini, 2011). The technical viability and process efficiency of CPM prototypes must however be investigated in detail, in order to ensure successful pilot-scale demonstrations (Mascia et al., 2013) which can be used toward reliable production-scale implementations keenly pursued by pharmaceutical corporations (Poechlauer et al., 2013).

Organic flow synthesis literature must thus be (and indeed, has been) extensively surveyed to identify a wide range of Active Pharmaceutical Ingredients (APIs) whose production has already been accomplished in continuous mode at laboratory scale. A wide range of promising API candidates have thus been identified as suitable for CPM implementation (Jolliffe and Gerogiorgis, 2015a), including ibuprofen and artemisinin, which have already been examined in recent publications (Jolliffe and Gerogiorgis, 2015b). Systematic flowsheet synthesis, process modelling and simulation are essential towards rapidly assessment of CPM potential, and are already widely applied for CPM case studies (Gernaey et al., 2012).

The present paper focuses on a plantwide process modeling and simulation study for the production of diphenhydramine, a first-generation antihistamine which has been developed and has anticholinergic, antitussive, antiemetic and sedative properties and is mainly used to treat allergies, drug-induced parkinsonism and other neurological symptoms; moreover, it has a strong hypnotic effect and is and FDA-approved nonprescription sleep aid, especially in the form of diphenhydramine citrate. Our process model is developed on the basis of a published continuous flow synthesis (Snead and Jamison, 2013), which demonstrates that this economically and societally important API can be produced using Plug Flow Reactors (PFRs) with molten ammonium salts, thereby allowing for heating heat well above the boiling point of all reaction components. Moreover, this method also achieves solvent and waste minimization, thereby reducing operating costs and suppressing hazards associated with excess and disposal of chemicals.

Process modelling and steady-state simulations have been conducted toward process design for the respective CPM routes, with novel kinetic expressions developed on the basis of published kinetic data from experimental campaigns and thermodynamic property estimations essential for process design. Economic evaluations of the respective plant construction and operation venture can accordingly be performed on the basis of plantwide mass and molar balances, in order to quantitatively evaluate the viability of a production-scale CPM process.

LITERATURE REFERENCES

1. Calabrese, G.S., Pissavini, S., 2011. From batch to continuous flow processing in chemicals manufacturing. *AIChE J.*, **57**(4): 828-834.

2. Gernaey, K.V., Cervera-Padrell, A.E., Woodley, J.M., 2012. A perspective on PSE in pharmaceutical process development and innovation, *Comput. Chem. Eng.*, **42**(1): 15-29.

3. Jolliffe, H.G., Gerogiorgis, D.I., 2015a. Process modelling and simulation for continuous pharmaceutical manufacturing of ibuprofen, *Chem. Eng. Res. Des.*, **97**: 175-191.

4. Jolliffe, H.G., Gerogiorgis, D.I., 2015b. Plantwide design and economic evaluation of two Continuous Pharmaceutical Manufacturing (CPM) cases: Ibuprofen and artemisinin, *Comp.-Aided Chem. Eng.* (in press).

5. Mascia, S., Heider, P.L., Zhang, H. et al., 2013. End-to-end continuous manufacturing of pharmaceuticals: Integrated synthesis, purification and final dosage formation, *Angew. Chem. Int. Ed.* **52**(47): 12359-12363.

6. Poechlauer, P., Colberg, J., Fisher, E., et al., 2013. Pharmaceutical roundtable study demonstrates the value of continuous manufacturing in the design of greener processes, *Org. Proc. Res. Dev.* **17**(12): 1472-1478.

7. Schaber, S.D., Gerogiorgis, D.I. et al., 2011. Economic comparison of integrated continuous and batch pharmaceutical manufacturing: a case study, *Ind. Eng. Chem. Res.* **50**(17): 10083-10092.

8. Snead, D.R., Jamison, T.F., 2013. End-to-end continuous flow synthesis and purification of diphenhydramine hydrochloride featuring atom economy, in-line separation, and flow of molten ammonium salts, *Chem. Sci.* **4**(7): 2822-2827.

405063 Efficient Detection of Amyloidâusing PMMA-Tag Fused VHH Antibody

Abstract:

VHH domains represent the smallest antigen-binding unit with a molecular size of 15 kDa in comparison to scFv fragments (30 kDa), Fab fragments (60 tkDa), and whole antibodies (150 kDa) [1]. For this reason, VHH fragments have been considered to have great potential in various industrial applications. Furthermore, the genetically fused PMMA tag to VHH antibody allow the high rate of refolding and immobilization. Recently, Amyloid beta (Aβ) denotes peptides of 36–43 amino acids that are crucially involved in Alzheimer's disease as the main component of the amyloid plaques found in the brains of Alzheimer patients. In the present study, we examined whether PMMA-tag fused VHH antibody could offer the potential for detecting Amyloid β antigen. To do it, the $(His)_6$-tagged recombinant protein expressed as inclusion bodies in *Escherichia coli* Rosseta (DE3 component cell) were purified by affinity chromatograph on an Histrap Ni^{2+}-Chealate column using a linear gradient of imidazole. His-tag-VHHs was purified as a monomer, eluting on the range from 0 to 400 mM imidazole under pH 6.0. As a results, the highly purified VHH-PM antibodies showed different recovery rate and antigen-binding activity, according to the different strains of VHH-PM antibodies. The recovery rate of Anti-Aβ-VHH-PM (L1-3) and Anti-Aβ-VHH-PM (61-3) was lower in pH 8.5. However, Anti-Aβ-VHH-PM (61-3) was highly recovered at pH 6.0. Furthermore Anti-Aβ-VHH-PM (L1-3) refolded in pH 8.5 showed high antigen binding activity, compared to that Anti-Aβ-VHH-PM (61-3). It was demonstrated that the optimal condition of refolding in Anti-Aβ-VHH-PM was changeable with pH and strain.

425853 Comparative Proteomic Analysis of Host and High Biopharmaceuticals Producing CHO Cell Lines

Abstract:

One important target in biopharmaceutical industry is to achieve a high production of therapeutic protein. However, the effect of different host cells on protein expression has not been fully understood, so it is very challengeable to rationally design host cells and construct production cell lines. In this study, comparative proteomics was performed to investigate the host cell proteins that could correlate with the expression of recombinant therapeutic proteins. The global protein maps and biopharmaceutical production of four cell lines were analyzed and compared, including adapted ATCC CHO/dhfr-, CHO/dhfr-/EPO, Life Tech CHO DG44, and CHO DG44/IgG. All the host cells and production cell lines were cultivated in suspension culture using chemical defined medium. The productivity analysis in a simple fed-bath cell culture with glucose feeding was performed in shaker flask, showing that the ATCC CHO/dhfr-/EPO produced erythropoietin with titer of 421 mg/L and the CHO DG44/IgG produced IgG with titer of 1,385 mg/L. The LC/MS/MS was used to create the profiles of the global intracellular proteins of these four cell lines, and more than 1,000 host proteins were analyzed. The host proteins associated with metabolism of carbon and energy in glycolysis pathway, TCA cycle and amino acids were investigated. Other proteins involved in, such as protein processing (e.g. protein disulfide-isomerase A3 and endoplasmin), cytoskeletal structure (e.g. alpha-actinin-4 and vimentin) and cell cycle control (e.g. heat shock protein HSP 90-alpha and cyclin-dependent kinase 1), were also analyzed. The possible host cell regulators that correlate with different dhfr- host cells and different therapeutic proteins were identified although further investigation is needed. This study provided a deep insight into the regulation mechanism of biopharmaceutical productions by CHO cells. The results could be used to direct the rational cell engineering for new-generation host cells in the future.

432085 Proteomic Time Profiling of Mab Producing CHO Fed-Batch Bioreactor Cultures

Abstract:

The production of recombinant proteins such as monoclonal antibodies (mAbs) is usually based on complex mammalian cell processes. Rising omics technologies have enabled the characterization of manufacturing processes on a molecular level. Getting a deeper insight into cellular mechanisms is crucial for the understanding and optimization of manufacturing processes, which is also highly promoted by regulatory agencies through the Quality by Design (QbD) initiative. In particular, the investigation of proteomic profiling can be helpful for the identification of important up- and down regulated pathways which have an influence on the process performance and quality of the produced biopharmaceuticals.

The time evolution of the intracellular protein abundance of a monoclonal antibody (mAb) producing Chinese hamster ovary cell line during fed-batch cultivation was investigated. Two different reactor scales were compared distinguishing possible scale-dependent differences. Multivariate data analysis (MVDA) such as principal component analysis (PCA) was used to emphasize highly up- or down expressed proteins over time. A classification of important biological functions was performed by gene ontology annotation using the proteins with highest fold-change. Furthermore, profiling of the produced mAb and the corresponding quality attributes on a proteomic level is discussed targeting possible operational changes. Scale and time dependent proteomic profiling in combination with multivariate statistical analysis will be highlighted for future applications.

Malvern

Characterizing and Understanding Protein Stability at High Concentration

Wei Qi, Samiul Amin and E.Neil Lewis
Salt Lake City, AICHE
Nov. 2015

Outline

- Challenges from High Concentration Formulation
- Raman Spectroscopy and Dynamic Light Scattering
- Case studies
  - Model Proteins – Thermal/Reducing conditions
  - Thermal stability of mAb and its modified version
  - Other applications
- Summary

What A Difference a Few Years Makes

2011

1	Lipitor Pfizer Inc	7,429,953	Stats
2	Plavix Bristol-Myers Squibb Company	6,560,260	Stats
3	Nexium AstraZeneca Pharmaceuticals, LP	5,963,917	Stats
4	Abilify Otsuka Pharmaceutical Co., Ltd	5,032,032	Stats
5	Advair Diskus GlaxoSmithKline	4,492,486	Stats
6	Seroquel AstraZeneca Pharmaceuticals, LP	4,492,001	Stats
7	Singulair Merck & Co., Inc.	4,449,747	Stats
8	Crestor AstraZeneca Pharmaceuticals, LP	4,266,282	Stats
9	Cymbalta Eli Lilly and Company	3,552,147	Stats
10	Humira AbbVie, Inc.	3,421,540	Stats

www.drugs.com

2014

Rank in 2013 (in 2012)	Product	Company	To treat	2013 sales ($US m)	2012 sales ($US m)	
1 (1)	Humira	AbbVie	Rheumatoid arthritis / autoimmune diseases	10,659	9,616	← Protein
2 (2)	Enbrel	Pfizer/Amgen	Rheumatoid arthritis / autoimmune diseases	8,776	8,496	← Protein
3 (4)	Remicade	Johnson & Johnson/ Merck & Co	Rheumatoid arthritis / autoimmune diseases	8,386	7,990	← Protein
4 (3)	Seretide/Advair	GlaxoSmithKline	Asthma	8,251	7,634	
5 (6)	Lantus	Sanofi	Diabetes	7,592	7,155	← Protein
6 (5)	Rituxan	Roche	Rheumatoid arthritis	7,503	6,377	← Protein
7 (9)	Avastin	Roche	Bowel cancer	6,751	6,282	← Protein
8 (7)	Herceptin	Roche	Breast cancer	6,562	6,253	← Protein
9 (8)	Crestor	AstraZeneca	Cholesterol	5,622	6,149	
10 (10)	Abilify	Otsuka Holdings	Depression / schizophrenia	5,500	5,304	

The Guardian, March 2014

Protein Research- Academic and Industry Interests !

Malvern www.malvern.com

Commercial Antibody Formulations

Product name	Dosage form	Concentration	pH	Formulation
ILARIS® (canakinumab)	Lyophilized	150 mg/mL	NA	92.4 mg/mL sucrose, L-histidine and L-histidine HCl, 0.6 mg/mL polysorbate-80
XOLAIR® (omalizumab)	Lyophilized	125 mg/mL	NA	125 mg/mL omalizumab, 90 mg/mL sucrose, 1.7 mg/mL L-histidine hydrochloride monohydrate, 1.1 mg/mL L-histidine, 0.3 mg/mL polysorbate-20
RAPTIVA® (efalizumab)	Lyophilized	100 mg/mL	~6.2	Approximately 82 mg/mL sucrose, 4.5 mg/mL L-histidine hydrochloride monohydrate, 2.9 mg/mL L-histidine, 2 mg/mL polysorbate-20
SYNAGIS® (palivizumab)	Lyophilized	100 mg/mL	NA	47 mM histidine, 3 mM Glycine, 5.6% mannitol
SIMPONI® (golimumab)	Liquid	100 mg/mL	~5.5	0.88 mg/mL histidine and histidine-HCl monohydrate, 41 mg/mL sorbitol, 0.16 mg/mL polysorbate-80
STELARA® (ustekinumab)	Liquid	90 mg/mL	5.7–6.3	1 mg/mL L-histidine and L-histidine-HCl, 76 mg/mL sucrose, 0.04 mg/mL polysorbate-80
HUMIRA® (adalimumumab)	Liquid	50 mg/mL	~5.2	6.2 mg/mL sodium chloride, 0.86 mg/mL sodium citrate, 1.3 mg/mL citric acid monohydrate, 12 mg/mL mannitol, 1 mg/mL polysorbate-80
CAMPATH® (alemtuzumab)	Liquid	30 mg/mL	NA	8 mg/mL sodium chloride, 1.44 mg/mL dibasic sodium phosphate, 0.2 mg potassium chloride, 0.2 mg/mL monobasic potassium phosphate, 0.1 mg/mL polysorbate-80, 0.0187 mg/mL disodium edentate dihydrate
AVASTIN® (bevacizumab)	Liquid	25 mg/mL	6.2	60 mg/mL trehalose, 5.8 mg/mL sodium phosphate monobasic monohydrate, 1.2 mg/mL sodium phosphate dibasic anhydrous, 0.4 mg/mL polysorbate-20
HERCEPTIN® (trastuzumab)	Lyophilized	21 mg/mL	~6	Approximately 20 mg/mL α,α-trehalose dehydrate, 0.5 mg/mL L-histidine HCl, 0.32 mg/mL L-histidine, 0.09 mg/mL polysorbate-20, 1.1% bacteriostatic water
VECTIBIX® (panitumumab)	Liquid	20 mg/mL	5.6–6.0	5.8 mg/mL sodium chloride, 6.8 mg/mL sodium acetate
ROACTEMRA® (tocilizumab)	Liquid	20 mg/mL	6.5	15 mM phosphate, 50 mg/mL sucrose, 0.5 mg/mL polysorbate-80
ARZERRA® (ofatumumab)	Liquid	20 mg/mL	6.5	8.55 mg/mL sodium citrate, 0.195 citric acid monohydrate, 5.85 mg/mL sodium chloride
SOLIRIS® (eculizumab)	Liquid	10 mg/mL	NA	Chloride, phosphate dibasic, phosphate monobasic, polysorbate-80
LUCENTIS® (ranibizumab)	Liquid	10 mg/mL	5.5	10 mM histidine–HCl, 10% trehalose, 0.01% polysorbate-80
REMICADE® (infliximab)	Lyophilized	10 mg/mL	~7.2	50 mg/mL sucrose, 0.05 mg/mL polysorbate-80, 0.22 mg/mL monobasic sodium phosphate monohydrate, 0.61 mg/mL dibasic sodium phosphate dihydrate
RITUXAN® (rituximab)	Liquid	10 mg/mL	~6.5	9 mg/mL sodium chloride, 7.35 mg/mL sodium citrate dihydrate, 0.7 mg/mL polysorbate-80
ZENAPAX® (daclizumab)	Liquid	5 mg/mL	~6.9	3.6 mg/mL sodium phosphate monobasic monohydrate, 11 mg/mL sodium phosphate dibasic heptahydrate, 4.6 mg/mL sodium chloride, 0.2 mg/mL polysorbate-80
SIMULECT® (basiliximab)	Lyophilized	4 mg/mL	NA	1.4 monobasic potassium phosphate, 0.20 mg/mL disodium hydrogen phosphate (anhydrous), 0.32 mg/mL sodium chloride, 4 mg/mL sucrose, 16 mg/mL mannitol, 8 mg/mL glycine
ERBITUX® (cetuximab)	Liquid	2 mg/mL	7.0–7.4	8.48 mg/mL sodium chloride, 1.88 mg/mL sodium phosphate dibasic heptahydrate, 0.42 mg/mL sodium phosphate
REOPRO® (abciximab)	Liquid	2 mg/mL	pH 7.2	10 mM sodium phosphate, 150 mM sodium chloride, 0.001% polysorbate-80

[a] All formulations are expressed as mg/mL after reconstitution. Because of fill volume variability, some concentrations are approximate.

European Journal of Pharmaceutics and Biopharmaceutics 78 (2011) 208–212

Malvern www.malvern.com

Analytical Challenges from High Concentration

- Characterization of proteins at the formulated conditions, with excipients.
- Stability profile, either storage conditions or forced degradation.
- Understanding of high concentration formulation, i.e viscosity.
- Conventional spectroscopic technologies suffer from low concentration requirement, excipients effect, and provide limited information.

Far UV CD --- Secondary Structure
Low concentration, interference from excipients.
Intrinsic Fluorescence --- Tertiary Structure
Low concentration, fluorescence quenching.
FTIR --- Secondary Structure
High concentration, water subtraction.

Protein Characterization --- Scales

Molecular Heterogeneity | Nano/Micro Heterogeneity

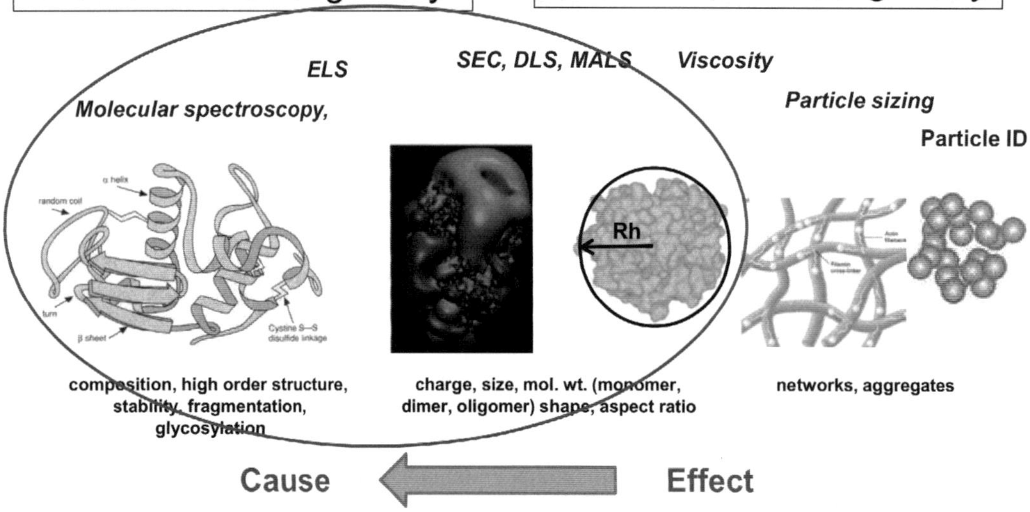

Structure, Function, Dimension, Behavior

Combine Elastic and Inelastic Scattering

Zetasizer ZS Helix – Hybrid DLS-Raman

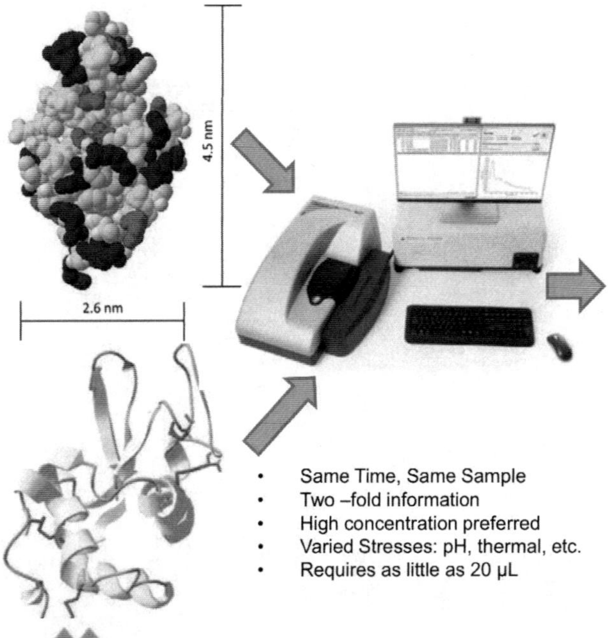

- Same Time, Same Sample
- Two –fold information
- High concentration preferred
- Varied Stresses: pH, thermal, etc.
- Requires as little as 20 µL

Hydrodynamic Size
Interaction Parameters (k_D)
Colloid Stability

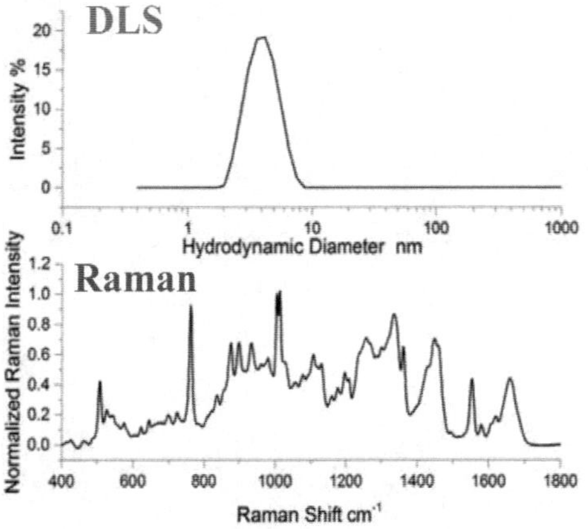

Secondary and Tertiary Structure
Aromatic Side Chain
Disulfide Bond
Conformational Stability

Lewis, US Patent Pending

Dynamic Light Scattering (DLS)

Correlation of short time scale (μs) intensity fluctuations leads to the diffusion coefficient (D), hydrodynamic radius (R_H), & polydispersity index (PDI).

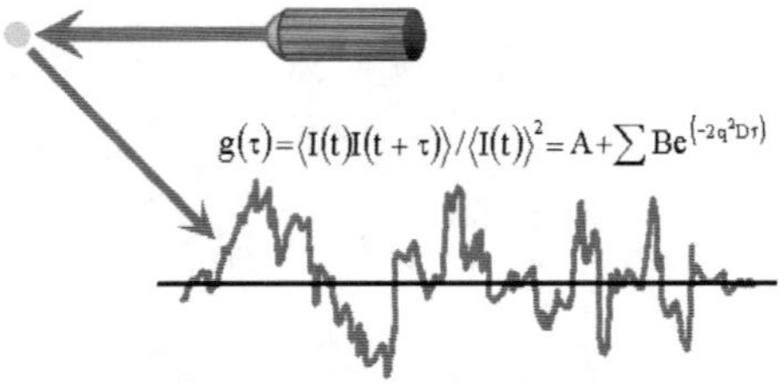

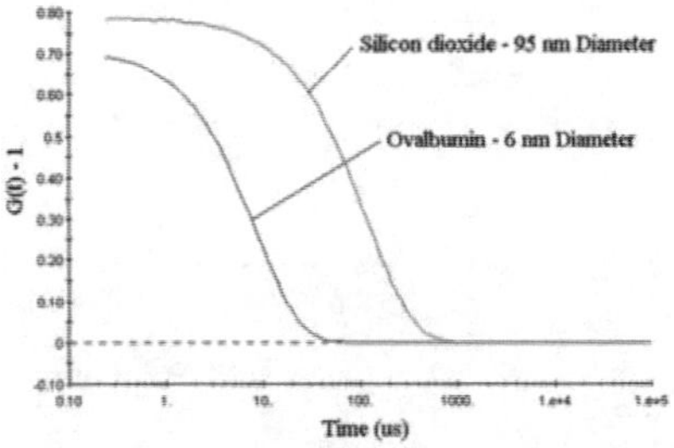

Stokes-Einstein Equation

$$R_H = \frac{kT}{6\pi\eta D}$$

q = Scattering vector
D = Diffusion coefficient
R_H = Hydrodynamic radius
k = Boltzmann constant
T = Temperature
η = Solvent viscosity

Raman Spectroscopy

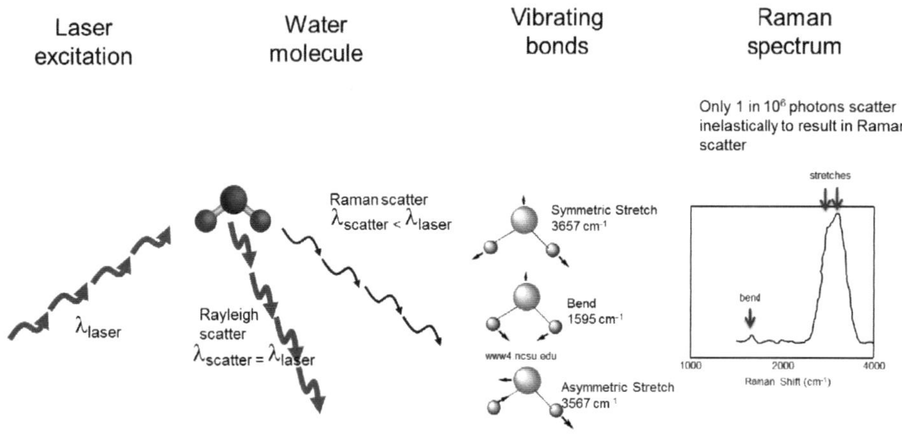

- A Non-Invasive, Non-destructive inelastic light scattering/vibrational spectroscopic technique.
- Different selection rules – polarizability (Raman) vs dipole moment (IR)
- Water weak to Raman spectrum

Typical Protein Raman Spectrum

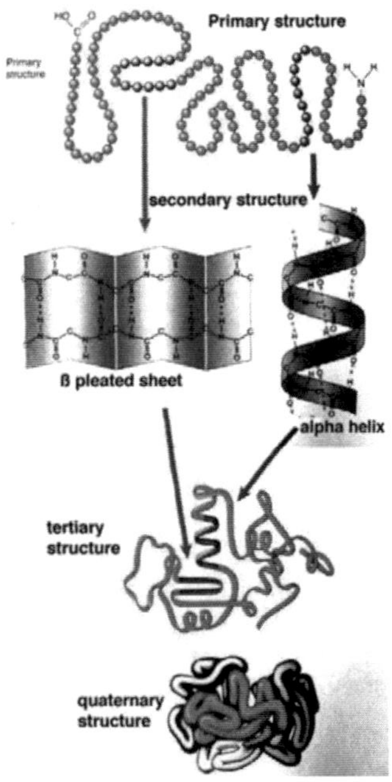

http://medicinenewbie.blogspot.com/2011/05/professional-i-exam-2-may-2011-part-4.html

Raman spectrum of proteins provides

- Backbone (Amide I, III) → helical/sheet
- Side chain orientation (Trp)
- Hydrophobic or hydrophilic environment (Trp)
- Hydrogen bonding patterns (Trp, Tyr)
- Disulfide bond (*g-g-g, g-g-t, t-g-t*)

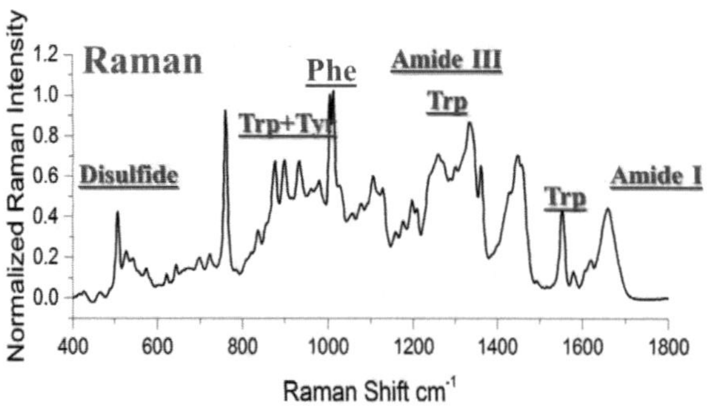

Malvern © 2015 Malvern Instruments Limited | Company Confidential www.malvern.com

Lysozyme --- Thermal induced Unfolding

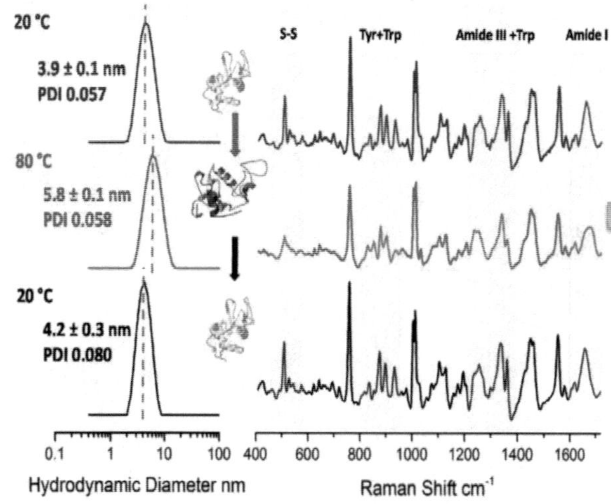

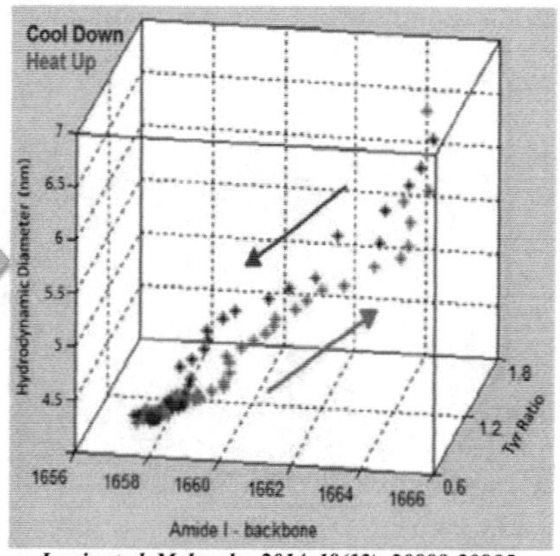

Lewis et al. Molecules 2014, 19(12), 20888-20905

- Real time profile --- full reversible thermal event
- Simultaneous characterization --- structure and size
- A thermal induced unfolding event without aggregation associated

Lysozyme --- TCEP induced Reducing

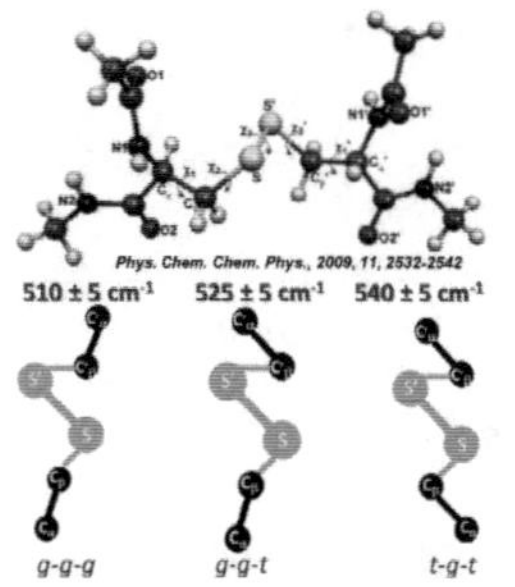

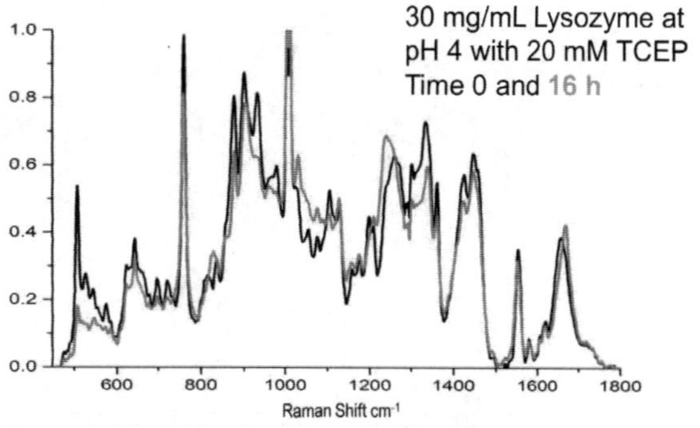

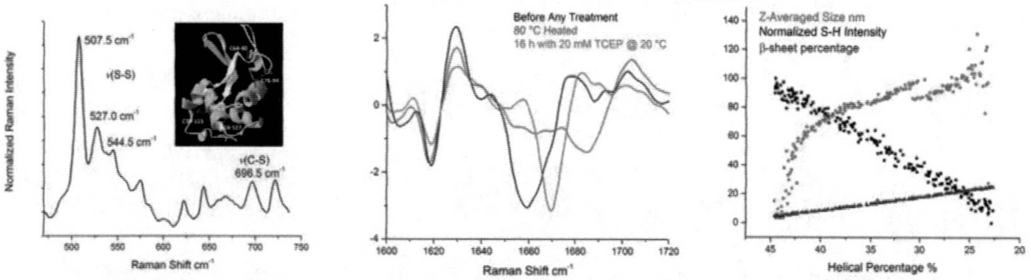

- A conformation change with aggregation/gelation process.
- Different denatured states for reducing and thermal stress.

RNase A --- Thermal induced Unfolding

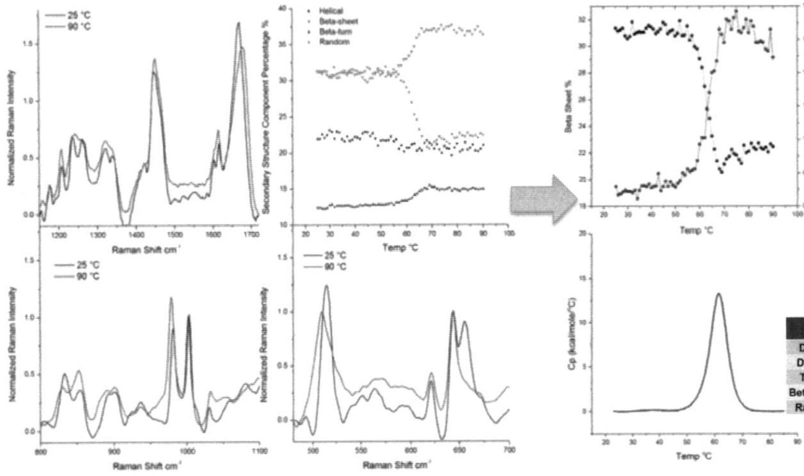

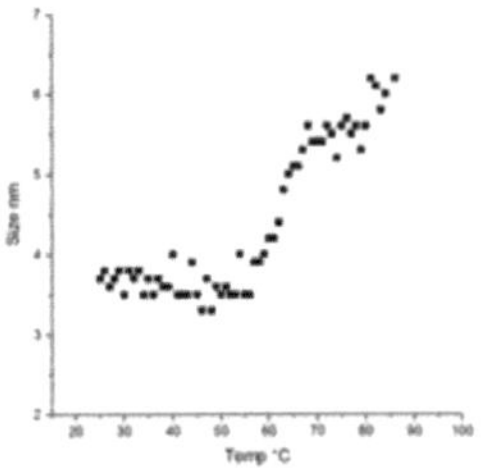

	T_m (°C)	ΔH_v (kJ/mol)	$\Delta H_v / \Delta H_{cal}$
DSC (Cp)	61.5 ± 0.5	457.8 ± 45.8 (ΔH_{cal})	1.00
DLS (size)	59.5 ± 0.4	436 ± 41	0.95
Tyr Ratio	61.0 ± 0.3	436 ± 6	0.95
Beta Sheet %	60.5 ± 0.2	569 ± 85.6	1.24
Random %	59.0 ± 0.4	441 ± 41	0.96

- Clear change of secondary structure (Amide I, III), Tyr and disulfide region.
- Simultaneous/cooperative transitions for tertiary, secondary structure and size.
- The van't Hoff enthalpy and calorimetry enthalpy ratio indicates a highly cooperative two state transition.

RNase A --- TCEP induced Reducing

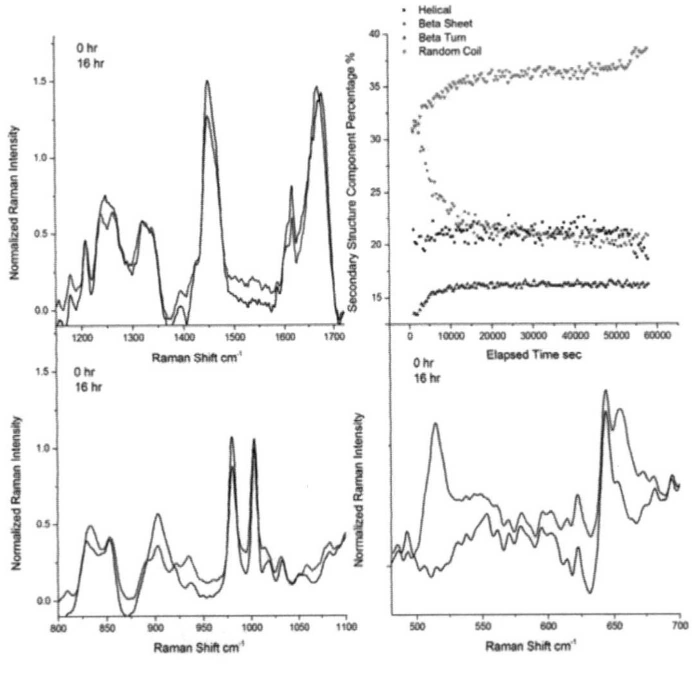

- 30 mg/mL RNase A pH 5.5, 20 mM TCEP
- Clear transition from beta-sheet to random coil/beta turn upon TCEP reduction.
- The Tyr ratio at 850/830 also changed, indicating a modification of the local hydrogen bonding environment, but less change than for thermal stress.
- The disulfide S-S mode at 515 cm^{-1} disappeared and as before (ref. heating) there is a similar decrease for the C-S mode at 655 cm^{-1}.

RNase A --- TCEP induced Reducing Kinetics

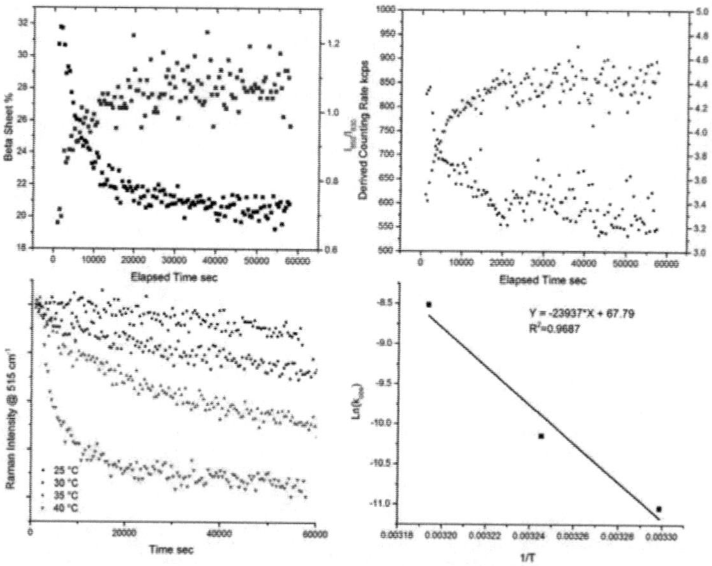

- Upon reduction, there is a decrease in the derived count rate and a small size increase, indicating a gradual expanding structure.
- Reduction experiments at multiple temperatures can be used to investigate the kinetics and activation energy of reduction.

- A reducing induced conformation change, but not associated with aggregation !

mAb X --- Thermal Induced S-S Region Change

25 °C

63 °C

67 °C

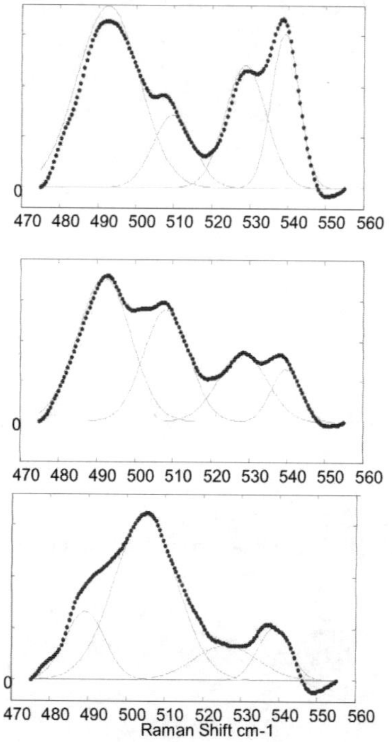

Disulfide bond conformers

Intensity at 539 cm^{-1}
trans-gauche-trans

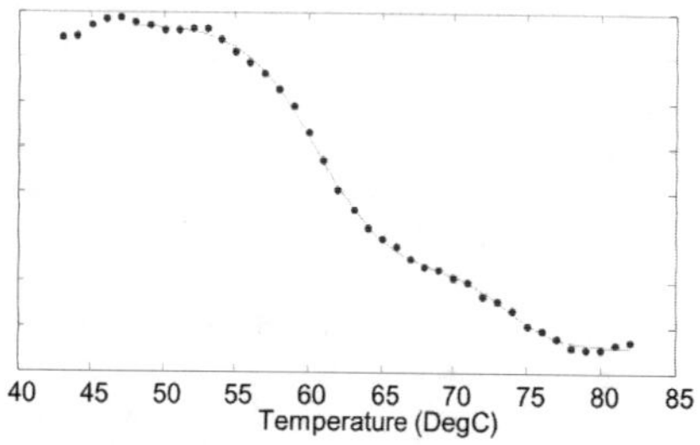

mAb X at 150 mg/mL Temp Ramp 25-90 °C
Disulfide bond conformers – double transition @ 65.0 & 60.2 °C

Malvern www.malvern.com

mAb Y and modified mAb Y --- Thermal Stability

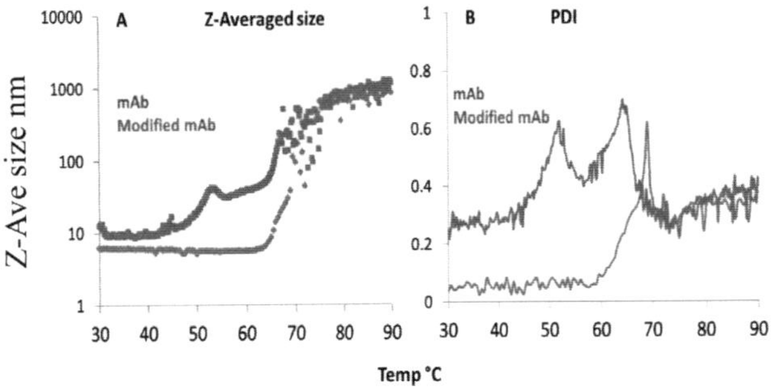

Sizing results
- mmAb double transition: ~53 & 65°C
- mAb single transition: ~ 69 °C
- PDI trend for mmAb – oligomerization followed by aggregation

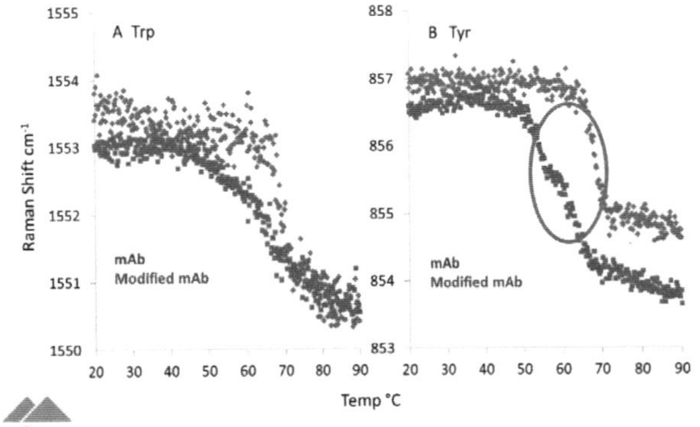

Raman results

- Minimum secondary structure change
- mmAb double transition: ~54 & 63 °C
- mAb single transition: ~ 69 °C

Tyrosine H-bonding environment most perturbed during low temperature transition

Oligomerization of modified mAbY
Isothermal Incubation at 46, 53 and 60 °C

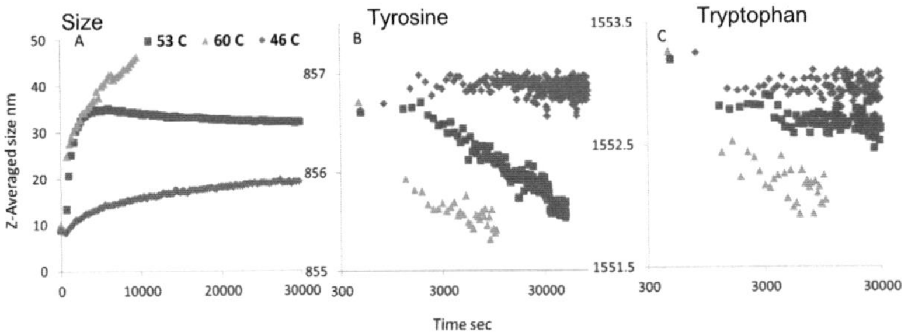

- 46 °C - pre-transition
  - size slowly increases
  - Raman maker bands show no change
- 53 °C – in-transition
  - size quickly increases from 10 to ~35 nm and then stabilizes
  - Tyr peak position drops, while the Trp peak position shifts slightly
  - Tyr peak position shifts from 856.8 to 855.5 cm^{-1}, the value seen at the first transition in the T ramps
- 60 °C – post-transition
  - size quickly increases to over 40 nm within 3 hours
  - Trp and Tyr peak positions quickly shift to lower frequencies
  - Tyr peak position shifts to ~855.5 cm^{-1}.

Oligomerization of mmAb

Isothermal Incubation at 46, 53 and 60 °C

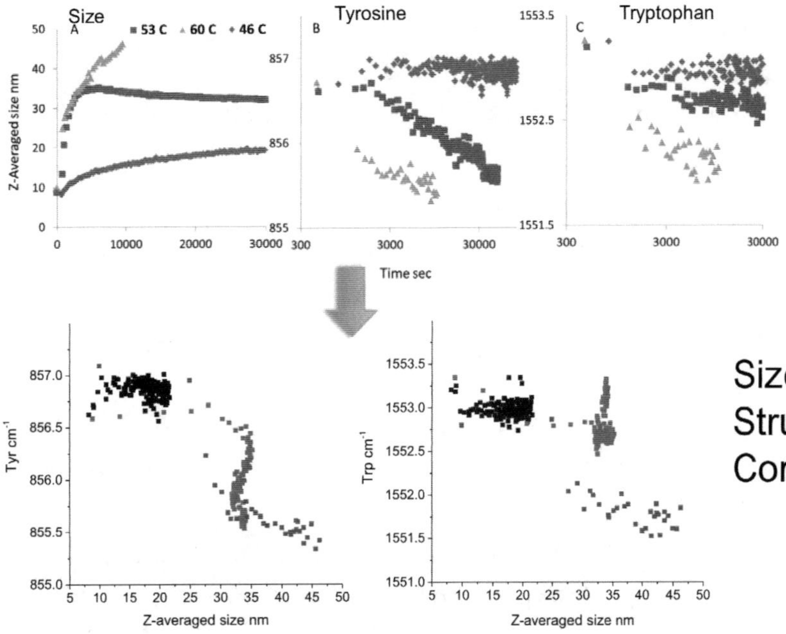

Size and Structure Correlation !

Protein Aggregation in Solution: Conformational and Colloidal Stability

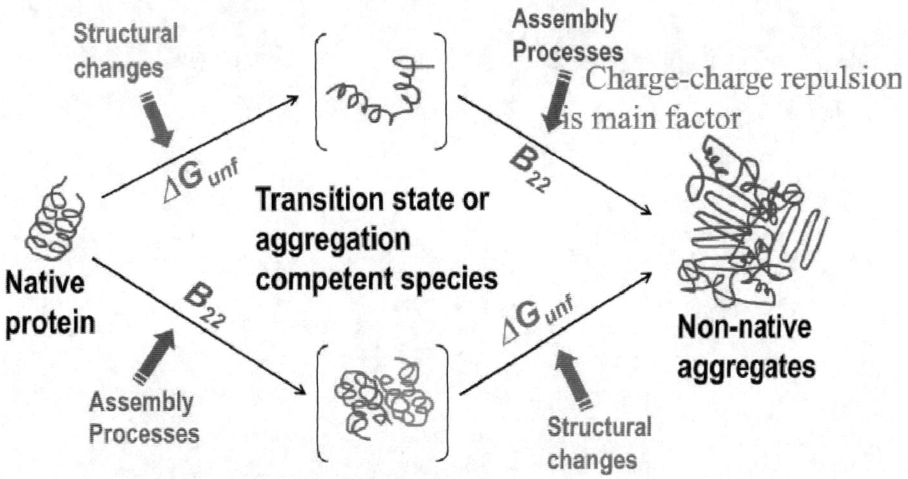

High protein concentration put additional challenges for formulation development and real time stability characterization.

Chi et al., 2003. Pharm. Res. 20:1325

Other Applications

- Hydrogen-deuterium exchange.
- Biosimilars Characterizations (Insulins, mAbs, etc.).
- Coupled with Micro-rheology with low frequency marker.
  Steve Blake's Talk on β-lactoglobulin at 2:20 pm in Rm. 206B
- Chemical denaturation (GdnHCl/Urea/H_2O_2)/Chemical modifications.
- Lyophilization.

Summary

- High concentration formulation brings new analytical challenges.
- Raman spectroscopy could benefit the user by providing abundant structure information simultaneously. Like all the spectroscopic techniques, it provides an ensemble information.
- Coupled with dynamic light scattering for size distribution, the platform could be versatile to probe the protein at the formulation conditions with selected stresses, in the real time, if preferred.
- Could be widely applied to many potential application areas.
- Samples at high concentration and want a lot of information.... DLS/Raman !

Malvern www.malvern.com

Malvern

Acknowledgements

Malvern Instruments
Bioscience Development Initiative, Columbia MD.

University of Colorado, Denver
Prof. John Carpenter

Questions?

wei.qi@mavlern.com
443-878-1319

427902 Design of Dendrimer-Doxorubicin Conjugates for Transport Modulation Across in Vitro Pulmonary Epithelium and Their Solution Formulation in Pressurized-Metered Dose Inhalers

Abstract:

Lung cancer is the leading cause of cancer death among both men and women in the United States. Adenocarcinoma accounts for more than 40% of all lung cancer cases. Doxorubicin (DOX) is an FDA-approved chemotherapeutic that has been widely used as primary anticancer drug in the treatment of a variety of cancers including lung cancers. However, rapid elimination, uncontrolled release, and life-threatening cardiotoxicity, has hindered the applicability of DOX and other potent anti-cancer agents. The potential use of DOX in the treatment of lung adenocarcinoma with DOX is also limited by its low accumulation in the lungs, which is generally administered through systemic route. The development of polymeric carriers for the modulation of transport, targeting and controlled release of potent anti-cancer agents and their aerosol formulation for local lung delivery is, therefore, of great relevance in the treatment of lung adenocarcinoma.

In this work, a series of PEGylated poly(amidoamine) dendrimer nanocarriers with acid-labile and acid-nonlabile DOX conjugates were synthesized and characterized. We investigated the effect of PEGylation density (low, medium and high) and number of DOX conjugates (low and medium) on the release of DOX from the dendrimer nanocarrier, *in vitro* cytotoxicity, and transport of dendrimer nanocarriers across a model of the lung epithelium (Calu-3 cells). PEGylation retards the release of DOX in acidic medium. PEGylation density also affects cytotoxicity as seen by a slight increase in IC_{50} for DOX-conjugated dendrimer compared to free DOX, due to the controlled release of the therapeutic. PEGylation degree significantly helped dendrimer nancarriers escape mucus trapping compared to bare dendrimer, aiding their transport to the basolateral side. Reduced transepithelial electrical resistance (TEER) and limited cellular internalization demonstrated that paracellular transport plays a vital role in transporting dendrimer-DOX conjugates across airway epithelium. The PEGylated dendrimer-DOX conjugates were directly formulated in the propellant-based metered-dose inhalers (pMDIs) without the aid of any surfactants and co-solvents, forming a solution aerosol formulation. The solution formulation tends to deliver dendrimer-DOX conjugates with high efficiency to the alveolar region (up to 80%). The results suggest the targeting, transport, release and formulation of the dendrimers can be controlled by careful design of the nanocarriers.

430373 Controlling Mechanical Properties and Dissolution Rate of Strip Films Loaded with Griseofulvin Nanoparticles

Abstract:

Polymer strip films have been demonstrated to be an ideal delivery form for poorly water-soluble drugs. Specifically for active pharmaceutical ingredient (API) particles that have been reduced in size to enhance their solubility and dissolution rate, polymer strip films provide a means by which these nano/micro-size particles can be delivered relatively quickly while retaining their nano/micro size upon delivery. However, the ability to intelligently manipulate the mechanical properties and dissolution rate of such films for the delivery of poorly water-soluble API nanoparticles has yet to be investigated. To this end, hydroxypropyl methylcellulose (HPMC) films were prepared, loaded with model API, griseofulvin, nanoparticles produced via wet stirred media milling in aqueous suspension. These films consisted of various plasticizers and HPMC molecular weights. Addition of glycerin, triacetin, and polyethylene glycol as plasticizers successfully reduced film strength and increased film elasticity with no significant impact on dissolution rate. Increasing the molecular weight of the polymeric film-former generally led to slower film dissolution rate, accompanied by a small increase in tensile strength. All films exhibited good uniformity of content in terms of film thickness and drug distribution (<6% RSD for 0.7 cm^2 samples). Results demonstrate that, when used in tandem, adjustment of plasticizer content and polymer molecular weight can be used to successfully tailor the mechanical properties and dissolution rate of polymer films loaded with poorly water-soluble drug nanoparticles without sacrificing the advantages or integrity of the strip film format, allowing for intelligent film formulation design for targeted applications.

406800 Lipid-Polymeric Hybrid Nanoparticle Based Nicotine Vaccine: Size Matters

Abstract:

Lipid-polymeric hybrid nanoparticle based nicotine vaccine: size matters

Zongmin Zhao , Yun Hu, Reece Hoerle, Meaghan Devine, Chenming Zhang

Department of Biological Systems Engineering, Virginia Tech, Blacksburg, VA 24061

Presenting author

Email addresses:

Zongmin Zhao: zmzhao@vt.edu

Yun Hu: hyun86@vt.edu

Reece Hoerle: rhoerle@vt.edu

Meaghan Devine: meaghd1@vt.edu

Chenming Zhang: chzhang2@vt.edu

Abstract

Tobacco use is prevalent worldwide and has consistently been the top preventable cause of many serious diseases, resulting in huge mortality, morbidity, and economic loss in recent decades. Due to the limited success of FDA approved pharmacological therapies, it is necessary to develop more effective treatment for smoking cessation. Currently, nicotine vaccine that can induce nicotine specific antibodies is considered as a promising approach against tobacco addiction. However, traditional protein-hapten conjugate nicotine vaccines have shown less than desired immunological efficacy due to their poor recognition and internalization by immune cells. Previous study in our group showed that negatively charged nanohorn supported liposome could effectively enhance the immunogenicity of nicotine protein conjugate based vaccine. In this study, nanohorn was substituted with Poly (lactic-co-glycolic acid) (PLGA) nanoparticle (NP) and a novel lipid-polymeric hybrid NP based nicotine vaccine was developed. In particular, the effects of particle size on the immunogenicity of hybrid NP based vaccine was investigated. 100 nm hybrid NPs were found to be taken up by dendritic cells more efficiently than 500 nm hybrid NPs. Mice immunization results demonstrated that both 100 nm and 500 nm nanovaccines could induce strong immune response against nicotine. However, a significantly higher titer of antibody against nicotine was achieved by 100 nm nanovaccine than 500 nm nanovaccine. Moreover, no evidence of systemic toxicity was observed on mice immunized with all the hybrid NP based nicotine vaccines. These results from this study suggest that lipid-polymeric NP based nicotine vaccine is a promising candidate for treating nicotine dependence.

434194 High Throughput Synthesis of Nanovaccine Formulations

Abstract:

Using nanoscale vehicles to deliver protein antigens in the development of vaccines has been a well-studied area of research. High throughput techniques utilizing combinatorial libraries have the potential to be a valuable resource in rapidly synthesizing and screening nanoscale biomaterial-based carriers for drug and vaccine delivery. In this study, we highlight the development of an automated high-throughput robot for synthesizing polyanhydride nanoparticles containing protein antigens. Polyanhydrides are a class of safe and biodegradable polymers that have been widely used as a vaccine adjuvant/carrier. The method we have developed was based on a multiplexed homogenizer and has the capacity to handle parallel streams of monomer or polymer solutions to synthesize polymers and/or nanoparticles. Copolymers were synthesized from sebacic acid (SA), 1,6-bis(*p*-carboxyphenoxy)hexane (CPH), and 1,8-bis(*p*-carboxyphenoxy)-3,6-dioxactane (CPTEG). Nanoparticles loaded with the model antigen, ovalbumin, were synthesized using anti-solvent nanoprecipitation. Concentrations of surfactant were varied to test its effect on protein encapsulation efficiency as well as protein release kinetics. We observed that surfactant concentration did not significantly affect protein release rate, but greatly enhanced protein encapsulation efficiency. Particles synthesized with the high throughput method were compared to particles made with conventional anti-solvent nanoprecipitation to validate the high throughput approach. Protein burst and release rates between the two methods were similar, even though particles synthesized using the high-throughput technique were observed to be smaller. Finally, we demonstrated that the high-throughput method could be adapted to functionalize the surface of the particles to aid in the design of targeted drug and vaccine delivery systems. Our results showed that the developed high-throughput system is a viable alternative to conventional methods for studying vaccine and drug delivery vehicles, with the potential to more rapidly optimize the design of these systems.

434315 Preparation and Characterization of Pegylated, Acid-Labile Dendrimer-Doxorubicin Conjugates and Their Aerosol Formulation in Propellant-Based Inhalers

Abstract:

Lung cancer is the leading cause of cancer death among both men and women in the United States. Adenocarcinoma accounts for more than 40% of all lung cancer cases. Doxorubicin (DOX) is an FDA-approved chemotherapeutic that has been widely used as primary anticancer drug in the treatment of a variety of cancers including lung cancers. However, rapid elimination, uncontrolled release, and life-threatening cardiotoxicity, has hindered the applicability of DOX and other potent anti-cancer agents. The potential efficacy of DOX in the treatment of lung adenocarcinoma is also limited by low accumulation in the lung tissue, which is usually administered through systemic route. The development of polymeric carriers for the modulation of transport, targeting and controlled release of potent anti-cancer agents and their aerosol formulation for local lung delivery is of great potential relevance in the treatment of lung adenocarcinoma.

In this work, a series of PEGylated poly(amidoamine) dendrimer nanocarriers with acid-labile DOX conjugates were synthesized and characterized. We employed a two-step PEGylation strategy to increase the payload of the hydrophobic DOX. We investigated the impact of pH (neutral and acidic pH), PEGylation density (low, medium and high) and number of DOX conjugates (low and medium) on the release of DOX from the dendrimer nanocarrier, the kinetics of carrier uptake, intracellular release kinetics of DOX from the nanocarrier, and toxicity in an alveolar adenocarcinoma cell line (A549). PEGylation retards the release of DOX in acidic medium, and also intracellularly, as determined by co-localization studies with confocal microscopy. We also observed that the kinetics of cellular entry of the nanocarrier with DOX increased significantly compared to free DOX, as determined by flow cytometry. At the highest PEGylation density, the rate of internalization of the nanocarrier containing DOX was even higher than that of free DOX. PEGylation density also affects cytotoxicity as seen by an increase in IC_{50} for DOX-conjugated dendrimer compared to free DOX, due to the controlled release of the therapeutic. PEGylated dendrimer-DOX conjugates were directly formulated in the propellant-based metered-dose inhalers (pMDIs) with a trace amount of ethanol (ca. 0.2%), forming a solution aerosol formulation. The stability and aerosol characteristics increased significantly as PEG density increased and DOX payload decreased. The formulation of highly PEGylated dendrimer-DOX conjugates resulted in aerosols with high deep lung deposition (up to 80%). These results are rationalized based on the phase behavior of the polymer in propellant HFA.

414380 A Drug-Sparing Approach to the Preparation of Drug Nanosuspensions: High-Intensity Vibratory Milling

Abstract:

Almost 40% of the new chemical entities in drug pipeline of pharmaceutical companies have low aqueous solubility, which leads to poor bioavailability *in vitro* and *in vivo*. Reduction of drug particle size down to nano-scale has been an effective way of increasing the dissolution rate of poorly water-soluble drugs. In this study, we aim to elucidate the impact of the process parameters on the breakage kinetics and particle size during the preparation of drug nanosuspensions via high-intensity vibratory milling, which can be used as a drug-sparing approach in early development. Griseofulvin (GF) particles were wet-milled in the presence of two stabilizers, hydroxypropyl cellulose (HPC) and sodium dodecyl sulfate (SDS), to prepare stable GF suspensions under various processing conditions. Laser diffraction (LD), scanning electron microscopy (SEM), and X-ray powder diffraction (XRPD) were used to characterize the suspensions. Several milling experiments were performed to determine the optimal bead size at low power density. Then, the process was intensified with the optimal bead size and operated at a much higher power density by increasing the vibration intensity and bead loading step-wise with the objective of increasing the breakage rate. After examining the breakage kinetics upon intensification, additional experiments were performed with various bead sizes to explore if and how the optimal bead size depends on the power density. Finally, a comparative assessment of the performance of vibratory milling vs. wet stirred media milling has been carried out in view of the power density, resultant breakage kinetics, and final drug particle size attained. This study provides a systematic process intensification approach as well as guidance for bead size selection and suggests process/design improvement strategies for fast preparation of drug nanosuspensions via vibratory milling.

The Use of GastroPlus™ Modeling in Enabling Drug Discovery Research - A Case Study

Ching Kim Tye, PhD (Bristol-Myers Squibb)

AIChE ANNUAL CONFERENCE, NOV 10, 2015

Bristol-Myers Squibb

2

Co-authors

- John Crison
- Zhanbin Wang
- Ching Chiang Su
- Michael J Hageman

Bristol-Myers Squibb

Outline

3

- Background – GastroPlus™ and its general usage
- How pH affects drug absorption
- Case study
  - The use of GastroPlus™ modeling in pH-dependent absorption risk assessment to enable decision making at preclinical stages

Bristol-Myers Squibb

GastroPlus™

4

- GastroPlus
  - Based on Advanced Compartmental Absorption and Transit (ACAT) model
  - Originally developed for drug absorption for oral administration
  - Physiological Based Pharmacokinetic (PBPK) Model
    - **Mechanistic-based** by mathematically transcribing anatomical and physiological descriptions of the phenomena involved in ADME processes
  - Significant publications since late 1990 - main usage during early clinical development and formulation optimization
- GastroPlus peer-reviewed publications
  - 159 compiled on Simulation Plus website (as of 10/19/2015)

Bristol-Myers Squibb

5

Advanced Compartmental Absorption and Transit Model (ACAT™)

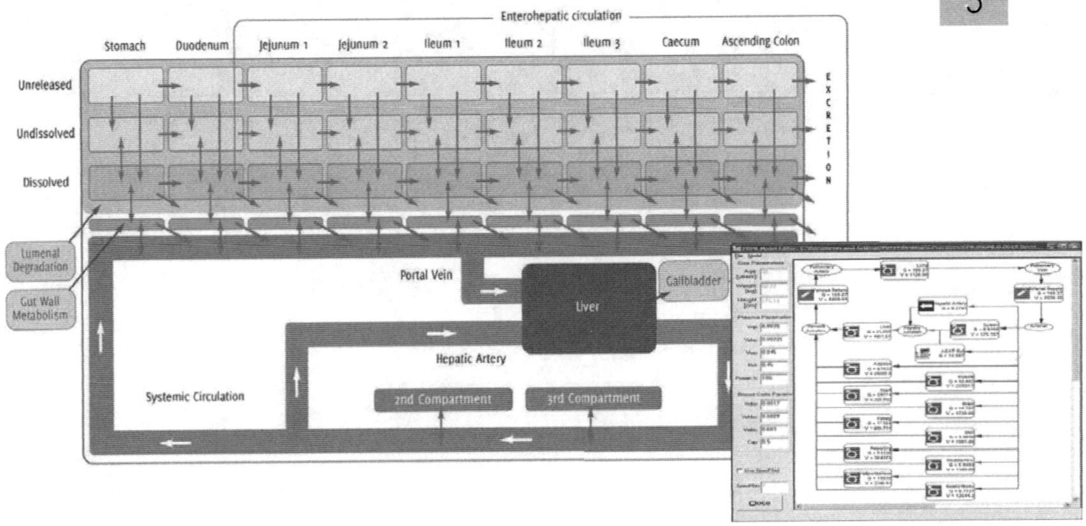

6

Applications of Absorption Modeling in Drug Discovery and Development

Drug Discovery	Pre-clinical Development		Clinical Development
Lead Optimisation	**Clinical Lead Selection**	**Entry Into Human**	**Phase1 Phase2 Phase3**
Predict absorption from *in vitro* and *in silico* data Explore limitations on absorption via parameter sensitivity analysis	Integrate *in vitro* and *in vivo* data to explore formulation possibilities Assist development of formulations for toxicological studies	Predict oral pharmacokinetics in human Extended biopharmaceutical evaluation to define clinical formulation strategy	Predict food effect Design extended release formulations Develop In Vitro In Vivo Correlations

Neil Parrott and Thierry Lave, Applications of physiologically based absorption models in drug discovery and development, Molecular Pharmaceutics, Volume 5(5), 2008.

How pH Affects Drug Absorption?

7

- Drug absorption is mainly through passive diffusion across cells - passive lipid diffusion
- The Henderson-Hasselbalch equation
  - Acid: pH = pKa + log $[A^-]/[HA]$
  - Base: pH = pKa + log $[B]/[HB^+]$
- Weak acids and weak bases can be ionized or un-ionized as the pH of the media changes. Ionized drugs are less likely to participate in the lipid diffusion across the biological membrane.

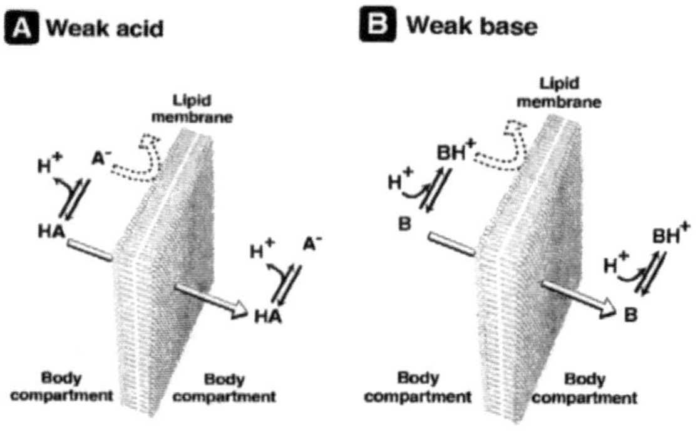

Bristol-Myers Squibb

The Relationships Between pH, pKa and Ionization

8

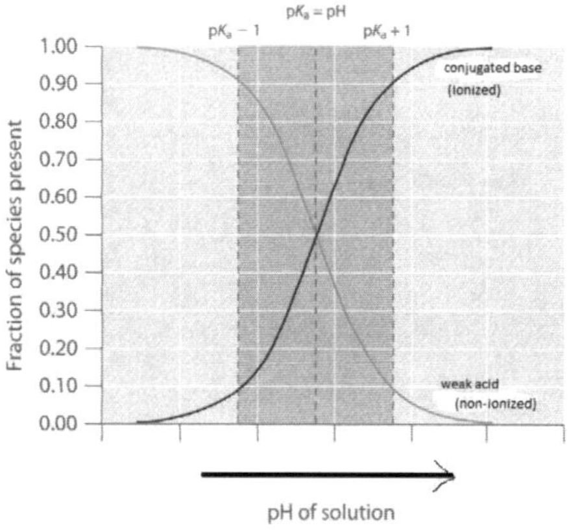

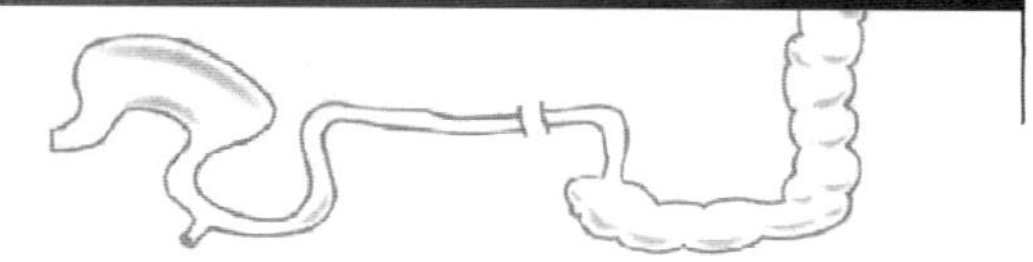

	Stomach	Jejunum	Ileum	Colon	
pH	1.4-2.1	4.4-6.6	6.8-8.6	5-8	fasting
	3-7	5.2-6.2	6.8-8.0	5-8	fed

Bhal, S. K. Lipophilicity Descriptors: Understanding When to Use Log*P* & Log*D*., *2012*

- Weak acid ionized at
  - 50% at its pKa
  - 91% at pKa +1
  - 99% at pKa + 2
- Opposite for weak base

Bristol-Myers Squibb

Advanced Compartmental Absorption and Transit Model (ACAT™)

9

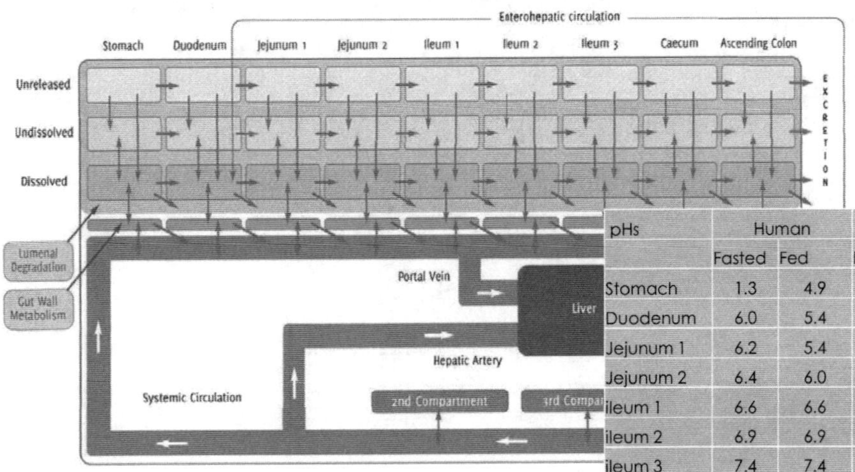

pHs	Human		Rat	
	Fasted	Fed	Fasted	Fed
Stomach	1.3	4.9	3.9	3.2
Duodenum	6.0	5.4	5.9	5.0
Jejunum 1	6.2	5.4	6.1	5.1
Jejunum 2	6.4	6.0	6.1	5.1
ileum 1	6.6	6.6	5.9	5.9
ileum 2	6.9	6.9	5.9	5.9
ileum 3	7.4	7.4	5.9	5.9
Caecum	6.4	6.4	6.6	5.9
Asc colon	6.8	6.8	6.2	5.5

Opt LogD Model SA/V6.1

SimulationsPlus
SCIENCE + SOFTWARE = SUCCESS

10

Case Study:
The Use of GastroPlus™ to Enable Drug Discovery

Compound BMS-1

Physicochemical properties

- Molecular weight ~ 500
- Log D (at pH 6.5) = 3.4
- High permeability (Caco-2)
- Weak base, pKa = 2.5
- Solubility
  - Low solubility in aqueous (<0.001 mg/ml)
  - pH dependent solubility
  - Good solubility in biorelevant media: FaSSIF and FeSSIF
  - Very soluble in polyethylene glycol (PEG)

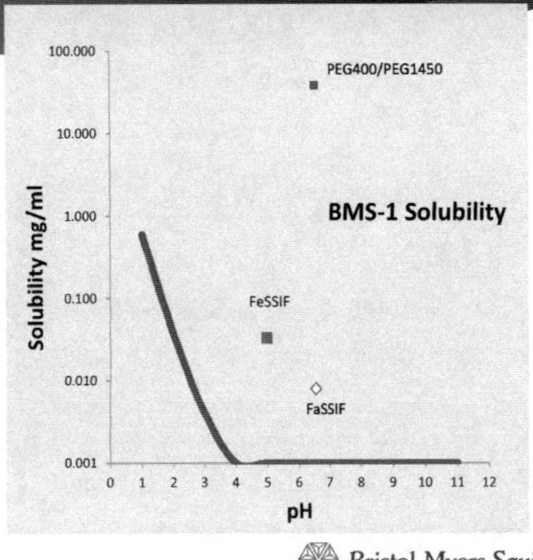

FaSSIF = Fasted State Simulated Intestinal Fluid, pH 6.5
FeSSIF = Fed State Simulated Intestinal Fluid, pH 5

Bristol-Myers Squibb

Preclinical Bioavailability

11

- Based on a solution formulation
- Variable bioavailability across species
- High fraction absorbed

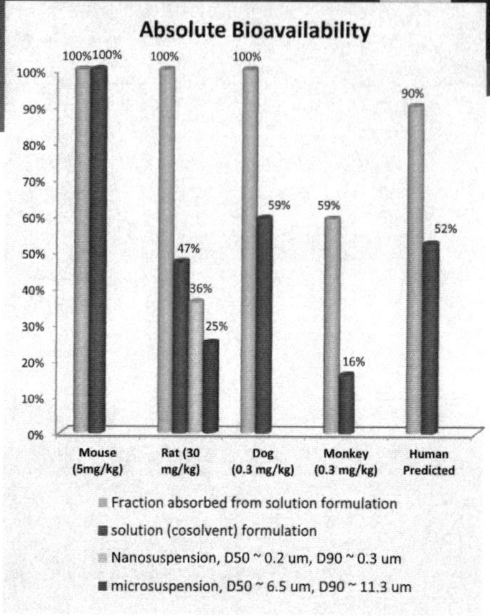

The Question
How Much pH-dependent Absorption Risk?

12

BMS-1 properties

- 600x higher solubility at pH 1 than at pH ≥ 4, a potential for pH-dependent absorption
- A highly potent oncologic compound with a low therapeutic index

A significant pH-dependent effect will be problematic for development.

Therefore, here are the questions:

- Is this a viable drug candidate?
- How much is the risk of significant pH-dependent absorption in human?
- If there is a pH-dependent absorption, can this be mitigated?

Bristol-Myers Squibb

Processes Needed to be Understood for pH-Dependent Absorption

13

- pKa and ionization, acid/base
- Solubility
- Supersaturation
- Precipitation timing and ease of redissolve
- Permeability
- stability
- Particle size
- Formulation factors (dissolution & release kinetics)
- And the interplay of these factors

Bristol-Myers Squibb

pH-Dependent Absorption Risk Assessment Methods

14

Noted that during discovery research, the dose is uncertain and API availability is very low

	Method Type	Method Details	Benefits and Limitations
1	In vitro	pH shift transfer dissolution tests	▪ Simple ▪ Set up to mimic human conditions ▪ High demand of API
2	In vivo	Pre-dosed rats with pentagastrin or famotidine	▪ Most resource intensive ▪ *In vivo* system can be more representative ▪ Not human
3	In silico	GastroPlus modeling	▪ No API needed ▪ Allow prediction in different doses ▪ Require physicochemical & biopharmaceutical property data ▪ Based on animal model

Bristol-Myers Squibb

In Vitro Microdissolution: pH-Shift Transfer Method*

15

API Suspension or powder:
Target: Clinical Dose/250 ml

FaSSIF Transfer:
14 ml

Dilution

Non-sink condition

Assumption: Dissolution is critical

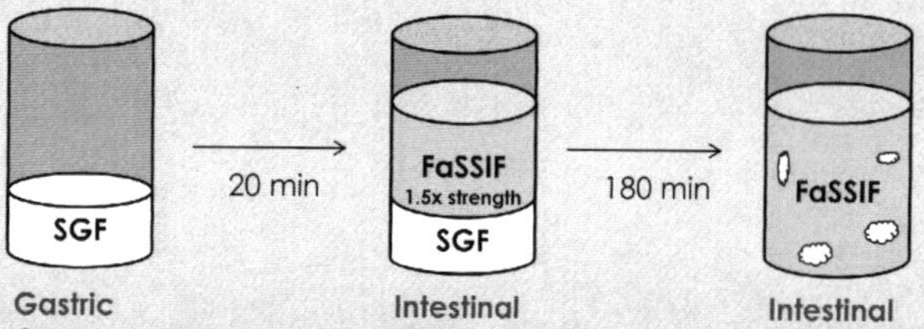

Gastric Compartment
Vol: 7 ml

Intestinal Compartment
Vol: 21 ml

Intestinal Compartment
Final pH ~6.5

Test A: $SGF_{pH2} \rightarrow FaSSIF_{pH6.5}$ (equivalent to un-treated or normal)
Test B: $SGF_{pH6} \rightarrow FaSSIF_{pH6.5}$ (equivalent to famotidine-treated or achlorhydria)

* Neil Mathias *et. al.* Assessing the risk of pH-dependent absorption for new molecular entities: a novel in vitro dissolution test, physicochemical analysis, and risk assessment strategy. Molecular Pharmaceutics 2013, 10, 4063-4073

In Vitro MicroDissolution Tests Indicated Moderate pH-Effect (at dose = 175 mg)

16

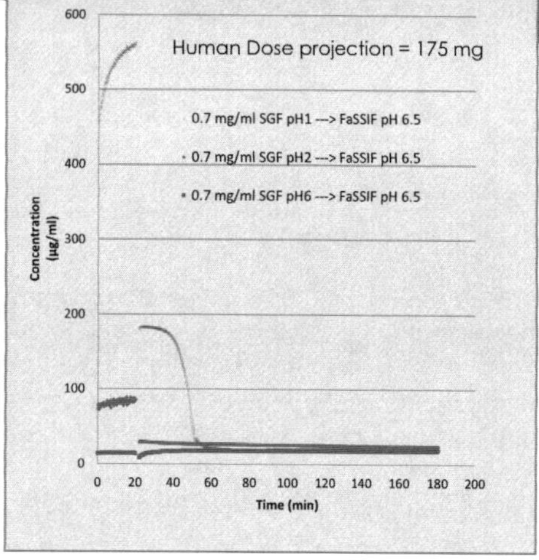

Table 4. *In Vitro*–*in Vivo* Correlation: pH-Effect Risk Categories*

risk category	*in vitro* AUC Ratio	clinical AUC or C_{max} ratio
high	<0.2	<0.5
moderate	0.2–0.5	0.5–0.8
low	>0.5	>0.8

* Neil Mathias et al. Assessing the risk of pH-Dependent Absorption for New Molecular Entities. Mol. Pharmaceutics 2013, 10, 4071

Test 1: pH effect = AUC (SGF pH 6 to FaSSIF pH 6.5) / AUC (SGF pH 2 to FaSSIF pH 6.5) = 0.50

Test 2: pH effect = AUC (SGF pH 6 to FaSSIF pH 6.5) / AUC (SGF pH 1 to FaSSIF pH 6.5) = 0.16

- Result indicated potential moderate pH-effect

Bristol-Myers Squibb

In vivo (Rat) Evaluation Showed Significant pH-Effect at 20mg/kg

17

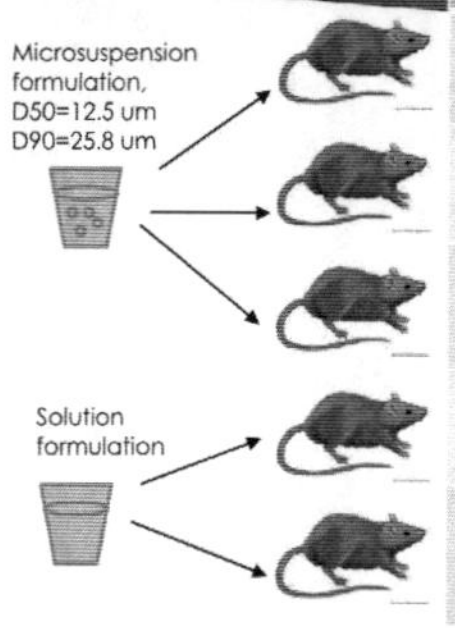

Treatment	Dose (mg/kg)	Cmax (μM)	Tmax (h)	T half (h)	AUC (μM*h)	Bioavailability (%)	AUC changes (%)
control	20	0.61 ± 0.15	4 ± 0	8.9 ± 4.4	4.9 ± 1.3	30	100%
Pretreatment: pentagastrin 0.25 mg/kg, SC 30 min prior	20	0.68 ± 0.21	3.3 ± 1.5	5.0 ± 1.5	4.5 ± 1.0	27	92%
Pretreatment: famotidine 25 mg/kg, PO 60 min prior	20	0.28 ± 0.08	4.8 ± 1.5	14 ± 11	2.7 ± 0.4	16	55%
control	30	2.1 ± 0.1	0.7 ± 0.3	13 ± 11	11.9 ± 1.3	48	100%
Pretreatment: famotidine 25 mg/kg, PO 60 min prior	20	1.53 ± 0.26	1.8 ± 1.9	5.1	9.5 ± 0.4	57	120% (dose normalized)

- Microsuspension formulation – pretreatment with famotidine resulted in 55% AUC reduction. This indicates a potential pH-effect at 20 mg/kg.
- Solution formulation – No significant AUC changes with famotidine pretreatment. This indicates that a solution formulation could be a strategy for pH-dependent risk mitigation.

Bristol-Myers Squibb

GastroPlus Modeling: Parameters Inputs and Model Building

18

N = 3, resource constraints in Discovery but data fits fairly well

- Physicochemical
  - MW, compound structure, partition coefficient, pKa, pH-solubility profile, biorelevant solubility, particle size, precipitation time, permeability*
- Pharmacokinetics
  - Use PKPlus model by fitting experimentally determined mean plasma concentration-time profile following intravenous administration to calculate clearance
- Physiological: Rat - physiological - Fasted
- A model was developed based on good fits across
  - Intravenous administration
  - Oral solution formulations at 5 mg/kg and 30 mg/kg
  - Oral suspension formulations at 30 mg/kg

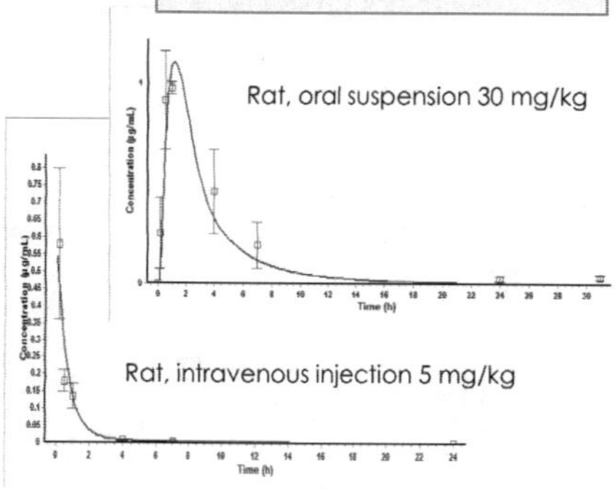

* rat intestine permeability was based on an internal correlation of Caco-2 permeability and intestinal permeability established for literature compounds.

Bristol-Myers Squibb

Parameter Sensitivity Analysis (PSA)

19

- With the validated rat PK model, based on the microsuspension data, PSA was run for
  - stomach pH between 1 to 8
  - Dose between 0.5 & 100 mg/rat

Bristol-Myers Squibb

GastroPlus Simulation for Stomach pH Sensitivity Effect in Rats (microsuspension, 30 mg/kg)

20

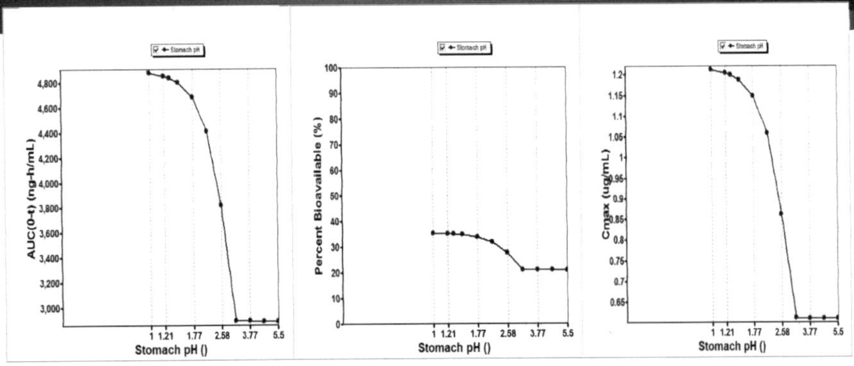

- Based on available rat data, it is predicted that there may be ~ 50% more drug available in a stomach with pH 1.5 in compare to pH > 3

Bristol-Myers Squibb

GastroPlus Output - The Effect of pH-Effect Absorption is Dose-Dependent

21

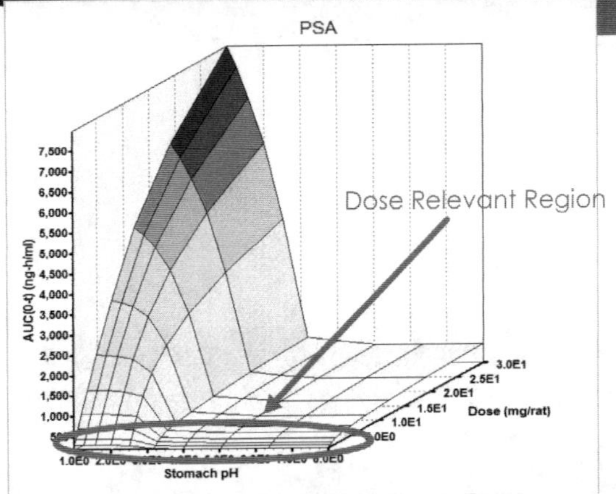

- As the program progressed, it was determined that the dose requirement will be low.
- The model predicted that the risk would be low
- A risk mitigation option available: solubilized formulations
- This allowed the program to move forward into early development quickly

Bristol-Myers Squibb

The No Effect Prediction was Later Confirmed in a Dog Study

22

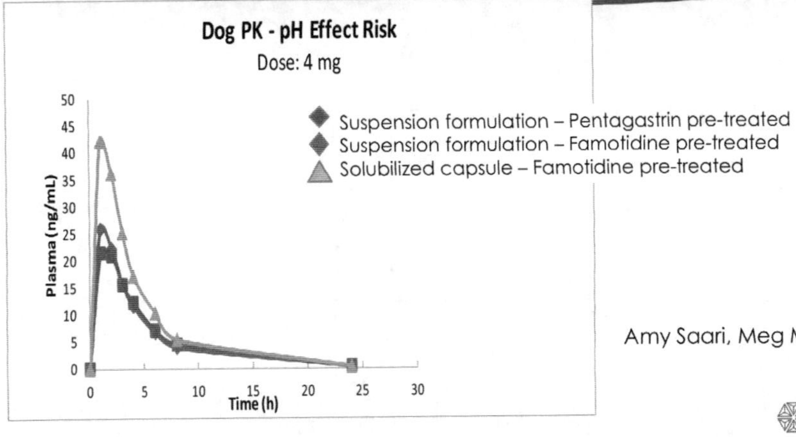

Amy Saari, Meg McCallum, Yan Xu

Bristol-Myers Squibb

23

Summary

The use mechanistic-based modeling (GastroPlus™) in combination with *in vitro* and *in vivo* data was successfully used to evaluate the risk of pH-dependent absorption across a wide range of doses to enable decision making in Discovery Research.

The benefits of mechanistic-based GastroPlus™ modeling in Discovery

- ACAT physiological model available for commonly used preclinical toxicology and pharmacology species including, mouse, rat, rabbit, cat, dog, etc
- Use animals to assess formulation performance and recommend formulations strategy for human testing - parameter sensitivity analysis
- Dose is a critical factor in the risk equation, however dose is usually not well defined during preclinical stage - modeling & simulation allow information to be gathered and projected for different doses.

Bristol-Myers Squibb

430408 Modeling of Drug Delivery Across the Blood-BRAIN Barrier for the Treatment of Autism Spectrum Disorder

Abstract:

Modeling of Drug Delivery Across the Blood-Brain BARRIER FOR the Treatment of Autism Spectrum Disorder

(Submitted to AIChE Session: Recent Challenges in the Biopharmaceutical and Pharmaceutical Fields)

Jamelle Simmons[1], Luke Achenie[2,3], and Yong W. Lee[1,3]

1. Virginia Tech, School of Biomedical Engineering and Sciences (SBES)
2. Virginia Tech, Department of Chemical Engineering
3. Virginia Tech Center for Autism Research (VTCAR)

Introduction: Autism spectrum disorder (ASD) is classified as a developmental disability with deficits in areas of communication and interaction. Signs are noticeable during early childhood and are lifelong. Between 2000 and 2010, ASD prevalence rose from 1 in 150 to 1 in 68 [1]. Previous research in our lab has found elevated levels of xanthine oxidase (XO) in male autism mouse brains along with increased neuronal loss and microglial activation. We hypothesize that inhibition of XO will reduce brain injury in ASD. In this presentation we will discuss the modeling of two mechanisms for transport of drugs across the blood-brain-barrier (BBB). We will also discuss our initial attempts to do multiscale modeling of transport using agent-based modeling under the Netlogo modeling environment.

Materials and Methods: Sodium fluorescein salt (Na-Fl, MW=376.27, Sigma-Aldrich) was used as a fluorescent substitute for our xanthine oxidase inhibitor febuxostat (MW=316.374). C6 (rat glioma) cells were cultured in DMEM supplemented with fetal bovine serum (FBS) and penicillin-streptomycin (Penn-Strep) on the basal side of Corning¨ Transwell¨ polyester membrane inserts (12mm with 0.4 micron pores). After C6 cells reached confluence, RBE4 (rat brain capillary) cells were seeded on the apical side of the inserts and cultured in DMEM/Ham's F-12 supplemented with FBS, Penn-Strep, basic fibroblast growth factor (bFGF), and hydrocortisone (Hydro). Four groups were studied, (1) uncoated inserts, (2) collagen coated inserts, (3) RBE4+Hydro mono-culture, and (4) RBE4+C6+Hydro co-culture. Transport experiments were run with apical chambers containing Na-Fl in RBE4 media with basal chambers containing only RBE4 media. Spectrophotometric analysis was conducted on basal chamber samples taken at 0, 3, 6, 9, and 12 hrs. Effective diffusion (D_{eff}) constants were calculated for each transport group and used to construct partial differential equation (PDE) diffusion models.

Results and Discussion: Effective diffusion constants (cm^2/hr) were calculated from the average of 3 and 6hr time points for each well type with (1) $D_{eff} = 1.02x10^{-4}$ for blanks, (2) $D_{eff} = 1.43x10^{-4}$ for collagen coated wells, (3) $D_{eff} = 6.85x10^{-5}$ for RBE4+Hydro wells, and (4) $D_{eff} = 7.54x10^{-4}$ for RBE4+C6+Hydro wells. Diffusion models were plotted for each well type (Figure 1).

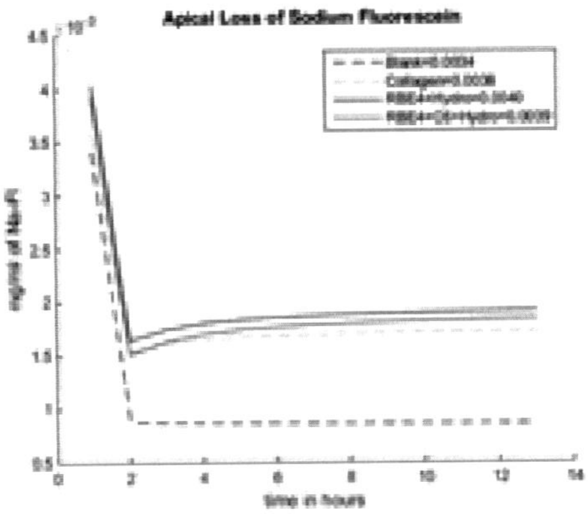

Figure 1. Loss of tracer drug from apical layer to basal layer over 12 hours. More sodium fluorescein is loss to the basal layer in the blank inserts compared to other Transwell models. The loss of tracer to the basal layer in the collagen coated inserts occurs slower than the blank insets but faster than the RBE4 models. Loss of tracer between RBE4+Hydrocortisone and RBE4+C6+Hydrocortisone models are similar. Experimental cumulative transport of Na-Fl across *in vitro* models were (1) 51.26% (SD=6.088, N=4) in blank wells, (2) 50.25% (SD=8.595, N=4) in collagen coated wells, (3) 37.304% (SD=2.957, N=5) in RBE4+Hydrocortisone wells, and (4) 37.648% (SD=3.625, N=5) in RBE4+C6+Hydrocortisone wells.

Conclusions: Current results indicate that our *in vitro* BBB models are capable of inhibiting the free diffusion of molecules compared to controls resulting in lower effective diffusion constants. Future studies will involve validation of the predictive capabilities of each diffusion model followed by microfluidic and animal studies to study the effects of diffusion across the BBB when flow dynamics are being considered *in vitro* and *in vivo*.

References: [1] (2015, March 25). Retrieved from http://www.cdc.gov/ncbddd/autism/index.html

422560 From in-Vitro Data to Virtual Patients: Development of an in-Silico Model of Acne Pathogenesis for Prioritizing Targets for Novel Acne Therapies and Prediction of Late Stage Clinical Outcomes

Abstract:

Quantitative Systems Pharmacology (QSP) is an integrated systems-level modeling approach that combines the fields of pharmacology and systems biology. It quantitatively links the pathogenesis of the disease to the mechanism of action of drugs and clinical outcomes by utilizing data from a number of different sources like in-vitro target engagement assays, in-vivo studies as well as patient data from clinical trials. It thus allows mechanistic modeling of the effect of different drugs on disease progression and late stage clinical outcomes and has emerged as an important tool to advance the discovery and development of therapeutic drugs.

In this talk, the application of QSP for modeling the pathogenesis of a skin disease, acne, will be presented. Acne is a multifactorial disease with limited understanding of its pathogenesis. Therefore, a QSP based approach was used to understand the interplay between the processes involved in acne such as sebum production, P. acnes growth and inflammation etc. and their role in disease progression. Also, the effect of modulating these acne processes on disease severity has been evaluated to guide the development of new acne treatments. In addition, different virtual patients have been simulated which can help in identifying the patient subsets that can benefit most from new acne treatments under development within GSK. We have also evaluated the efficacy of three GSK compounds in reducing acne severity as well as compared them to currently available treatments using this model. The talk thus presents an example of how a QSP model of a disease, developed by utilizing current literature knowledge and in-silico tools, can help guide decisions in different stages of drug development.

422856 Quantitative Systems Pharmacology (QSP) Modeling and Simulation in Drug Discovery to Support Pharmacology: Strategy and Applications

Abstract:

Significant challenges are facing the pharmaceutical industry in developing new drugs, and the high rate of attrition demands new approaches in research and development. Advances in modeling and simulations (M&S) have the promise to enable modern drug discovery and development in accelerating decision making. Model based drug development has shown value in increasing probability of success and M&S tools has become integrated in drug discovery and development processes. In this presentation an overview of the applications of M&S to support drug discovery and development projects will be discussed. Examples of modeling drug pharmacokinetics processes including absorption, distribution, metabolism and elimination, as well as modeling systems pharmacology concepts will be presented. Strategies for implementation of M&S tools in drug discovery will be discussed and recommendations for a recipe of success will be proposed.

435605 How Modeling Tools Benefit Early to Late Stage Formulation Development and the Design of Clinical Studies

Abstract:

Pharmacokinetic absorption modeling and simulation is widely used through many drug development stages as a powerful tool for assessing biopharmaceutical risk to, and in silico data. PK absorption modeling tools leverage in vitro, and in vivo to greatly impact formulation development, clinical study design, and regulatory submission. The aim of this presentation is to demonstrate the applications of modeling tools and how they affect drug development decisions.

Four case studies of absorption modeling and simulations combined with conventional analysis tools will be presented. Commercially available software (GastroPlus, v8.5) was used to build PBPK models for the dog and human physiologies. These models were parameterized based on the compound's physicochemical properties, in vitro biorelevant dissolution in fasted state simulated intestinal fluid (FaSSIF), fed state simulated intestinal fluid (FeSSIF) and in vivo data (clearance, plasma time profile, etc.) for each drug.

Rodent and human physiology models in the GastroPlus program were tested, modified, and optimized to obtain a best reliable model to enable decision making based on certain stage of the project. Parameter sensitivity analysis was used to analyze the impact of formulation factors (dose level, particle size, dissolution rate, etc.) to AUG and Cmax in simulated human clinical studies. The case studies cover:

1. Comparing the exposures and pharmacokinetic profile of crystalline material and amorphous dispersions.
2. The impact of drug dose and administration with food on exposure.
3. The impact of particle size on expsoure
4. The impact of patient variability on single and multiple doses using virtual clinical trial simulation

An exploratory clinical study was conducted with larger particle size material, higher dose levels and a standard to evaluate the food effect, dose linearity and particle size risk, and the results aligned with the GastroPlus predictions. The PBPK model output was used to provide guidance on evaluating the biopharmaceutical risk in humans and served as enable decision on formulation development and influencing the clinical development plan.

427792 Effects of Operating Parameters on Tablet Quality in a Continuous Manufacturing Line

Abstract:

As continuous manufacturing is gaining increasing importance within the pharmaceutical industry there still remains a knowledge gap in building quality into the final product. In the context of establishing continuous manufacturing technology, characterization of these processes becomes important. As there are many parameters impacting the product quality, it is important to understand the interplay between these process parameters and tablet performance. Hence, the present study evaluates experimentally the effect of variation in critical process parameters (CPP) on the critical quality attributes (CQA) of tablets produced by direct compression continuous manufacturing. The study was performed on a model API, acetaminophen (APAP), using lactose monohydrate and magnesium stearate (MgSt) as excipients. Multivariate effects were analyzed using Quality-by-design principles. A fractional factorial design with 30 experiments, including 3 repeated center points, was used to study the influence of four process variables – drug concentration (%), compaction force (KN), blending speed (rpm), and feed frame speed (rpm) – on the final product. Design of experiments was executed on a continuous tablet manufacturing line consisting of three feeders, a mill, a blender, and a tablet press. Tablets were characterized for hardness, dissolution, content uniformity, and weight variability. Additionally, they were subjected to NIR transmission spectroscopy analysis and ultrasound testing. The processed NIR spectral datasets were analyzed using PCA, and it was determined that the first and second principal components track tablet drug content and compaction force. Analyses of designed experiments also revealed that drug concentration and compaction force significantly affect tablet dissolution. The results were further used to optimize the process and establish boundary limits for critical process parameters that influence the product to achieve desirable tablet quality.

416476 Enhanced Spray Drying Throughput with High Productivity Hypromellose Acetate Succinate

Abstract:

New drug entities frequently suffer from low solubility and present significant formulation challenges. One of the leading technologies utilized to overcome the low solubility of drugs is the formation of amorphous solid dispersions via spray drying of the API with an excipient. While a number of excipients have been utilized to create amorphous solid dispersions, hypromellose acetate succinate (HPMCAS) has proven to be highly effective and versatile in creating dispersions that demonstrate enhanced free drug levels in solution and inhibit recrystallization. To date, there are no excipients that have been specifically developed for spray drying applications and the excipients used in this application space were all developed for other applications such as enteric coatings and are not optimized for the spray drying process.

The present study compares AFFINISOL™ High Productivity HPMCAS, a new grade of HPMCAS designed specifically for spray drying, to standard HPMCAS. The High Productivity HPMCAS has reduced viscosity compared to standard commercial HPMCAS, allowing for increased throughput and reduced use of organic solvent in the spray drying process. Spray drying performance of the two HPMCAS materials was compared by measuring solution mass flow rate as a function of applied pressure during nozzle atomization testing. A head pressure was applied to the solution container and the resulting mass flow through the nozzle was measured using a Micromotion Coriolis mass flow meter. The size of the spray dried particles were measured using a Malvern Mastersizer 2000 Scirocco particle size analyzer with a variable dispersive air pressure. It was found that to match the atomization performance of standard HPMCAS, the concentration of High Productivity HPMCAS in solution had to be increased by 1.7x relative to standard HPMCAS. In addition, viscosity vs. shear rate at constant temperature of the HP HMPCAS was measured and compared to that of standard HPMCAS. Furthermore, solutions at low concentration show a Newtonian viscosity behavior as a function of the shear rate, without any onset of pseudoplasticity in the detectable shear rate region. With an increase in concentration a small onset of pseudoplasticity becomes apparent and is more pronounced for the conventional HPMCAS than for the High Productivity HPMCAS. Finally, this study will discuss the in vitro drug release rate and crystallization inhibition of amorphous spray dried dispersions formed with itraconazole and phenytoin.

425187 Measuring Blend Hydrophobicity By Single Droplet Penetration Technique

Abstract:

Hydrophobicity is recognized as a critical property for pharmaceutical intermediate blends. For lubricated blends, the hydrophobicity is especially important as it affects the drug release in later *in vitro* dissolution tests. Hydrophobicity also affects the characteristic wetting time during wet granulation processes. In this work, a method was developed to monitor the penetration process of a droplet on powder surface. Using a high-speed camera and video processing technique, the droplet penetration profile was obtained. As a case study, blends with varying hydrophobicity were prepared using a modified Couette shear cell. Comparison of droplet penetration profiles showed that blends with higher hydrophobicity had longer penetration time. Interestingly, a "waiting phase" was observed before the droplet initiates the penetration for less hydrophobic blends. A non-wetting liquid was used in addition to DI water. The penetration model was used to describe the droplet penetration process with the assumption of constant drawing area. The method to compare blend hydrophobicity using penetration profiles from two liquids was demonstrated. This study presented a potential technique to measure blend hydrophobicity to indicated process changes.

427913 Amorphous Solid Dispersion of Poorly Soluble API By Hot Melt Extrusion (HME) and Spray Drying (SD): A Rational Screening Approach

Abstract:

Amorphous solid dispersion of poorly soluble API by hot melt extrusion (HME) and spray drying (SD): A rational screening approach

A.Paudel[1], A. Mercuri[1], S. Mohr[1], M. Bresciani[1], J.G. Khinast[1]

[1]Research Center Pharmaceutical Engineering, Graz, Austria

Polymeric amorphous solid dispersion (ASD) of poorly water soluble active pharmaceutical ingredients (APIs) are one of the promising strategies of solubility and dissolution rate enhancement, and hence, tackling the growing challenges in oral bioavailability (1). Hot melt extrusion (HME) and spray drying (SD) are two industrial technologies for manufacturing ASD (2). Major hurdle associated to the ASD is the possible risk of crystallization of amorphous form during storage and/or administration, thus negating the claimed solubility advantage. Since solid-state API-carrier molecular miscibility and carrier-induced supersaturation play a critical role towards stabilization of ASD and in vivo performance, selection of suitable carrier (s), optimal drug loading and appropriate manufacturing process/parameters is crucial for the success of this formulation strategy (3). HME and SD both exhibit different pros and cons associated with processability of the range of carriers, available solvent range, thermal degradation and molecular, particulate and bulk-level properties of resulting amorphous intermediates. Thus, there is a different degree of downstream processability for finished dosage development.

This work aims towards a multi-methodological approach, using a model drug candidate and a wide diversity of ASD carriers, such as Soluplus, HPMC, HPMC-AS, PVPVA 64 and their blends with Eudragit E or HPC: (i) solution-state excipient-induced biorelevant supersaturation, (ii) film-based mini-formulation preparation via solution casting (SC) and melt-quench-cooling (MC) using selected carriers, (iii) solid-state characterization and mini-dissolution studies, (iv) formulating ASD formulation via HME and SD and (v) solid-state characterization, in vitro biorelevant dynamic dissolution and accelerated short-term physical stability of formulations prepared by HME and SD.

The first set of carriers was selected based on the extent of API supersaturation generation and its maintenance in fasted-state simulated intestinal fluid (FaSSIF). These were Soluplus, HPMC and HPMC-AS/HPC blends. SC and MC films were prepared of API with the varying ratios of latter polymers using HME relevant temperatures and SD relevant solvents, respectively. Drug-carrier miscibility and interaction, glass transition, crystallinity and wettability of film formulations were evaluated using several methods, such as DSC, pXRD, ATR-FTIR spectroscopy, contact angle etc. Based on the outcome, supersaturating formulations with a homogenous amorphous phase were generated by HME and/or SD using different carriers (Soluplus, HPMC and HPMC-AS/HPC). Solid-state characterization, *in vitro* biorelevant dynamic dissolution and accelerated short term stability study were performed on formulations prepared by HME and SD.

It was found that supersaturation screening and small-scale dissolution tests can partially predict the *in vitro* behaviour of ASDs. Film based screening (SC vs MC) aids in predicting the performance rank order of ASD formulations. The results showed that strong functional interactions like API-carrier H-bonding is not always a prerequisite for stabilizing miscible API-polymer systems. Comparable *in vitro* dissolution behaviour of amorphous and partially crystalline Soluplus based dispersion implied that complete amorphicity of API in dispersion is not essential for enhancing *in vitro* performance of solid dispersion. ASD derived from HME and SD can behave completely different when exposed to elevated T/%RH, even though identical carrier with identical composition and physical states are used.

References

1. Van den Mooter, G. (2012) The use of amorphous solid dispersions: A formulation strategy to overcome poor solubility and dissolution rate. *Drug Discovery Today: Technologies 9*, e79–e85.

2. Paudel, A., Worku, Z.A., Meeus, J., Guns, S., Van den Mooter, G. (2013) Manufacturing of solid dispersions of poorly water soluble drugs by spray drying: formulation and process considerations, *International journal of pharmaceutics 453, 253-284.*

3. Paudel, A., Van Humbeeck, J., Van den Mooter, G. (2010) Theoretical and experimental investigation on the solid solubility and miscibility of naproxen in poly (vinylpyrrolidone), *Molecular pharmaceutics* 7, 1133-1148.

GlaxoSmithKline

Pharmaceutical Processes Selection from Shear Cell Analysis and Compaction Simulation

Don Clancy
Modeling Center of Excellence

Presented to AIChE
Wednesday, November 11, 2015: 10:10 AM
Ballroom D (Salt Palace Convention Center)

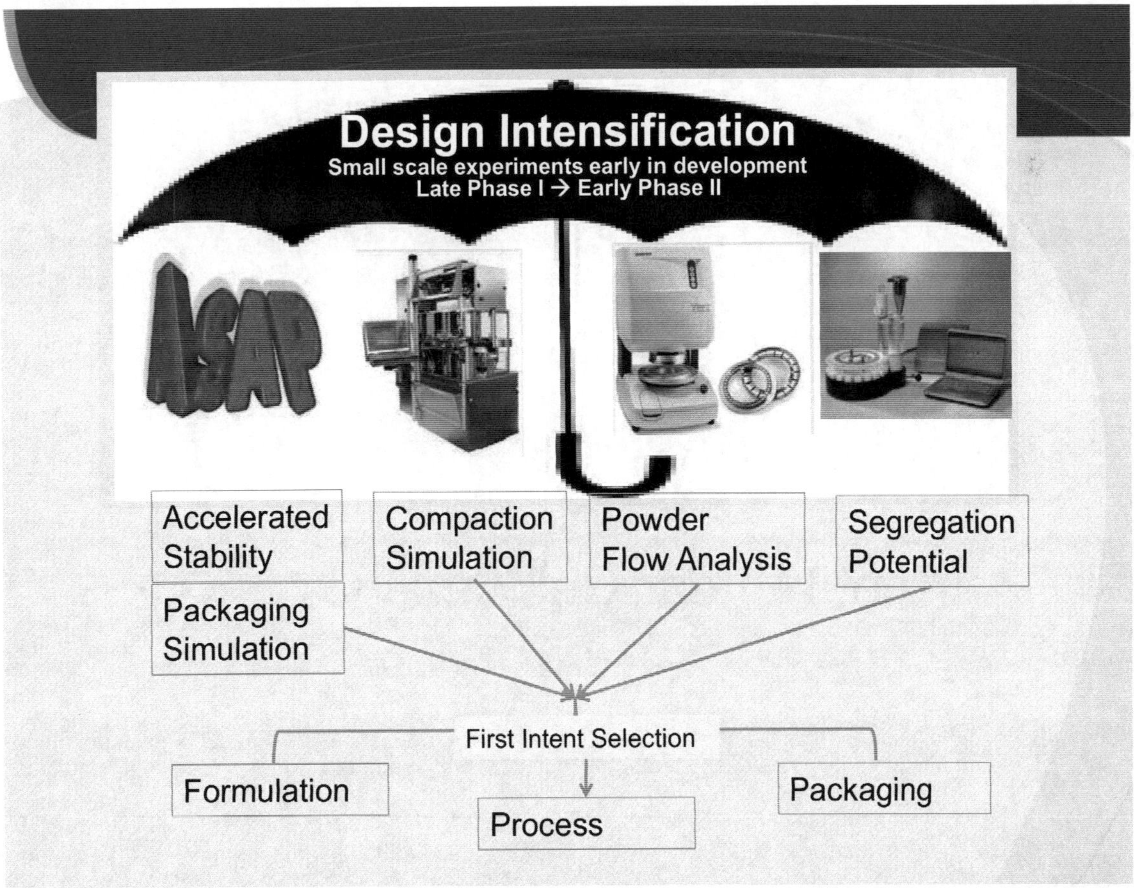

First Intent Selection of Secondary Processing Routes

Secondary Continuous Processing Design Space

Powder Flow	Poor Compressibility	Good Compressibility
Good Flow	Continuous Blend and Capsule Fill (2)	Continuous Blend and Tablet Compression (1)
Moderate	Continuous Wet Granulation (4)	Continuous Dry Granulation (3)
Poor Flow	Batch High Shear Wet Granulation (5)	Batch High Shear Wet Granulation (5)

Powder Compressibility

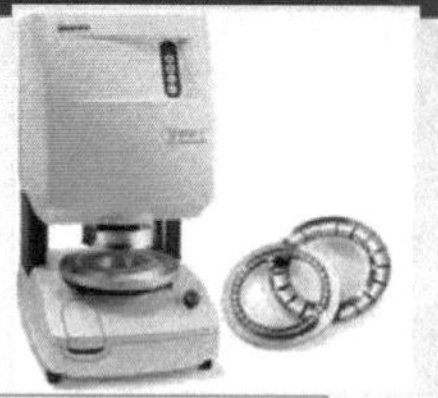

Shear Cell

Compaction Simulation

- This research focus started with the concept that small scale testing in a shear cell and a compaction simulator can provide the information necessary for first intent selection of secondary process routes.

gsk GlaxoSmithKline

About Shear Cell Analysis

Method: Annular Shear Cell - Brookfield Powder Flow Tester

Principle: Tester Cell Lid

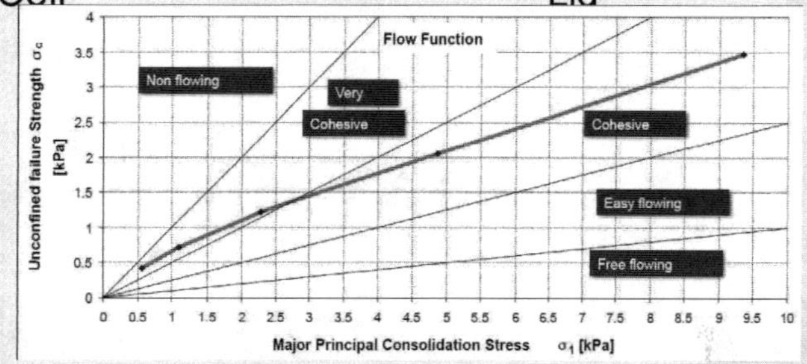

- Cell is filled with the powder being poured loosely and excess material is scraped off
- Powder is compressed by the lid and sheared rotationally
- Shear stress values are recorded by the tester indicating the force required to get the particles to flow against each other
  - This step is repeated for various consolidation loads allowing to build a flow function

Simplified Explanation of How "Flow Index" is Calculated

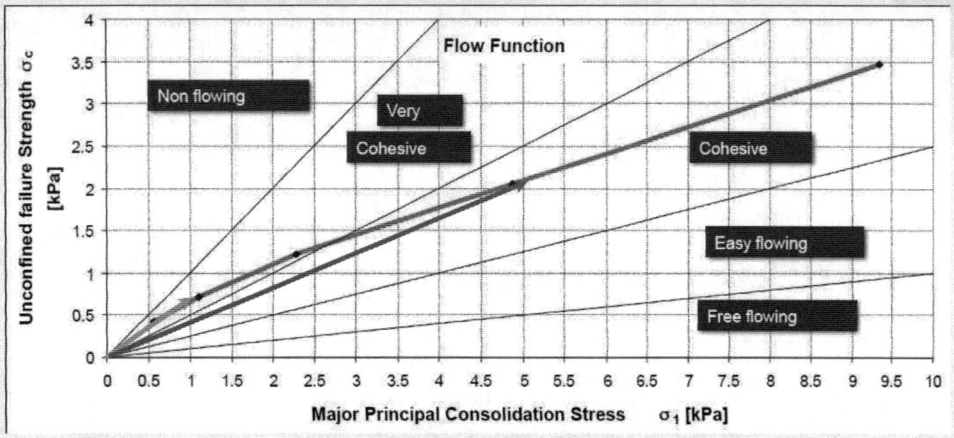

- Two shear cell terms that are used extensively are the "Flow Function", and the "Flow Index."
- "Flow Index" (FI) is the slope of the line that goes from the origin through the measured failure strength at a given major principal consolidation stress.
  - Flow Index at 1-KPa is slope of green line or approximately 0.65.
  - Flow Index at 5-Kpa is slope of blue line, or approximately 0.42.
- The "Flow Function" (FF) is the inverse of the "Flow Index." FF = 1/FI
- A lower Flow Index is indicative of better flow.

GlaxoSmithKline

Tablet Compression

Don Clancy and Laura Wehmeyer

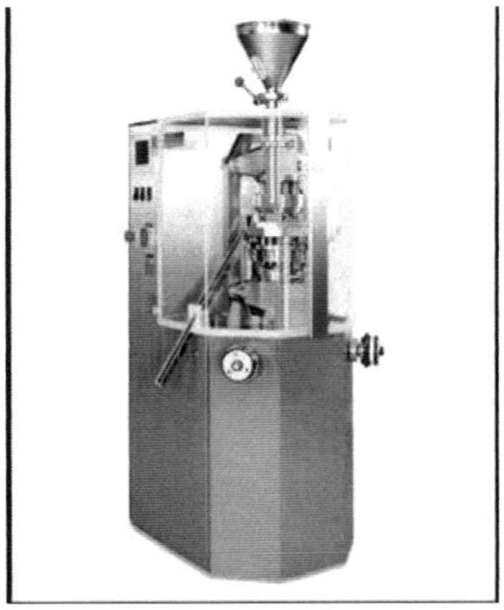

Rational for Defining a Powder Flow Limit for Tablet Compression

- Powder flow increasingly affects the tablet weight variability as the time given for die fill decreases.
- Tablet weight variability affects tablet content uniformity, which has a typical limit of 6% RSD.
- Conservatively, the tablet weight uniformity should not exceed 3% RSD, as both weight variability and blending API content uniformity affect overall tablet content variability.

Typical Experimental Design for evaluating Compression

Blend #	Formulation	Flow Index at 1 KPA	RPM	Tablet Weight (mg)	RSD% tablet weight
1	79.2% Fastflo Lactose, 19.8% Avicel PH102, 1% MgSt	0.10	13	202.28	1.10%
2	19.8% Fastflow Lactose 79.2% Avicel PH102, 1% MgSt	0.17	13	200.82	1.51%
3	99% Avicel PH102, 1% MgSt	0.21	13	186.98	1.70%
4	99% Avicel PH101, 1% MgSt	0.28	13	169.3	2.05%
1	79.2% Fastflo Lactose, 19.8% Avicel PH102, 1% MgSt	0.10	66	195.6	1.26%
2	19.8% Fastflow Lactose 79.2% Avicel PH102, 1% MgSt	0.17	66	155.9	2.66%
3	99% Avicel PH102, 1% MgSt	0.21	66	162.94	2.37%
4	99% Avicel PH101, 1% MgSt	0.28	66	138.98	5.96%
1	79.2% Fastflo Lactose, 19.8% Avicel PH102, 1% MgSt	0.10	120	169.4	1.31%
2	19.8% Fastflow Lactose 79.2% Avicel PH102, 1% MgSt	0.17	120	133.62	2.61%
3	99% Avicel PH102, 1% MgSt	0.21	120	148.72	3.10%
4	99% Avicel PH101, 1% MgSt	0.28	120	146.2	4.32%

- This is an example experiment design for a single size die cavity on the Korsch XL300.

Tablet Compression – Gravity Fed Analysis

Actual by Predicted Plot

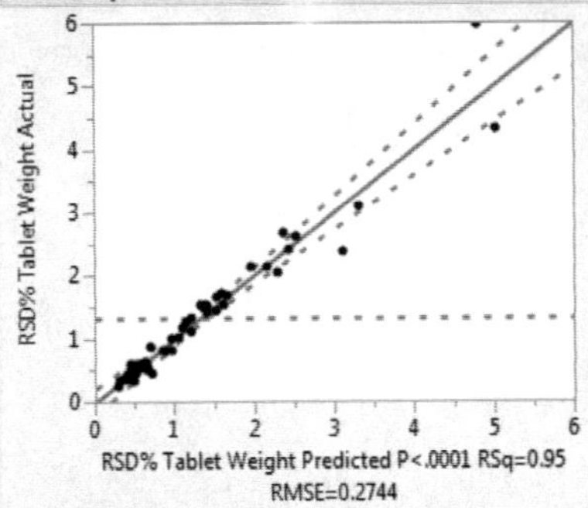

Summary of Fit

RSquare	0.945261
RSquare Adj	0.935916
Root Mean Square Error	0.274434
Mean of Response	1.321794
Observations (or Sum Wgts)	49

Analysis of Variance

Source	DF	Sum of Squares	Mean Square	F Ratio
Model	7	53.323252	7.61761	101.1446
Error	41	3.087875	0.07531	Prob > F
C. Total	48	56.411127		<.0001*

Analysis of Variance

Source	DF	Sum of Squares	Mean Square	F Ratio
Model	7	53.323252	7.61761	101.1446
Error	41	3.087875	0.07531	Prob > F
C. Total	48	56.411127		<.0001*

Parameter Estimates

Term	Estimate	Std Error	t Ratio	Prob>|t|
Intercept	0.46523	0.168395	2.76	0.0085*
Flow Index at 1 KPA	11.078386	0.799136	13.86	<.0001*
Tablet Weight (mg)	-0.001521	0.000163	-9.32	<.0001*
1/RPM	-9.125136	2.848993	-3.20	0.0026*
(Flow Index at 1 KPA-0.14642)*(Flow Index at 1 KPA-0.14642)	31.193683	12.82952	2.43	0.0195*
(Flow Index at 1 KPA-0.14642)*(Tablet Weight (mg)-538.497)	-0.010197	0.00299	-3.41	0.0015*
(Tablet Weight (mg)-538.497)*(Tablet Weight (mg)-538.497)	1.6606e-6	7.439e-7	2.23	0.0311*
(Flow Index at 1 KPA-0.14642)*(1/RPM-0.02267)	-218.5304	37.43267	-5.84	<.0001*

Contour Plot for RSD% versus Tablet Weight and Flow Index at 1 KPA – Gravity Fed Presses

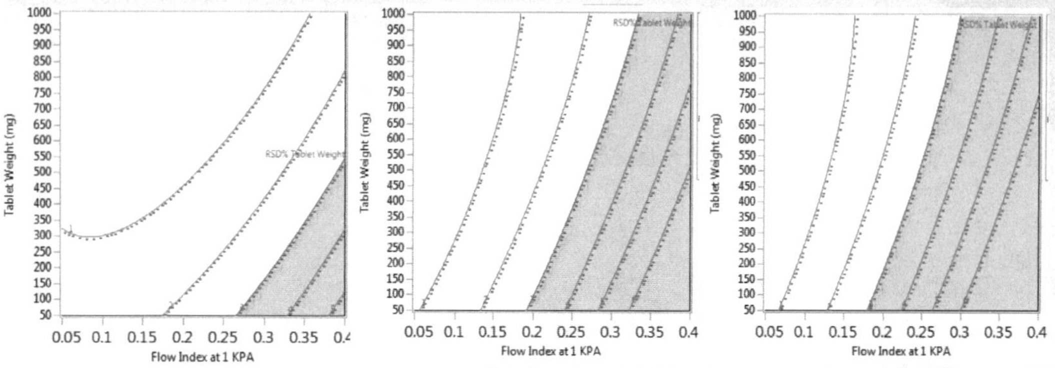

15 RPM | 45 RPM | 120 RPM

- 200 mg tablets require a flow index of 0.2 at 1 KPA for 3% RSD at 120 RPM, which is where the design space upper limit is set.
- The limit is dependent on tablet weight, as 3% RSD upper limit, shows possible flow index for larger tablets could reach 0.3 for a 500 mg tablet.
- Model is extrapolated above 0.28 flow index due to problems with arching in the feed hopper preventing further material with worse flow properties than 0.28 from being processed. Larger presses with larger hopper feed outlets will have less problems with arching in the feed hopper.

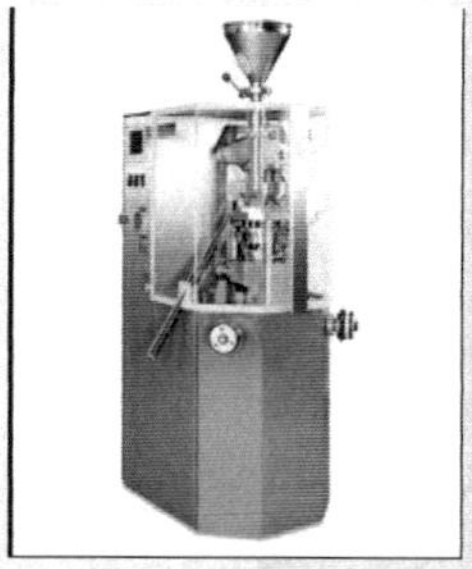

High Speed Compression - Fette 102i with Paddle Feeder

Whole Model

Actual by Predicted Plot

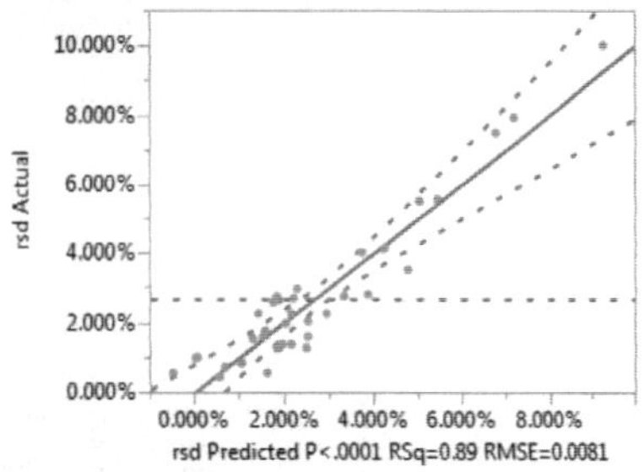

Summary of Fit

RSquare	0.894036
RSquare Adj	0.862639
Root Mean Square Error	0.008119
Mean of Response	0.026702
Observations (or Sum Wgts)	36

Analysis of Variance

Source	DF	Sum of Squares	Mean Square	F Ratio
Model	8	0.01501471	0.001877	28.4755
Error	27	0.00177959	0.000066	Prob > F
C. Total	35	0.01679430		<.0001*

Parameter Estimates

Term	Estimate	Std Error	t Ratio	Prob>|t|
Intercept	-0.027731	0.004961	-5.59	<.0001*
Flow Index	0.0873655	0.011679	7.48	<.0001*
RPM	0.000275	3.36e-5	8.18	<.0001*
Tablet wt.	5.1275e-5	6.561e-6	7.82	<.0001*
(Flow Index-0.235)*(Flow Index-0.235)	-0.289955	0.115668	-2.51	0.0185*
(Flow Index-0.235)*(RPM-65)	0.0013241	0.000284	4.66	<.0001*
(RPM-65)*(RPM-65)	2.0141e-6	1.011e-6	1.99	0.0566
(Flow Index-0.235)*(Tablet wt.-356.161)	0.0001756	4.878e-5	3.60	0.0013*
(RPM-65)*(Tablet wt.-356.161)	1.0893e-6	1.579e-7	6.90	<.0001*

Fette 102i – 24 Station Paddle Fed up to 172,800 tablets per hour

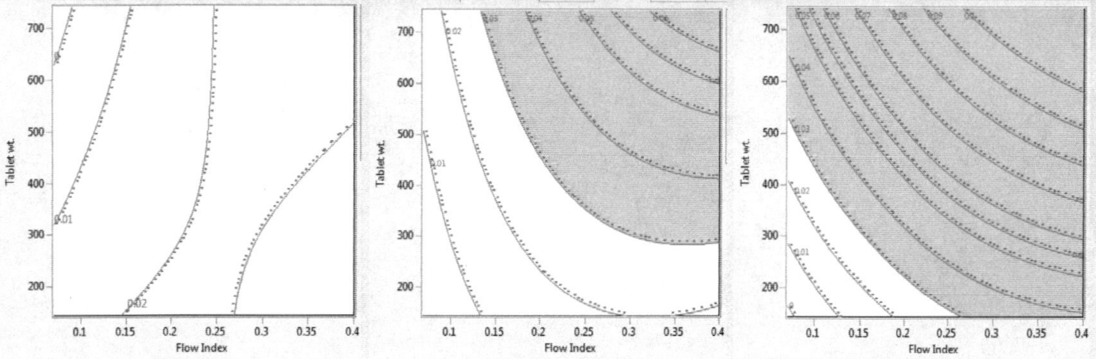

10 RPM 65 RPM 120 RPM

- The Fette 102i with a paddle feeder is far more capable of handling poorly flowing materials at lower speeds. Material with flow index as high as 0.44 was successfully processed even at 65 rpm for 200 mg tablets.
- The same approximate limitation of a flow index of 0.23 for a 200 mg tablet weight is observed at 120 rpm, which is not much different than the observation on the Korsch.
- The extreme limitation for larger tablets >400 mg might be due to limitations of the feed rate through this particular paddle feeder.

gsk GlaxoSmithKline

Part II:

Powder Flow Requirements for Encapsulation

With thanks to:
Fernando Muzzio, et.al. - Rutgers

Predicting Capsule Weight Variation

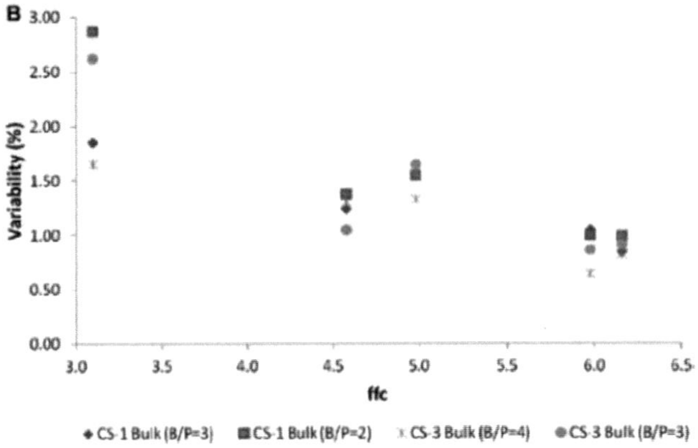

- Capsule weight variation can be reasonably predicted from the Powder Flow Function (ff) (left graph) measured by shear cell at 6 KPa.
- For capsule RSD<3%, a Flow Function greater than 3 is needed. This corresponds to a Flow Index of 0.33 at 6 Kpa. The approximate corresponding flow index at 1 Kpa is 0.4.

Reference Link

First Intent Selection of Secondary Processing Routes Using Powder Flow and Compressibility

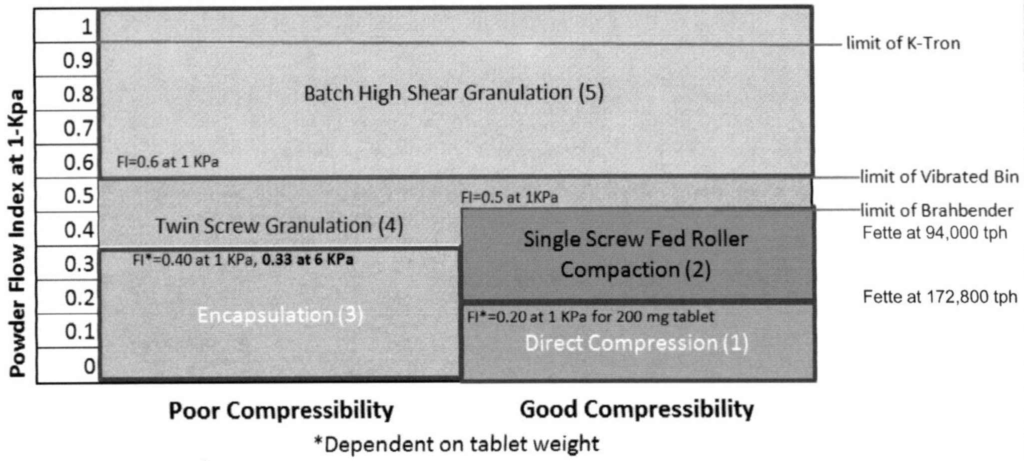

- The resulting "Design Space" confirmed by experimentation is shown in the above figure.
- Further information on segregation potential and melting point can also affect this process selection.

Part III: Powder Flow Requirements for Twin Screw Granulation

Don Clancy

Significant Contributions:
Vicky He, Matt Bahr, Scott Croy, Sayantan Chattoraj

"K-Tron" Twin Screw Hopper

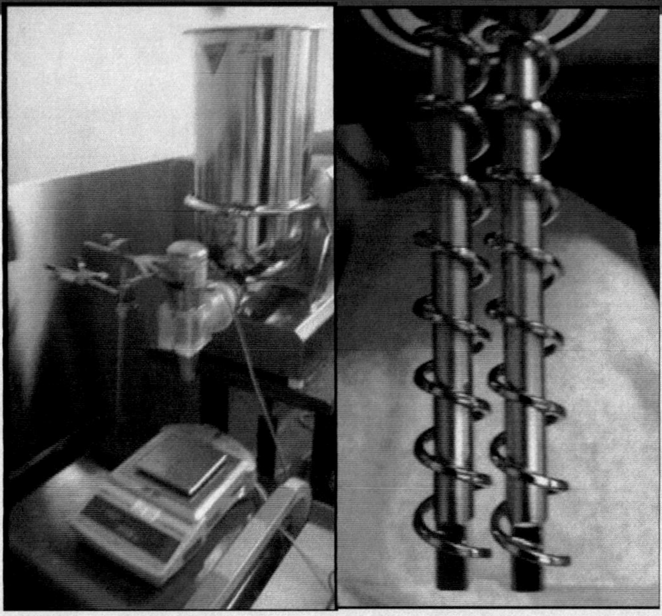

- K-Tron is the main hopper used for Twin Screw granulation.
- The hopper features twin helical screw conveyors.
- Outlet of the twin screw conveyors leads to a vertical "Cramming" unit to push powder into the twin screw granulator.

Good Versus Poor Powder Flow From Screw Conveyed Hoppers

Good powder flow (movie link) Poor flowing powder (movie link)

- Powders cannot flow continuously like water. When a pressure is put on a powder, it will not move until it reaching point exceeding failure pressure. Then it will move as a sudden avalanche.
- Good flowing material will appear to flow near continuously as shown on left, but poor powder flow will discharge as packets of material as shown on right.
- Variation can be observed by weighing the powder output at rapid time intervals. In this study, the fastest our scale can weigh dynamically is once every 3 seconds.

K-Tron Hopper 3-Second Flow Variability Related to Shear Cell Flow Index at 1 KPa

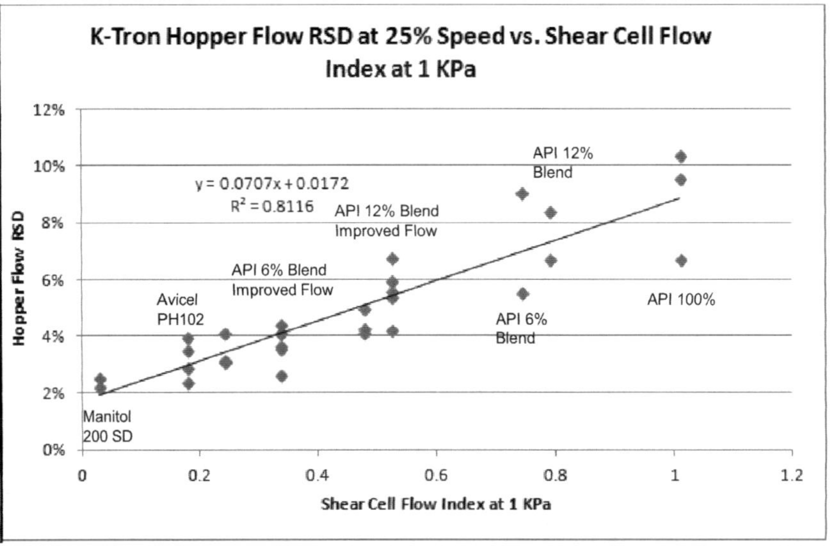

- The K-Tron hopper feeding twin screw granulation is remarkably robust, and has successfully fed pure PHI API with flow index up to 1.0, though with increasing levels of instantaneous variation in feed rate.

Dual Helix Hoppers (K-Tron) –Discharge Variability

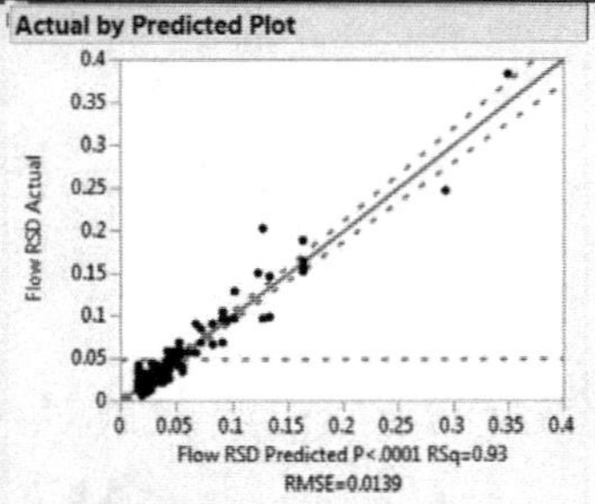

Summary of Fit

RSquare	0.928332
RSquare Adj	0.925963
Root Mean Square Error	0.013944
Mean of Response	0.048511
Observations (or Sum Wgts)	126

Analysis of Variance

Source	DF	Sum of Squares	Mean Square	F Ratio
Model	4	0.30473011	0.076183	391.8343
Error	121	0.02352547	0.000194	Prob > F
C. Total	125	0.32825558		<.0001*

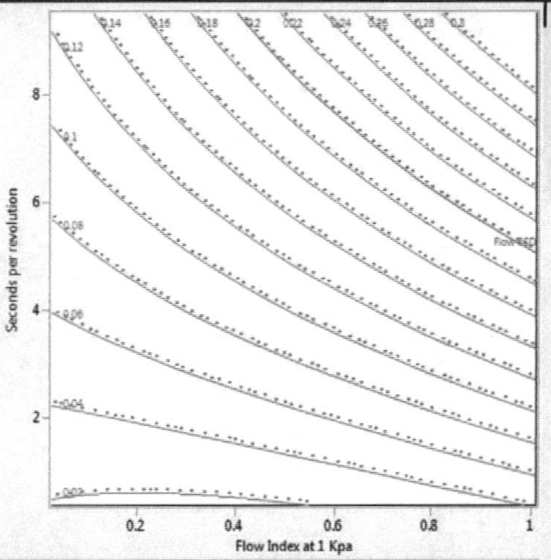

Parameter Estimates

Term	Estimate	Std Error	t Ratio	Prob>|t|
Intercept	-0.011143	0.002509	-4.44	<.0001*
Flow Index at 1 Kpa	0.0449248	0.004573	9.82	<.0001*
Seconds per revolution	0.0217191	0.000776	27.99	<.0001*
(Flow Index at 1 Kpa-0.47865)*(Flow Index at 1 Kpa-0.47865)	0.0449677	0.014787	3.04	0.0029*
(Flow Index at 1 Kpa-0.47865)*(Seconds per revolution-1.4832)	0.0228292	0.00197	11.59	<.0001*

- This is a remarkably resilient hopper, and will not be the real limitation.
- Discharge powder flow RSD is mostly impacted by time needed for discharge screw rotation (RPM), with some impact of flow properties at slow discharge rates.

Conclusions Regarding K-Tron Dual Helix Hopper Feed to Continuous Twin Screw Granulation

- The K-Tron hopper is a remarkable piece of equipment and unlikely to create a hard limit of flow index that can be fed. Material with a flow index as high as 0.94 at 1 Kpa consolidation pressure was successfully fed with no incidence. This is the worst flowing API, GSK has with a rod-like crystalline structure.

- Instead, there's a softer limit imposed by the variability in flow that's correlated to the flow index and the feeder RPM. A more variable inlet flow will lead to a more variable continuous granulation process, and a narrower design space for the granulation.

GlaxoSmithKline

Part IV: Requirements for Single Screw Hopper fed Unit Operations (Roller Compaction)

Don Clancy
June 2014

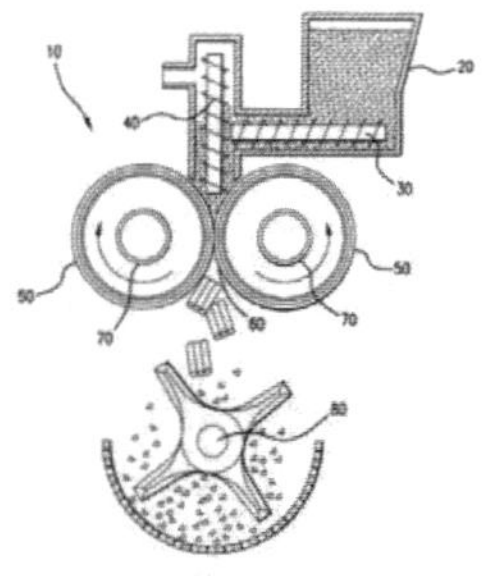

FIG. 1

Part IV - Flow Requirements for Single Screw Hoppers - Brahbender

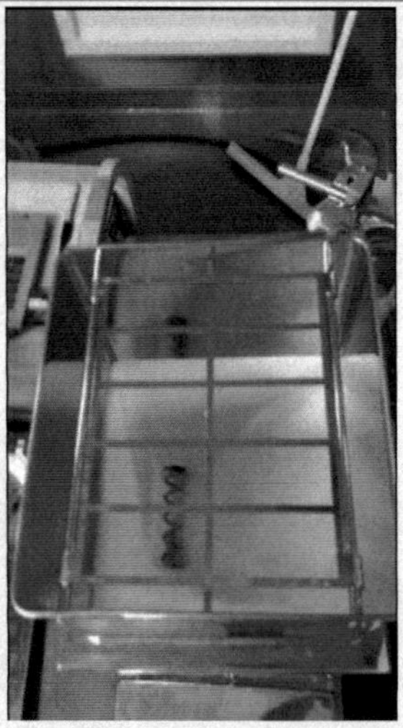

View from Top

- This Brahbender hopper features a single screw, and has a useful advantage of being able to work with very little powder. It's able to deliver powder with only 30 to 60 grams of powder retention left in the hopper.
- Single screw hoppers such as this are not able to feed as wide a variety of materials as the dual screw K-Tron.

Single Helix Hopper (Brahbender) –Discharge Variability

Actual by Predicted Plot

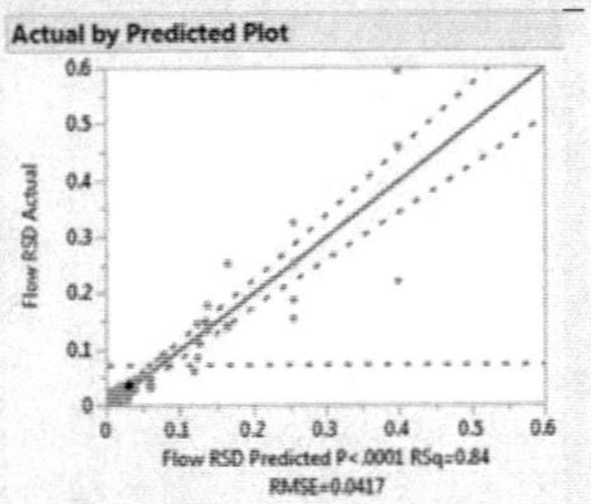

Summary of Fit

RSquare	0.842022
RSquare Adj	0.83259
Root Mean Square Error	0.041742
Mean of Response	0.072676
Observations (or Sum Wgts)	72

Analysis of Variance

Source	DF	Sum of Squares	Mean Square	F Ratio
Model	4	0.62222723	0.155557	89.2771
Error	67	0.11674108	0.001742	Prob > F
C. Total	71	0.73896832		<.0001*

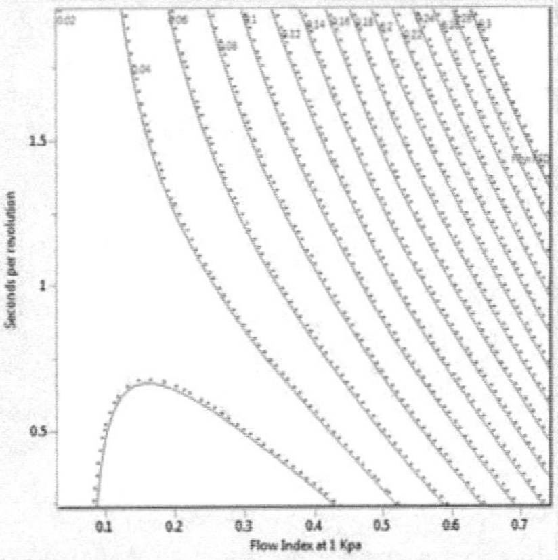

Parameter Estimates

Term	Estimate	Std Error	t Ratio	Prob>|t|
Intercept	-0.064341	0.010601	-6.07	<.0001*
Flow Index at 1 Kpa	0.1889273	0.027612	6.84	<.0001*
Seconds per revolution	0.0564204	0.008212	6.87	<.0001*
(Flow Index at 1 Kpa-0.30083)*(Flow Index at 1 Kpa-0.30083)	0.5057237	0.094609	5.35	<.0001*
(Flow Index at 1 Kpa-0.30083)*(Seconds per revolution-0.86368)	0.233865	0.030841	7.58	<.0001*

- Heavily impacted by powder flow properties, only suitable for flow index <0.5 at 1 Kpa.
- Problems with arching and unpredictability at >0.5 flow index at 1 Kpa.
- Same helix impeller design used for feeding roller compaction and twin screw

Beware of Bins Upstream of Continuous Feeders!

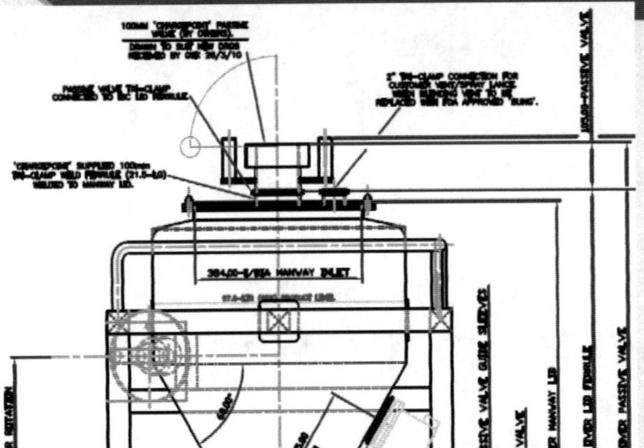

- This bin is used to blend excipients and API together prior to twin screw granulation or roller compaction
- The limitation is the narrow outlet, which is 15 cm in diameter.
- Powders arch over the outlet and powder flow can be intermittent. The arching dimension needs to be less than 15-cm, which corresponds to a flow index of approximately 0.6 at 1 KPa for most powders.

First Intent Selection of Secondary Processing Routes Using Powder Flow and Compressibility

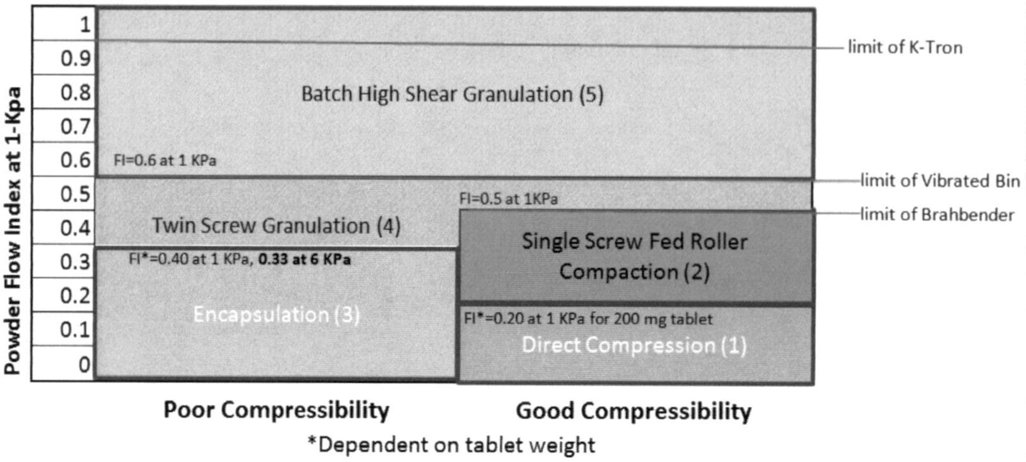

Final mapping of flow requirements for processing.

GlaxoSmithKline

Example Case Study:

Formulation Development of An Expensive Drug Product for Roller Compaction

Don Clancy, Maria Barnett and Francis Ignatious

Nov 2014

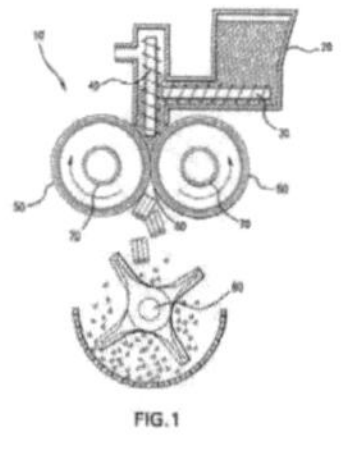

FIG.1

Background to Case Study

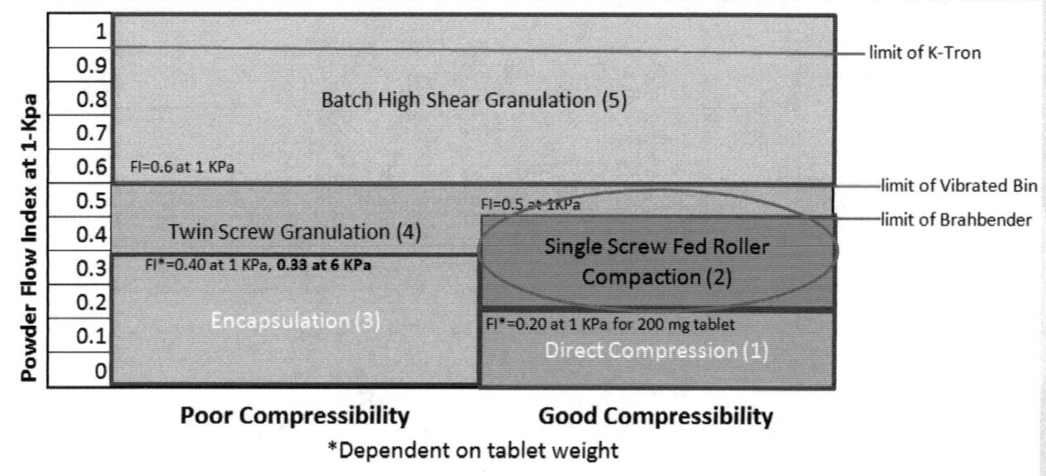

- Formulation exhibits good compressibility characteristics.
- Desire is to minimize tablet size by increasing drug content to the limit of what can be robustly processed by roller compaction.

API Content in Blends Versus Flow Index at 1 KPa

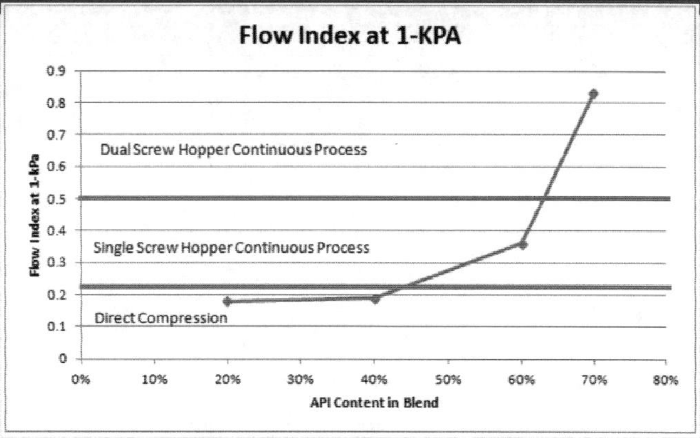

Flow Function Results:

Data Set #	Flow Function Slope	Flow Function Intercept (kPa)	Flow Index	Batch Number
1	0.06	0.259	0.18	EE771611D_20%Blend
2	0.08	0.521	0.19	EE771611C_40%Blend
3	0.09	0.839	0.36	EE771611B_60%Blend
4	0.09	0.767	0.83	EE771611A_70%Blend
5	0.10	0.102	0.22	Avicel PH102_142385515

- To qualify for direct compression, the flow index needs to be better (smaller) than Avicel PH102 at 1 KPa pressure. In this case, the 20% and 40% blends could be directly compressed without significant concerns of tablet weight variability problems.
- The 60%, and 70% blends have significantly poorer flow compared to Avicel PH102, and would need to be granulated prior to compression.

IBC Bin Requirements for formulation with flow aid

☑ **Rat-hole Diameter:** Bin Diameter 1 .000 m Height 1 .000 m ☑ **Arching Dimension:**

Results:

Data Set #	Arching Dimension (m)	Rat-hole Diameter (m)	Batch Number
1	0.030	0.216	EE771611D_20%Blend
2	free flow	0.296	EE771611C_40%Blend
3	0.070	0.528	EE771611B_60%Blend
4	0.301	1.000	EE771611A_70%Blend
5	0.041	0.213	Avicel PH102_142385515

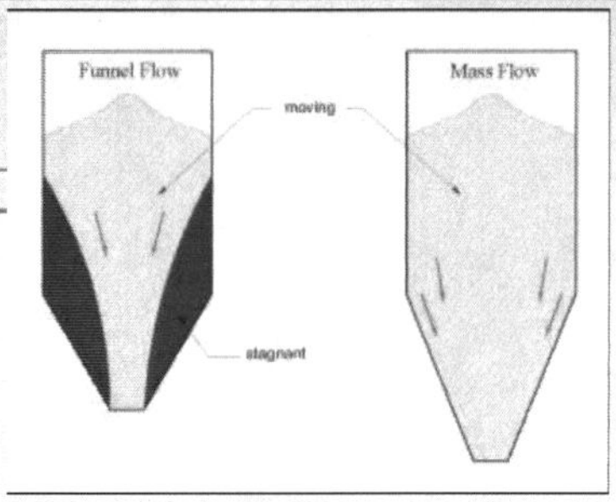

- Of interest is the 60% blend (data set#3), and whether the 60% and 70% formulations can be fed reliably out of an rectangular IBC bin to feeder into roller compaction.
- The 60% blend needs a bin opening outlet greater than 7-cm in diameter to prevent stable arch formation. Flow type will be funnel flow which will result in a rat hole that forms approximately 0.5 meter in diameter.
- The 70% blend will not feed reliably from an IBC bin. It will be able to arch over a 30-cm outlet. Assuming it's agitated to break the arch, the rat hole diameter equals the bin diameter. A rat hole will form and then no further flow would happen after that point. Approximately half the batch in the bin would be within the stagnant region.

Impact of 0.5% Flow Aids at 55% API Content, High Shear Blending

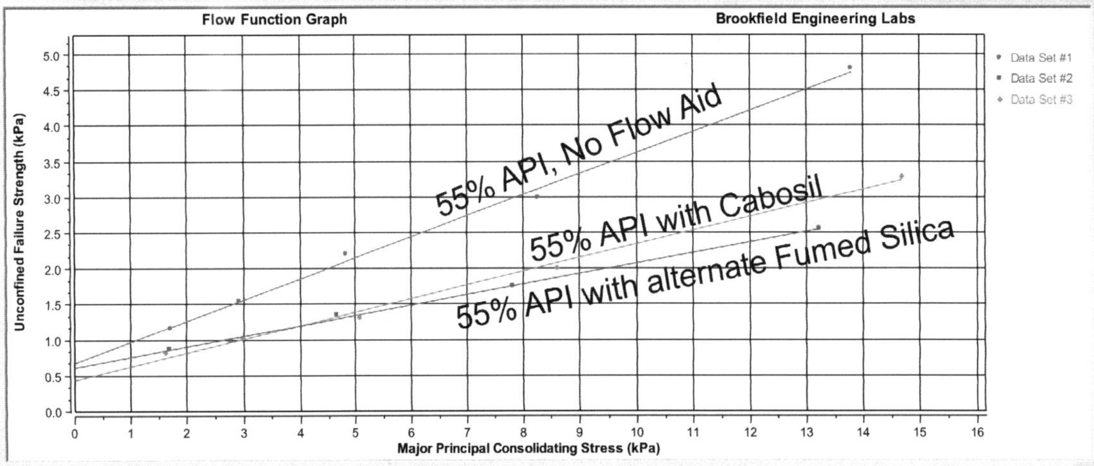

- 0.5% Fumed silica has a substantial impact of improving flow properties of this blend, and making it possible to feed into roller compaction.
- There's very little difference between Cabosil and the other type of fumed silica, though Cabosil is showing slightly improved flow at low pressure.

Conclusions from Case Study

- The 20% formulation has good flow properties, is classified as "easy flowing close to free flowing" and would be a candidate for direct compression with little tablet weight variation.
- The 40% formulation has good flow properties at low pressure that would be classified as "easy flowing." This formulation could be directly compressed with tablet weight variation similar to Avicel PH102.
- The 60% formulation would need to go through a granulation process to improve the flow properties before going on to compression. This formulation can be hopper fed from a typical single screw conveyed hopper without expecting problems. It can be fed from a bin with at least a 7-cm outlet.
- The 70% formulation can not be fed from an IBC bin, or a single screwed hopper. It can be fed on a double screw conveyed hopper, like the "K-Tron" hopper for the twin screw.
- A fumed silica flow aid is needed. There's little difference between the fumed silica suppliers on the resulting improvement in flow.

Conclusion

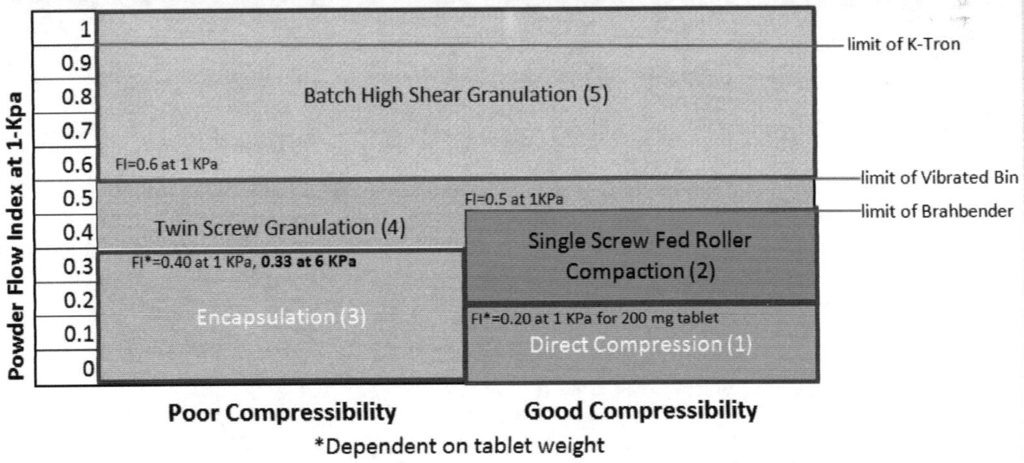

The use of a shear cell, and a compaction simulator, can provide product teams with useful information to designate the process needed, or alternatively provide feedback on formulation options to optimize the product for a given process.

Acknowledgements

- Robert Yule - GSK
- Sayantan Chattoraj - GSK
- Laura Wehmeyer - GSK
- Fran Muller - GSK
- Vicky He - GSK

437432 Continuous Chromatography: Breaking the "Bottleneck" in Downstream Processing

Abstract:

Multi-column continuous chromatography has recently become an enabling technology that will "break the bottleneck" in downstream processing of bio-pharmaceuticals. This development is not only driven by the tremendous increases in expression levels in the upstream fermentations, especially for monoclonal antibodies, but also by the introduction of biosimilars/biobetters the market. Continuous chromatography has shown promises in reduction of manufacturing cost. However, up to today there has not been a reported case at the production scale. What are the remaining technical barriers when implementing the technology in the GMP environment? This presentation highlights how to overcome the major barriers when developing this technology platform.

418599 Design and Simulation of a Continuous Bioreactor

Abstract:

A novel continuous bioreactor consists of an outer shell cylinder and inner core cylinder, each with adjustable rotational velocities to control the mixing in the reactor and provide various shear environments. Inlet flows are controlled to provide additional medium and oxygen to the reactor volume. A spiroid is attached to the inside of the bioreactor to increase the gas-liquid contact area and improve oxygen transfer. This bioreactor has dramatic advantages in cell cultivation because of the high cell survival rate, enhanced growth rate, and overall high productivity.

Computational fluid dynamics (CFD) simulations were successfully performed to describe the range of flow patterns possible in this bioreactor. Simulations are conducted for multiphase fluids with various operating conditions to simulate the actual reaction environment in the bioreactor. Velocity profiles, shear distributions and turbulent energy dissipation were shown to predict the high performance of the bioreactor. In addition, various experiments along with mathematical models were used to calculate the volumetric mass transfer coefficients at different operating conditions for bioreactors with and without the spiroid. The comparison of the results supported the improved performance of the bioreactor with the spiroid. Cell lines were selected for cultivation in the bioreactor.

Key words: computational fluid dynamics, continuous bioreactor, oxygen transfer, cell cultivation

422745 New Synthetic Adsorptive Depth Filtration Media for the Primary and Secondary Clarification of Mab from CHO-S Harvests

Abstract:

We report the design and performance characteristics of a synthetic depth filtration media that demonstrates a greatly improved filtration capacity and decreased effluent turbidity for the clarification of mAb from CHO-S harvests. This new depth filtration media is constructed entirely from synthetic components, which include a silica-based filter aid, acrylic fibers, and a charged polymeric binder resin. When compared to commercial depth filtration media comprising mixtures of diatomaceous earth and cellulose, the synthetic depth filter media is shown to have greatly decreased organic extractibles and reduced WFI flushing requirements by TOC analysis. The synthetic depth filter media also presents a significantly increased filtration capacity versus the commercial benchmark due to the incorporation of a pre-filtration layer. As an added benefit, the silica filter aid is shown to impart a modest HCP clearance capability through an adsorption mechanism. The impact of HCP clearance on downstream process purification steps is briefly explored.

418751 Strategies for Implementation of a New CHO Platform Process

Abstract:

Platform processes are integral for the rapid development of new therapeutic molecules within the biopharmaceutical space. Since platform processes are inherently limited in their capacity for change, there are serious strategic questions related to timing and implementation of modifications to a platform. Under discussion are the implications of changing a platform process with a focus on mammalian cell culture development. This includes necessity and timing for the change, the method of exploration of new platform setpoints and conditions, qualification of the platform, and characterization of platform robustness.

A case study is presented regarding application and implementation of a new platform process. Exploratory studies were used to identify incremental improvements to an existing platform. Changes included culture feed and feed timing, gassing conditions, and seed train. When evidence was generated to support use of the new platform, a series of qualification studies on multiple established molecules were conducted to support the use of the new platform. Finally, after the platform was accepted a collection of process robustness studies were performed based on a platform risk analysis. These studies will provide the process knowledge to support a quality by design approach through both early and late stage development.

431560 The Impact of Cell Culture Media Variability on the Process Development of Biosimilars

Abstract:

Sourcing limitations and the efforts to minimize introduction of adventitious viral or prion agents in the production of biologics have over the years accelerated the adoption of serum-free cell culture media. The subsequent use of animal component free media containing plant derived hydrolysates in cell culture processes has come with observed variability in cell growth, productivity and product quality of biologics. In its purest and supposedly most consistent form, cell culture media has evolved to chemically defined media. As part of a media vendor qualification study, four proprietary media formulations manufactured by five vendors were tested for composition, particle size distribution and solubility characteristics. Media from two vendors were then used in functional cell culture testing. There was a direct correlation between the media powder particle volume mean diameter (VMD), time to solution and final media turbidity. Differences in media turbidities before and after filtration indicated particulate removal during filtration step, suggesting that some components were not going into solution and therefore unavailable to cells in the bioreactor. Differences between expected and measured levels of media components such as trace metals suggested that there were other extraneous sources of trace elements. Functional cell culture testing results showed that these variations in concentrations of media components had some product quality impact that can affect comparability. Media quality, among other process variables, needs to be controlled for consistency in cell culture process performance.

437406 Correlation of Non-Baffled, Angled and Off-Center Agitated Process Vessel Blend Times

Abstract:

Non-baffled vessels are appealing for appropriate light duty mixing applications in the pharmaceutical industry because they may simplify the equipment cleaning challenges associated with baffles. The use of angled and off-center axial flow mixers in non-baffled dished bottom vessels can provide light duty mixing with good vertical mixing and minimal swirling. The blend time to achieve 95% homogeneity for a family of geometrically similar tanks from 100L to 5,000L from a manufacturing facility was characterized and fitted to multiple published engineering literature correlations which have been previously developed for baffled tanks. A comparison of the data fits to the correlations will be presented.

425170 Experimental Determination of Solids Suspension with Angled Impellers in Pharmaceutical Mixing Vessels

Abstract:

Angled-mounted impellers are commonly used in a number of pharmaceutical industry applications, from laboratory reactors to full-scale tanks. In these systems, the impeller is not centrally and vertically mounted, as in most stirred vessels, but it enters the vessel, and is therefore immersed in the fluid, at an angle from the vertical. This arrangement provides some baffling effects that would not otherwise exist if the impeller was centrally mounted.

One common requirement for such systems is that they are capable of suspending small solids dispersed in the liquid. Therefore, it is critical to know the value of the minimum agitation speed at which solids become just suspended, N_{js}. Despite their common use in industry, and especially the pharmaceutical industry, these systems have received little attention in the literature in general and as far as their ability to solid suspension is concerned in particular.

Therefore, the focus of this work was to determine under what conditions fine divided solids become suspended in a liquid-filled, unbaffled stirred vessel provided with angled-mounted impellers. A small laboratory vessel with a hemispherical bottom was used here. Different types of axial and radial impellers mounted at different angles (0°, 5° and 10°) at different off-bottom impeller clearances were tested. The value of N_{js} was experimentally obtained by visually inspecting the tank bottom and determining the impeller agitation speed at which the solids were observed to rest on the tank bottom for no more than 1-2 seconds before being swept away. In addition, a novel observer-independent criterion was developed, relying on the experimental determination of the fraction of solids remaining on the vessel bottom at increasing agitation speeds. This approach results in values of N_{js} that are close agreement with the visually determined N_{js}values.
This approach was used to determine N_{js} for different impellers under a number of operating conditions. In general, mounting the impeller at an angle resulted in N_{js} values lower than those obtained with centrally mounted impellers. In addition, it was found that lower impeller clearances resulted in lower N_{js} values, although the minimum for $N_{js}$was obtained when the impeller clearance was about 20% of the vessel diameter.

The results obtained in this work are expected to be applicable to a number of systems currently used in the pharmaceutical industry.

427057 Multiphase Bioreactor Simulation with GPUs

Abstract:

The effectiveness of a biopharmaceutical manufacturing process depends to a large extent on the efficiency of the bioreactor, especially in the field of generic drugs. The improvement of the productivity of industrial size bioreactors is, up to now, mainly driven by empirical knowledge. Computer simulation would help to understand the processes inside the reactor and hence lay the foundation for new reactor designs. However, despite the recent improvement of computational capabilities, only a few seconds of real time operation of an industrial scale reactor takes months to simulate.

Thus, the goal of this study is to use graphic cards to speed up this simulation. To take full advantage of the processor's parallelism, localized calculation algorithms are used, like the lattice Boltzmann method for modelling the fluid flow field. The calculation of the bubble movement is done by solving the Newton's equation of motion. For coalescence and breakup stochastic models are used. The phases are coupled with a two way approach.

To simulate large reactors, the code has a multi GPU functionality, hence it can distribute the workload of the simulation on several graphic processors. Validation is done with literature data as well as with results from holdup measurements of a 150l laboratory reactor. To show the usefulness of this code for scale up studies, a 40 m^3 reactor was simulated.

To mimic the behavior of real fermentation broth a non-Newtonian viscosity model was added using the power law model. An important process within the reactor is the transport of species (dissolved oxygen, nutrients, carbon dioxide, etc.) by the fluid flow field. To model the distribution of the species a particle based solver was implemented.

434243 Retreat Curve Impeller Flow and Heat Transfer Characteristics

Abstract:

Glass-lined vessels and impellers are often encountered in mixing operations, particularly in the pharmaceutical and fine chemicals industries. Retreat Curve Impellers are designed without sharp edges to make them easy to coat with glass. Although other more efficient glass-coated impeller designs are currently available, the retreat curve impeller is still in common use. A number of studies, both experimental and computational, have been conducted to characterize the flow patterns, mixing times, power requirements and just suspended speeds of mixing systems using these impellers. Very few studies have included the heat transfer characteristics. This talk will present the procedures and results of a previously unpublished experimental investigation to develop tables and correlations for the Power Number, Flow Number, and Nusselt Number associated with a dished bottom vessel fitted with a retreat curve impeller of various diameters and off-bottom clearances, as well as different baffle arrangements. The heat transfer data was obtained using a novel wall heat flux system with data for the vessel sides and bottom treated separately. These correlations have been used to design and optimize numerous stirred tank vessels and processes.

427476 Advances in Gas Sparging Simulation for Bioreactor Modeling to Create Comprehensive "Simulation Design Space"

Abstract:

Gas-liquid dispersion in stirred tanks is a commonly used industrial process, e.g., bio-pharmaceutical or fine-chemical manufacturing. While CFD models have been used to predict gas holdup, bubble size distribution and mass transfer coefficients, there have been fewer studies on deploying the CFD models to range of operating conditions. This study focuses CFD simulation at different rotational speed (rpm) and gas flow rates to create a comprehensive "Simulation Design Space" for mass transfer co-efficient (Kla). Mass transfer co-efficient is experimentally obtained by measuring Dissolved Oxygen (DO) concentration by absorption in pure water. In current study, simulation model is developed numerically to calculate DO through multiphase and multispecies CFD simulations. Population balance model is coupled with the Eulerian multiphase-multispecies model to predict the bubble size distribution and oxygen distribution. The study also looks into effects of turbulence on drag co-efficient which is a key factor in predicting an accurate bubble size distribution and gas hold-up. Simulation best practices are developed and then deployed for validation of the model against experimental data [1] on wide range of operating conditions.

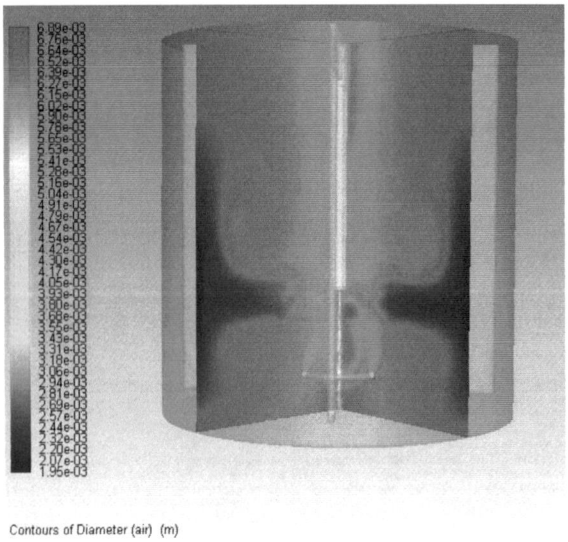

Bubble size distribution on planes between baffles

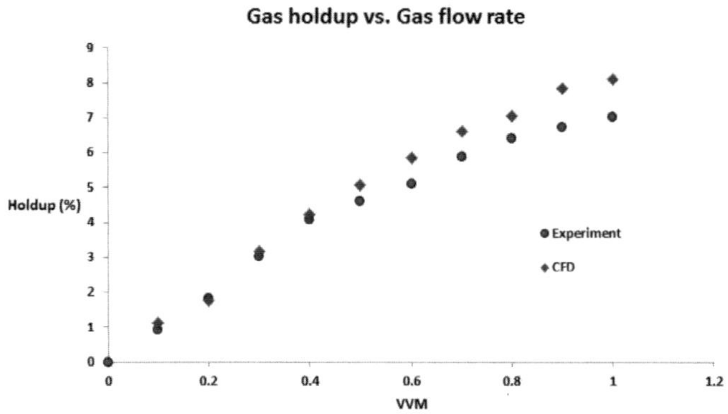

Variation of gas holdup with gas flow rate

1. M. Laakkonen et al., Chemical Engineering Science 62 (2007) 721 – 740

439687 Enhanced Dissolution of Itraconazole from Surfactant-Free Nanocomposite Microparticles

Abstract:

Bioavailability of a poorly soluble drug can be improved by preparing a drug nanosuspension and subsequently drying it into nanocomposite microparticles[1,2]. Unfortunately, drug nanoparticles aggregate during milling and drying, causing incomplete recovery and relatively slow dissolution[3,4]. Various dispersants can be used to mitigate the extent of nanoparticle aggregation. The aim of this study is to investigate the impact of various classes of dispersants on the dissolution of drug nanoparticles, with the ultimate goal of enhancing the bioavailability of poorly water-soluble drugs via high drug nanoparticle loaded, surfactant-free nanocomposites. Precursor suspensions of itraconazole (ITZ, a model poorly water-soluble BCS Class II drug) nanoparticles in the presence of various dispersants were prepared via wet stirred media milling and spray dried to form the nanocomposites. Hydroxypropyl cellulose (HPC, polymer) alone and with sodium dodecyl sulfate (SDS, surfactant) was used as base-line stabilizers/dispersants during milling. Three swellable crosslinked polymers known as superdisintegrants, croscarmellose sodium (CCS), sodium starch glycolate (SSG), and crospovidone (CP) as well as two conventional soluble matrix formers, sucrose and mannitol, were used in addition to HPC. Laser diffraction, scanning electron microscopy, powder X-ray diffraction, ultraviolet spectroscopy, and dissolution tests were used for the characterization. The results indicate that the enhanced wettability imparted by SDS was critical to the full recovery of drug nanoparticles and their dissolution, whereas HPC had relatively small positive effect. Use of co-milled superdisintegrants led to significant enhancement of the drug dissolution with SSG performing close to SDS, thus demonstrating the feasibility of spray drying to prepare high drug-loaded, surfactant-free nanocomposites. The positive impact of the superdisintegrants positively correlated with their swelling capacity (SSG > CCS >CP), signifying a swelling-induced erosion/disintegration mechanism for fast drug release from the nanocomposite particles.

References Cited

1. Müller RH, Peters K. Nanosuspensions for the formulation of poorly soluble drugs: I. Preparation by a size-reduction technique. *International Journal of Pharmaceutics.* 1998;160(2):229-237.

2. Lee J. Drug nano- and microparticles processed into solid dosage forms: physical properties. Journal of Pharmaceutical Sciences. 2003;92(10):2057-2068.

3. Bhakay A, Azad M, Vizzotti E, Dave RN, Bilgili E. Enhanced recovery and dissolution of griseofulvin nanoparticles from surfactant-free nanocomposite microparticles incorporating wet-milled swellable dispersants. Drug Development and Industrial Pharmacy. 2013(0):1-14.

4. Bhakay A, Davé R, Bilgili E. Recovery of BCS Class II drugs during aqueous redispersion of core–shell type nanocomposite particles produced via fluidized bed coating. Powder Technology. 2013;236(0):221-234.

421481 CFD Investigation of Nozzle Configuration and Interface Tension Effects on Core-Shell Droplet Formation in Coaxial Electrohydrodynamic Atomization Process

Abstract:

Core-shell microspheres have shown particular advantages in drug delivery application as drug carriers, such as preventing burst release of drug, improving drug stability and achieving synergistic therapeutic effects. Coaxial electrohydrodynamic atomization (CEHDA) is an emerging technique having several advantages in fabrication of core-shell micropaticles, such as uniform size distribution, effective protection of bioactivity and high encapsulation efficiency. However, the complex physics behind the process is still one of the challenges in the development of CEHDA. In addition, the productivity is limited by the flow rate especially for fine particles production. In order to better design the microparticle structure, improve the productivity of CEHDA process and further facilitate the scale-up of the fabrication process to industry, a computational fluid dynamic (CFD) model was developed to simulate and provide a better understanding of the CEHDA process. A volume of fluid model (VOF) was employed to simulate the liquid cone-jet mode and droplet formation. The electric body forces were included into the Navies-Stokes equations using user-defined function (UDF). The effects of nozzle configurations on the cone-jet formation and droplet formation which had not been systematically studied in literatures were investigated.. Furthermore, the effects of interface tension between core and shell fluids were investigated. The results showed that the configurations of nozzle tip have significant effect on the droplet production. Possible mechanisms were proposed to illustrate the differences of core-shell droplet formations with various nozzle configurations. The results also showed that using specific nozzle configuration, the limitation of flow rate on the productivity can be further overcome for the production of micro-/nano- particles.

423605 Synthesis and Stability Study of Polymorphic Transformed Mannitol/ LB Agar Microcarriers for Pulmonary Drug Delivery

Abstract:

Novel rough mannitol/LB Agar (lab-made) microparticles were synthesized by polymorphic transformation method. Spray-dried mannitol/LB Agar (lab-made) microparticles were immersed in hexane solution to initiate polymorphic transformation. After 5 to 15 days of reaction, rough mannitol/LB Agar (lab-made) microparticles were obtained. As-prepared microparticles were characterized by X-ray diffraction spectra (XRD), differential scanning calorimetry (DSC), scanning electron microscopy (SEM), thermal gravimetric analysis (TGA) and Andersen Cascade Impactor (ACI).

The XRD and DSC results indicate that δ-mannitol was completely transformed to α-mannitol after immersing the spray-dried mannitol/LB Agar (lab-made) microparticles in hexane. SEM shows that the spray-dried microparticles had smooth surfaces, while the surface of the transformed microparticles were rough. TGA result indicates that no hexane remained in the transformed microparticles. ACI result shows that the microparticles synthesized in the study had comparable fine particle fraction (FPF) with the best performing carriers reported in the literature. For the stability study, XRD indicates that the α-mannitol present in the transformed mannitol/LB Agar (lab-made) microparticles was stable for 33 months. SEM indicates that the small mannitol crystals in the transformed microparticles coaggulated with each other and self-assembled into nanorods. The transformed microparticles were able to maintain the rough-surface and spherical shape for 21 months. It is substantially longer than that of the mannitol/LB Agar (commercial) microparticles, which can only maintain a stable shape for 6 months.

425544 Detailed Simulation of Particle and Liquid Distribution in the Mixing Zone of a Twin-Screw Granulator

Abstract:

Wet granulation for continuous manufacturing in the pharmaceutical industry can be achieved using a twin-screw granulator, which is emerging as a promising technology. This is due to the high degree of flexibility through varying (i) the screw configuration which has a modular structure and (ii) the process parameters, allowing to achieve mixing and granulation in a short time (in the order of seconds). However, to optimally take advantage of this flexibility, a thorough understanding of the mixing and flow characteristics, which drive the wet granulation, in the twin-screw granulator is necessary. Despite the progress that has been made by experimental and meso-scale modelling studies in recent years, the description of the flow and mixing of the wet particulate materials in the twin-screw granulator is still challenging. In this study, we examined the time evolution of the spatial liquid distribution in the kneading section of a twin-screw wet granulation process. Particle scale, discrete element method (DEM) based simulations of a short quasi-two-dimensional simulation domain were performed. A capillary force model, as well a liquid transfer model (Mohan et al., Powder Technol 264: 377–395) were used to account for pendular liquid bridge formation, rupture and inter-particular forces. The analysis of liquid distribution between particles and liquid bridges showed that, if the liquid addition rate and source area are changed, and despite the fact that the same amount of liquid was added to the particles, the partitioning between the liquid present on the particle surface and the liquid bridges changes. For a larger spatial extent of the region in which liquid was added to the simulation, the number of liquid bridges increased. However, again the partitioning of the liquid in the bridges and on the particle surface changed. This suggests a higher aggregation level due to a higher number of liquid bridges, but also the possibility of increased breakage rates in case of high shear treatment due to the weak liquid bridge forces. Our results help identifying key parameters that impact liquid spreading (i.e., solid-liquid mixing) during particle processing in twin-screw granulators. Consequently, our results are an important guide in the future development of advanced closures for mesoscopic models, e.g., population balance models used in industrial applications.

425948 Production of Amorphous Nano-Solid Dispersions Using a Solvent Controlled Precipitation Process – Benchmarking with the Spray Drying Process

Wednesday, November 11, 2015: 10:00 AM
254A (Salt Palace Convention Center)
Íris Duarte[1,2], Pedro Serôdio[2], M. Luísa Corvo[1], João Vicente[2], João F. Pinto[1] and **Márcio Temtem**[2], (1)iMed.ULisboa, Faculty of Pharmacy, University of Lisbon, Lisbon, Portugal, (2)R&D Drug Product Development, Hovione FarmaCiência SA, Loures, Portugal

Production of amorphous nano-solid dispersions using a solvent controlled precipitation process – benchmarking with the spray drying process

Iris Duarte[1,2], Pedro Serodio[1], M. Luisa Corvo[2], Joao Vicente[1], Joao F. Pinto[2], Marcio Temtem[1*]

[1] Hovione FarmaCiencia SA, Sete Casas, 2674-506 Loures, Portugal;
[*]mtemtem@hovione.com or +351 219 847 569

[2] iMed.ULisboa, Faculdade de Farmácia da Universidade de Lisboa, Av. Prof. Gama Pinto, 1649-003, Lisboa, Portugal

Current pharmaceutical pipelines are highly populated with new molecules presenting poor physicochemical properties, which typically translate into solubility issues. Poor solubility is one of the major concerns in the oral-drug delivery field, mainly because it limits bioavailability. Among the different formulation strategies applied to address this issue, the use of amorphous solid dispersions (ASDs) is a popular technology, with a large number of amorphous-based medicines reaching the market.

Bioavailability enhancement may be achieved by improving the dissolution kinetics - applicable to Class IIa compounds, according to the recent *Developability Classification System* (DCS) - and/or by increasing the maximum concentration of the active compound in solution - applicable to DCS Class IIb compounds [1]. In the particular case of DCS class IIa drugs the particle size can play an important role in the enhancement of the dissolution rate. Although various methods are reported in literature for the preparation of solid dispersions (*e.g.* spray drying, freeze drying, hot melt extrusion, etc), the state of the art is scarce in technologies that enable both control of the particle size in the submicron range while maintaining the amorphous state.

In order to address this, the purpose of this work was to evaluate the feasibility of using a solvent controlled precipitation (SCP) process to produce nano-sized ASDs [2]. Moreover, the influence of varying the formulation and process conditions on the co-precipitated products was assessed.

<> Carbamazepine (anhydrous, Form III), a DSC Class IIa molecule, was the model drug selected to conduct this feasibility study. Solvent controlled precipitation experiments were undertaken using *Microfluidics Reaction Technology*, followed by a spray-drying step to isolate the particles [3]. Figure 1 represents the half-factorial design 2^{3-1} + 2 central points conducted to study the effect of critical formulation variables [*i.e.* type of polymer, drug load and feed solids concentration (C_feed)] on typical critical quality attributes (CQAs) of ASDs (*e.g.* physical stability, particle size and morphology, bulk powder rheology, dissolution/supersaturation potential, *in vivo* bioavailability). Among the API/Polymer systems tested, the ones that resulted in true amorphous solutions were also produced using spray-drying (SD), for benchmarking purposes.

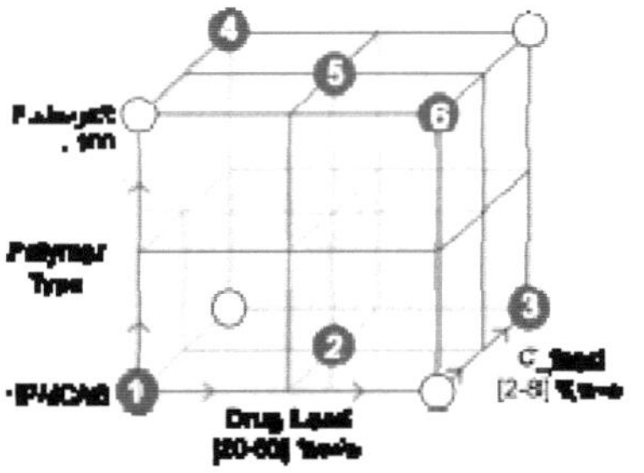

Figure 1. Half-factorial design (2^{3-1}) + 2 central points for the solvent controlled precipitation process study.

Figure 2, on the left, shows the XRPD diffractograms obtained for the different co-precipitated products. These results indicated that drug's solid-state and physical stability was mainly dependent on the type of stabilizing polymer and drug load in the formulation. For the CBZ:HPMCAS systems, a gradual increase in the relative intensity of the characteristic peaks of CBZ with increasing drug loading was observed, which indicated that glass solutions were not obtained. By opposition, for the CBZ:Eudragit® L100 systems, Test #4 resulted in a glass solution while Test #5 and #6 resulted in crystalline suspensions of CBZ within the polymer, indicated by the absence and presence of CBZ characteristic peaks, respectively. C_feed had no significant effect in the amorphization and physical stability of the products.

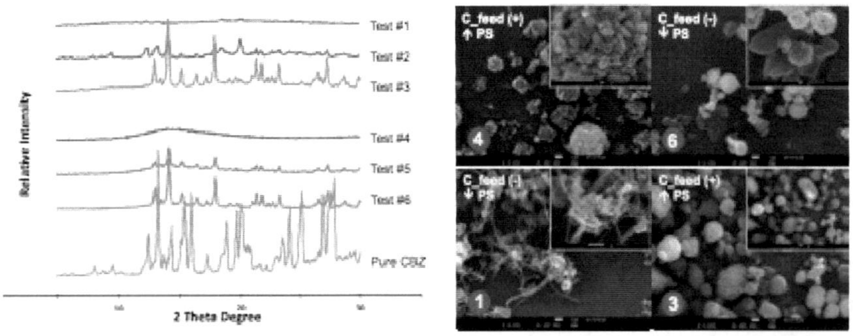

Figure 2. Left: XRPD diffractograms corresponding to Test#1 to Test#6. Right: SEM micrographs corresponding to Tests #4 and #6 and Tests #1 and #2.

On the right hand side of Figure 2 are represented the SEM results obtained for Tests #1 and #3 and Tests #4 and #6, at 5000x magnification. In relation to particle size and morphology, agglomerated and spherical particles were obtained among all tests, varying in a size range of 5-10μm, regardless the drug-polymer system or drug load. However, in terms of number distribution in each sample, a higher number of particles with a larger diameter were observed in the samples corresponding to high solids' concentration in solution, and vice-versa. These results indicated that bulk particle size was mainly dependent on the feed solids concentration, and both variables varied proportionately. Finally, the observation of the particles surface under high magnification revealed that the agglomerates consisted of individual particles, most of them with a diameter below 100nm (inserts). Moreover, these nano-particles were more individualized in case of low drug load formulations, and more entangled/fused with each other for higher drug load formulations. Consequently, formulations with lower drug load presented a higher superficial area, which is believed to have a positive influence on the dissolution profile of a BCS/DCS Class IIa drugs.

To assess the *in vitro* performance of the amorphous nanocomposite particles produced by SCP (Test #4) powder dissolution experiments were conducted (Figure 3). For benchamrking purposes, two additonal CBZ-based formulations were produced: (1) 60 wt% CBZ:Eudragit® L100 by SCP and (2) 20 wt% CBZ:Eudragit® L100 by SD, both at 8 wt% C_feed. The former formulation resulted in a crystalline suspension of CBZ within the polymer (similar to Test #6) and maintained its nano-features, while the latter resulted in a true glass solution, but in the form of microparticles. Figure 3, on the right, shows these differences, in terms of particle size/morphology.

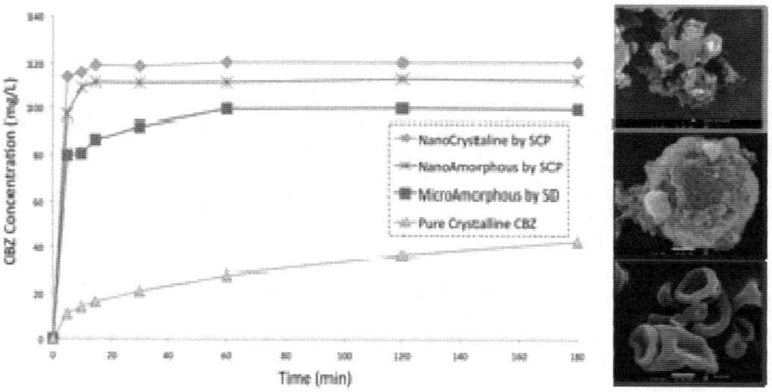

Figure 3. Left: Dissolution profiles of different CBZ:Eudragit® L100 formulations, either produced by SCP or SD and pure crystalline CBZ (n=1). Right: respective SEM micrographs of CBZ:Eudragit® L100 formulations tested.

When compared with spray-dried ASDs, the rapid dissolution of high-surface area nanoparticles produced by SCP favoured the creation and maintenance of higher supersaturated levels. Differences between the NanoCrystalline and NanoAmorphous formulations may be related with different powder wetting properties; nevertheless additional dissolution experiments will be conducted (n=3) to better understand these differences. Moreover, the ongoing *in vivo* studies will also support a greater understanding of this synergistic effect (nano + amorphous) in absorption and bioavailability of DCS Class IIa drugs.

To assess the bulk powder rheology of the amorphous nanocomposite particles produced by SCP (Test #4), a FT4 powder rheometer was used. Similarly, and for comparison, the respective amorphous spray-dried formulation (20 wt% CBZ:Eudragit® L100, C_feed at 8 wt%) was also tested.

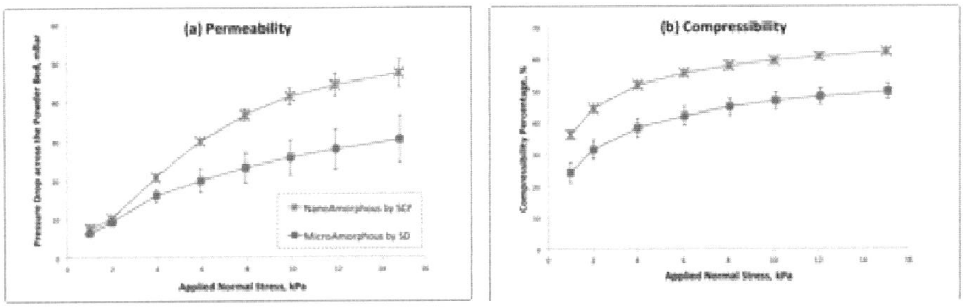

Figure 4. Left: Permeability as a function of normal stress at constant air velocity of 2 mm/s. Right: Compressibility (*i.e.* % of bulk density change) as a function of normal stress.

Figure 4 shows the results from the permeability and compressibility tests at 15 kPa. Permeability is a measure of how easily the powder can transmit air through its bulk, while compressibility is a measure of the volume change in a powder as a consequence of an applied consolidating stress [4]. The NanoAmorphous formulation showed to be less permeable, and consequently more compressible than the MicroAmorphous formulation. These differences are clearly related with the different particle properties of both powders, namely particle size, shape, surface properties, amount of fines, etc.

References:

[1] Butler J. and Dressman J., The Developability Classification System: Application of Biopharmaceutics Concepts to Formulation Development, *J Pharm Sci*, 2010, 99 (12), pp. 4940-54;

[2] Provisional patent application: PT107846;

[3] Panagiotou T., *PureNano™ and MRT Applications*, Microfluidics, 2012;

[4] Freeman Technology Limited, An Introduction to powders, 2014.

434832 Regulatory Science Perspective on Continuous Manufacturing

Abstract:

The continuous manufacturing of pharmaceutical drug products is gaining momentum within the industry. The successful implementation of continuous manufacturing represents the culmination of innovative quality and regulatory approaches from the previous decade, including PAT, increased product and process understanding, and advanced control strategies. The FDA's Office of Pharmaceutical Quality supports the development and implementation of control strategies that lead to increased quality assurance and availability of medicines to the public. This presentation will address the control strategy considerations for implementation of a continuous manufacturing process, focusing on assurance of quality aspects. The discussion will include concepts such as science-based justification for material traceability, sampling frequency and throughput/run time. The importance of understanding failure modes and implementing of controls at appropriate levels will be highlighted with emphasis on increased process understanding and demonstration of a state of control. In conclusion, design and operation of a continuous manufacturing process is based on science and engineering principles, but a control strategy that ensures product quality and consistency is the key to addressing the regulatory challenges for this technology.

430211 Key Considerations to Achieve Manufacture Excellence in Continuous Manufacturing Using Hot Melt Extrusion

Abstract:

Key Considerations to Achieve Manufacture Excellence in Continuous Manufacturing Using Hot Melt Extrusion

Hot melt extrusion is one of the most important enabling processing technology for manufacturing amorphous solid dispersions (ASDs) to enhance oral drug absorption. The extrusion process is inherently continuous. Back-to-back extrusion runs without shutdown have been demonstrated successfully in commercial manufacturing over the past decade with improved manufacture efficiency and flexible batch size to meet dynamic market supply needs. Manufacturing excellence is ensured through fundamental understanding of formulation degradation mechanism and impact of process/formulation variability on product quality as well as implementation of a robust quality control strategy including but not limited to PAT tools. Best practices to achieve manufacture excellence in continuous manufacturing using hot melt extrusion will be shared through a case study. Techniques and control strategies to enable fully continuous operation and real-time release will be reviewed and discussed.

414679 Considerations in Choosing Between the Use of Process Data and Analytical Data to Monitor and Control Cqa's in Continuous Drug Product Processes

Abstract:

Continuous drug product manufacturing introduces new risks and opportunities in providing assurance of quality. Many of these involve the ability to measure or predict product quality and reject/accept/adjust the process in real-time. The primary approach discussed in the literature for this is through the intensive use of on-line and at-line analytical testing (often referred to as PAT). A clear definition of the objectives and a first-principles approach to addressing them can result in a greater reliance on achieving routine quality though process monitoring and control and the focusing of analytical testing on those areas not being addressed with process data. An example is provided for the continuous direct compression platform where potency, blend uniformity and content uniformity can all be assured by the use of process data. This requires the process meet the criteria of being in a state of control and having good to excellent capability and also having acceptable risk of special cause upsets. When used in conjunction with process data this approach might also change validation requirements for on-line/at-line analytical methodologies to a less demanding level of rigor depending on its intended purpose.

430155 Nonlinear Steady-State Data Reconciliation for a Continuous Tableting Line

Abstract:

The robust operation of a continuous process requires a sufficient level and quality of online process measurements and an appropriate level of process control. As the pharmaceutical industry transitions from batch to continuous operation the questions of how many measurement points and what quality of online measurements are necessary or sufficient and the level of sophistication of the process control systems required are very much under discussion. Under the aegis of the NSF ERC on Structured Organic Particulate Systems, we have been investigating these questions in the context of continuous production of tablets. Tablets are the most common dosage form and thus the conversion from batch to continuous production of this drug product form has thus received quite a bit of attention [1].

In this paper we specifically focus on answering questions concerning the selection of and impact of online measurements. To address these questions, we first develop a nonlinear steady state data reconciliation model for a continuous dry granulation line. The data reconciliation model of such a process network consists of the relationship between all of the measured and relevant unmeasured variables of the process together with a description of the error distributions of these measurements [2, 3]. It is assumed that the number of measurements is larger than the degree of freedom of the network model, hence that there is redundancy. Given a set of measurements, an optimization problem is posed which has as its objective function a likelihood type function. This function is the minimization of the difference between the measured values and their best estimates, weighted by the standard deviation of the measurement error of the associated sensor model, with the above mentioned reconciliation model forming the constraint set [3, 4].

The solution of the maximum likelihood problem constitutes the best real time estimate of the state of a processing system taking into account all of the measurements and their error characteristics. Such an estimate can serve as basis for checking whether all process variables are within acceptable bounds, for identify gross measurement errors, for control action or for process optimization steps. Additionally the reconciliation formulation provides a framework for investigating the impact of adding or eliminating measurements, of replacing instruments with given error characteristics with those with tighter variances or of introducing measurement redundancy [5].

While the data reconciliation problem has received much attention in the process systems engineering literature, most of the attention has been directed at linear models, which consist of the component mass balances of the process [3]. Nonlinearities have principally been introduced through the addition of energy balances, which in their most common form involve equations that are bilinear in flow and composition or flow and temperature [3]. Additionally, it is generally assumed that measurement errors are normally distributed. In this paper we consider nonlinearities which arise because of the inclusion of online measurements of properties such as density and variables such as screw speed, roll speed, etc. The model nonlinearities have implications on computation performance, which is important for real time application. Finally, given the nature of some of the spectroscopic measurements used in the tableting application [6, 7], the assumption of normality of the measurement error comes into question [5]. We illustrate the approach using case studies carried out on a continuous tableting line.

References

[1] Reklaitis G. The Pharmaceutical Manufacturing Institute. **2013**, (White Paper) Purdue University, West Lafayette, IN.

[2] F. Carl Knopf. Modeling, Analysis and Optimization of Process and Energy Systems. First Edition. *John Wiley & Sons, Inc*. **2002.**

[3] Narasimhan S. & Jordache C. Data reconciliation & Gross Error detection. Houston Texas. *Gulf Publishing Company*. **2000.**

[4] Narasimhan S., Bhatt N. Deconstructing principal component analysis using a data reconciliation perspective. *Computers and Chemical Engineering*. **2015**, 77, 74–84

[5] Cencica O. and Frühwirth R. A general framework for data reconciliation. Part I: Linear constraints. *Computers and Chemical Engineering*. **2015**, 75, 196–208

[6] Austin, J.; Gupta, A.; McDonnell, R.; Reklaitis, G.V.; Harris, M.T. The use of Near-Infrared and Microwave Resonance Sensing to Monitor a Continuous Roller Compaction Process. *Wiley Online Library*. **2013**.

[7] Austin, J.; Gupta, A.; McDonnell, R.; Reklaitis, G.V.; Harris, M.T. A novel microwave sensor to determine particulate blend composition on-line. *Analytica Chimica Acta*. **2014,** 819, 82-93.

415000 Novel Use of Cognex Vision System for on-Line Level Monitoring/Control in a Continuous Direct Compression Manufacturing Process

Abstract:

The continuous direct compression manufacturing process consists of several loss-in-weight (LIW) feeders, a horizontal mixer, a surge hopper, and tablet press equipped with a near infrared (NIR) probe at the die-filling station of the tablet feed frame. The surge hopper atop the tablet feed frame helps both dampen upstream flow fluctuations and maintain a consistent head of powder above the feed frame so as not to starve the tablet press of powder. Use of a Cognex Vision system allowed for a novel means of measuring level in the surge hopper to gauge the mass balance of material flowing into and out of the surge hopper. With the Cognex Vision system (or comparable level sensor) measuring fluctuations in surge level, the tablet press rotor/turret speed can be adjusted and has been shown to have minimal impact to tablet properties through monitoring of the force control on the main compression stage. Additionally, measurement of the surge level enables application of a reduced system model which, in part, would forecast when an upstream disturbance in API concentration would reach the feed frame and when that disturbance would clear the feed frame.

433765 A Novel Tablet Coating Technology to Support a Continuous Tableting Line

Abstract:

Continuous Manufacturing is rapidly being developed and implemented by the Pharmaceutical Industry, driven mainly by the reduced Research and Development costs and improved manufacturing quality. For example, manufacturing lines for 10 kg/hr to 100 kg/hr have been implemented for Continuous Direct Compression, Continuous Wet Granulation and Drying, and Continuous Dry Granulation in a tablet manufacturing line. However, to support this throughput range of tablet manufacturing, continuous tablet coating technologies have been very limited and not attractive in supporting the plug flow manufacturing concept.

A novel tablet coating technology was developed to support the 10-100 kg/hr continuous tableting line which is able to coat small sub-batches (3-6 kg) very rapidly (7-15 minutes) with a high degree of accuracy. Using this method, each sub-batch is able to be tracked from the upstream tableting process and critical process parameters monitored during coating to ensure product quality of each sub-batch. Employing a small-footprint twin coating system, the coating process can be linked to the tablet manufacturing line uninterrupted and while maintaining the plug flow from the upstream processes. Additionally, this coating method enables in-line coating thickness and build rate measurement.

410319 PAT Data Management for Continuous Manufacturing – the Next Generation

Abstract:

The mind shift from batch production to continuous manufacturing has been made in the pharmaceutical industry for a couple of years, but the IT infrastructure to cover the analyzer integration, data management and advanced process control was put under pressure by this new paradigm. The current generations of PAT data management platforms are bridging the gap when it comes to full robust and reliable integration of analyzers, high throughput dataflows, and a further and deeper integration in the process automation layer.

Also optimized time alignment is critical to combine the right pieces of information. In this presentation we will dig a bit deeper in time alignment applied on hot melt extrusion lines with dynamic line speeds, and how this can be translated into optimized time alignment. Automated selection of PAT data collection methodology with specific time alignment settings will be discussed based on actual process parameters and decision trees. This will show the importance of a very close integration between the automation layer and the PAT data management on top. And we will go even a step further…

Also the user interface is an important aspect in smoothening the integration of PAT data management into continuous manufacturing. UX or user experience is the new trend to optimize the user interfaces for your operators into the IT platforms. Logic user-focused workflows are guiding the operator though the different necessary steps of operation. Some examples from the new SIPAT 4.1 operator dashboard will be shown after it was revamped by a team of UX specialists.

A last item to be discussed is data security. The IT landscape is more and more the backbone of all the decisions taken on the quality of your products. This makes your network a vulnerable source for hacking, sabotage and potential terrorism. In how far are you aware that data from your analyzers could be captured on the fly and slightly changed, so the wrong decisions are made in the end? Man-in-the-middle attacks have caused serious issues in the past and should be avoided at all times. We have gone that additional step by offering data solutions, fully ensuring and securing your data security.

425747 Particle Engineering of Cocrystals Using a Solvent-Free Approach By Spray Congealing

Wednesday, November 11, 2015: 12:30 PM
Ballroom D (Salt Palace Convention Center)
Íris Duarte[1,2], João F. Pinto[1] and **Márcio Temtem**[2], (1)iMed.ULisboa, Faculty of Pharmacy, University of Lisbon, Lisbon, Portugal, (2)R&D Drug Product Development, Hovione FarmaCiência SA, Loures, Portugal

Particle engineering of cocrystals using a solvent-free approach by spray congealing

Iris Duarte[1,2], Joao F. Pinto[2], Marcio Temtem[1*]

[1] Hovione Farmaciencia SA, Sete Casas, 2674-506 Loures, Portugal;
[*]mtemtem@hovione.com or +351 219 847 569

[2] iMed.ULisboa, Faculdade de Farmacia da Universidade de Lisboa, Av. Prof. Gama Pinto, 1649-003, Lisboa, Portugal

Around 90% of the new chemical entities emerging from the drug discovery phase present, at least, solubility constraints (BCS Class II+IV). Among the different solubilization strategies that have been applied to address this issue, the use of pharmaceutical cocrystals emerged as an alternative crystal-engineering platform to enhance the poor physicochemical properties of these challenging molecules [1].

Pharmaceutical cocrystals are stable multicomponent crystals, typically comprising two molecules, *e.g.* an API, or the salt of an API, and a coformer. The coformer can be an active or non-active ingredient (*e.g.* pharmaceutical excipients, vitamins, minerals, amino acids, other APIs) that has a favorable intermolecular interaction with the API, promoting hydrogen bonding, van der Wall forces or pi-stacking [2].

In general, and despite the great potential of cocrystallization, producing cocrystals with high efficiency at industrial scales is challenging due to the tight thermodynamic and kinetic crystallization windows necessary to achieve the desired crystalline form. Specifically referring to the pharmaceutical field, the wide application of cocrystals is still limited due, in part, to the scarcity of suitable large-scale production methods and lack of robust and cost-effective processes.

In order to address some of these challenges, spray congealing has recently emerged as new process for the preparation of cocrystals [3]. Spray congealing, which is a hybrid technology governed by similar principles of spray drying and hot melt extrusion, consists in feeding a molten mixture to an atomizer, which then breaks the liquid feed into small droplets, and those droplets are cooled thanks to a co-current stream of cooling gas (Figure 1).

The advantages of this method are the fact it can be conducted in the same apparatus of spray drying, with minor modifications, thus making the commercial scale of the process readily available; for being a solvent free technique, cocrystallization via spray congealing will also comply with green chemistry requirements and, most importantly, it can be used to produce "engineered" cocrystals (*i.e.* with the desired properties) in a particulate form, in a single stage operation without the need of any downstream processing.

The feasibility of using spray congealing to produce pharmaceutical cocrystals was successfully demonstrated using three model API:coformer systems (Table 1).

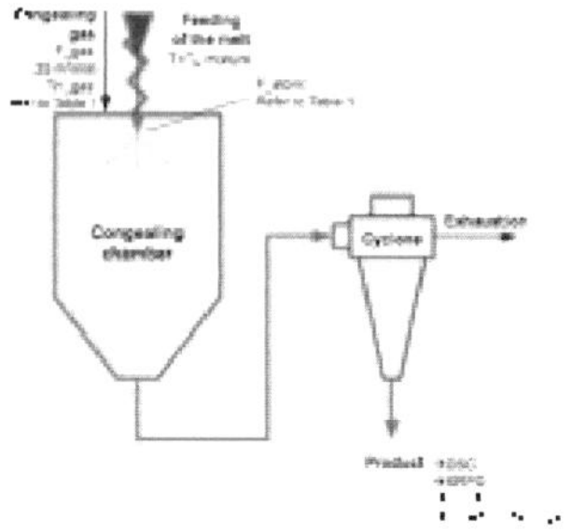

Figure 1. Schematic representation of the spray congealing set up used.

Table 1. Case-studies and respective process variables defined for the tests. The flow rate of the congealing (F_gas) was set at 0.35m^3/min.

System	Molar Proportion	Process Variables		Ref
		ΔT [°C]*	**F_atom [L/min]****	
Caffeine:Glutaric acid (CAF:GLU)	1:1	0-50	11-20	4,5
Caffeine:Salicylic acid (CAF:SAL)	1:1	50	9	5
Carbamazepine:Nicotinamide (CBZ:NIC)	1:1	125	12	5-7

* ΔT: difference between the mixture melting temperature observed inside the beaker ($T_{M,mix}$) and the inlet temperature of the congealing gas ($T_{in,\ gas}$);

**F_atom: flow of the atomization gas.

All the final products obtained with spray congealing presented DSC thermal profiles and XRPD diffractograms characteristic of the respective cocrystals and different from the pure components or physical mixtures. As an example, Figure 2 shows the results obtained for the 1:1 Caffeine: Salicylic Acid system.

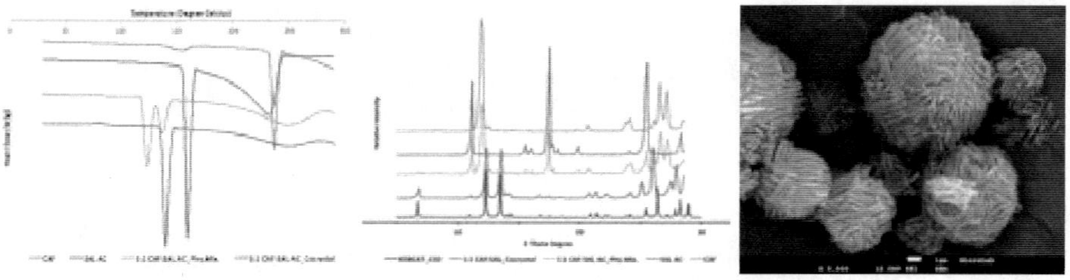

Figure 2. Results obtained for the 1:1 CAF:SAL system. From left to right: DSC, XRPD and SEM.

Moreover, and for one specific case-study, *i.e.* 1:1 Caffeine:Glutaric acid (CAF:GLU), a DoE with 2 parameters at 2 levels + 1 central point was conducted, to assess the effect of process variables on the final cocrystal particle and bulk powder properties. The parameters or process variables (*i.e.* ΔT and F_atom) were varied according to the ranges shown in Table 1.

According to Figure 3 it can be observed that particle properties (size, morphology, shape) can be adjusted, *in situ*, as part of the spray congealing process, without compromising cocrystal formation (DSC and XRPD data not shown).

Moreover, particle properties are known to affect the bulk behaviour of powders, which can impact subsequent downstream processing (*i.e.* blending, capsule filling, tableting, etc.) [8]. In order to evaluate this, bulk (*i.e*, compressibility and permeability) and shear properties of the powders produced were measured using a FT4 powder rheometer. Figure 4 shows the results from the permeability and compressibility tests at 15 kPa, and the shear cell test on samples that have been pre-consolidated at 9kPa normal stress.

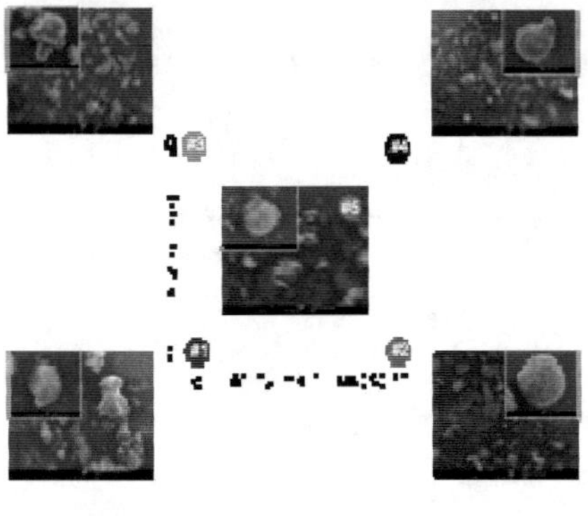

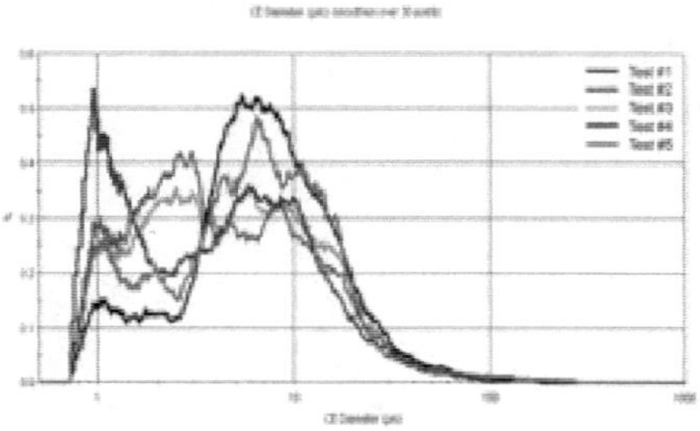

Figure 3. SEM results (right) and number-based circular equivalente diameter (CED) distributions (above) for the 1:1 CAF:GLU systems produced, when varying ΔT and F_atom.

Shear testing classified the powders as identical, whilst the permeability and compressibility tests show they are significantly different. The shear data suggest that the powders may behave similarly in a moderate to high stress environment, such as in a hopper, but it does not necessarily mean that they will behave similarly when aerated or flowing under low stress. The different tendencies of the permeability and compressibility results are clearly a consequence of the influence of multiple variables or particle properties (*e.g.* particle size, shape, surface properties, amount of fines, etc.) on the bulk powder rheology.

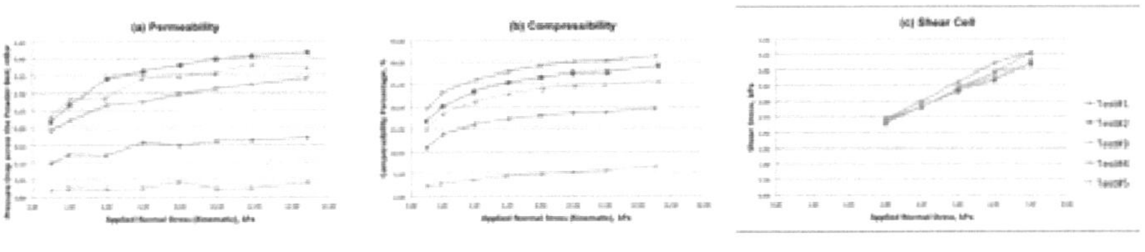

Figure 4. From left to right: permeability as a function of normal stress at constant air velocity of 2 mm/s, compressibility (*i.e.* % of bulk density change) as a function of normal stress and shear stress as a function of applied normal stress.

References:

[1] Blagden N. *et al.*, Pharmaceutical co-crystals - are we there yet?. *CrystEngComm*, 2014, 16, pp. 5753-5761;

[2] Qiao, N. *et al.*, Pharmaceutical cocrystals: An overview. *International Journal of Pharmaceutics*, 2011, 419, pp. 1-11;

[3] Duarte, I., *et al.* Synthesis and particle engineering of cocrystals. WO2015/036799A1 (2015);

[4] Trask *et al.*, Solvent-drop grinding: green polymorph control of cocrystallization. *Chem. Commun.*, 2004, 7, 890-891;

[5] Lu *et al.*, A rapid thermal method for cocrystal screening. *CrystEngComm*, 2008, 10, 665-668;

[6] Chieng *et al.*, Formation Kinetics and Stability of Carbamazepine - Nicotinamide Cocrystals Prepared by Mechanical Activation. *Cryst. Growth Des.,* 2009, 9(5), 2377-2386;

[7] Liu *et al.*, Improving the Chemical Stability of Amorphous Solid Dispersion with Cocrystal Technique by Hot Melt Extrusion. *Pharm. Res.*, 2012, 29, 806-817;

[8] Freeman R., Measuring the flow properties of consolidated, conditioned and aerated powders – a comparative study using a powder rheometer and rotational shear cell. *Powder Technology*, 2007, 174, pp. 25-33.

433246 A Case Study in Tuning API Powder Properties Via Crystallization to Meet Formulation Needs

Abstract:

Crystal form and powder properties are universal quality attributes which characterize every API. This presentation is a case study of one API that has the potential to form numerous polymorphs and a variety of crystal morphologies. Process development efforts were focused on obtaining the thermodynamically stable crystal form while targeting a narrow range of powder properties suitable for bioavailability and downstream drug product formulation. The work includes screening studies to select the appropriate solvent system and crystallization parameter optimization to control particle size and minimize agglomeration. Multiple technologies were employed to support this work, including FTIR, FBRM, wet milling, and dry milling. Understanding the impacts of crystallization parameters allowed for the generation of several lots of material with a variety of powder properties, which were evaluated for formulation behavior and bioavailability. The feedback and collaboration between drug substance and drug product process development teams enabled the selection of the desired target material properties (eg. particle size distribution, D90/D50 ratio) and formulation process (roller compaction or wet granulation) that would robustly lead to high quality drug to patients.

432520 Improving Dissolution Kinetics of Pharmaceuticals By Fluidized Bed Impregnation

Abstract:

According to the current Biopharmaceutics Classification System (BCS) adopted by the U.S. Federal Drug Administration (FDA), drug substances are categorized into four main classes, based on their solubility and permeability. Two of those classes (BCS classes II and IV) are characterized with low solubility. The trend in the past decade has been toward steady increase in the number of active pharmaceutical ingredients (APIs) belonging to BCS classes II and IV. It has been estimated that up to 60% of new chemical entities fall in the low solubility class, compared to 39% for marked drugs. This trend towards decrease in API solubility is undoubtedly one of the main hurdles for formulators today, as they continuously strive to develop novel formulations capable of effectively delivering the drug into the blood stream.

API solubility has a direct effect on the drug's bioavailability and therefore its efficacy. The absorption of BCS Class II and IV compounds and the resulting blood concentration-time profile depends on the rate-limiting step, which can be one of the following: permeability limited, dissolution limited and solubility limited. For poorly-soluble compounds, improving the rate (for dissolution limited) and the extent of dissolution (for solubility limited) will directly increase their bioavailability. There are two main techniques that are widely used today for improving dissolution rate and increasing solubility. The first technique involves particle size reduction as suggested by the Nernst-Brunner/Noyes-Whitney equation, where the dissolution rate for any API is proportional to the surface area available for dissolution. Solubility of any substance is a thermodynamic property, which among other environmental factors (solvent nature, pH and temperature) depends strongly on the chemical or physical structure of the substance itself (e.g. salts, crystalline or amorphous solids, solvates, co-crystals). Many variations of these two main techniques for improving dissolution rate and increasing solubility are employed into today's development or marketed drug formulations. The most common include various bottom up or top down approaches for size reduction and numerous means for producing amorphous materials (e.g. spray drying, freeze drying, hot-melt extrusion, polymer additives). Often these methods have major challenges and limitations associated with: difficulty in control or scale up, API stability, applicability to wide range of APIs, high cost, etc.

In this work we present a drug formulation method that is capable of increasing significantly the dissolution kinetics of a poorly soluble API (Fenofibrate), which shows a high degree of simplicity compared to many of the traditional methods. The method involves impregnation of the API into highly porous excipient in a fluidized bed (FB). Loadings of up to 40% were successfully formulated, showing significant improvement in the dissolution profile. Additional improvement in dissolution was achieved by mild milling of the impregnated drug product. Moreover, co-impregnation of the API with an FDA approved additive showed significant further improvement in dissolution profile, making our formulation comparable to one of Fenofibrate's patented market formulations (TriCor, 45mg). Other benefits of our impregnation method include: high blend uniformity independent on API nature or loading, excellent flow properties of the impregnated powder (no need for more additives), elimination of several unit operations from drug substance and drug product development.

423877 A Drop-on-Demand Manufacturing System for the Production of Amorphous Drug Products

Abstract:

Innovation is the key to success within the pharmaceutical industry. However, it has been largely sought through new drug discovery and development, while the development of more efficient manufacturing methods has received inadequate attention. Recently, as a result of encouragement from the regulatory authorities, the advent of globalization and the increasing awareness of environmental impact, the pharmaceutical industry has been reconsidering the way drug products are developed and manufactured. The US Food and Drug Administration (FDA) has promoted the Quality by Design (QbD) approach to increase process understanding and improve quality and efficiency while minimizing risk [1]. The Process Analytical Technology (PAT) guidance introduced by the FDA encouraged the monitoring of critical quality and performance attributes during processing with the goal of ensuring final product quality [2,3].

As part of this renewed emphasis on improvement of manufacturing and under the US National Science Foundation supported Engineering Research Center for Structured Organic Particulate Systems; we have developed a dropwise additive manufacturing process for solid oral dosage production. The process utilizes the drop-on-demand (DoD) inkjet printing technology for predictable and highly controllable deposition of active pharmaceutical ingredients (API) onto an edible substrate, such as a polymeric film or placebo tablet [4, 5]. This process uses fluid operations suitable for low volume production of personalized dosage forms. The advantages of liquid processing and reproducible production of small droplets create the special opportunity for the production of high potency, low dose drugs, which are difficult to produce with consistent quality via conventional powder processing methods.

Using the developed drop on demand manufacturing process for pharmaceuticals, different drug formulations including solvent based systems [4], i.e. solvent-polymer-API solutions, as well as melt based systems [5], i.e. polymer-API melts, can be printed. Melt-based printing applications eliminate the solvent evaporation step after drop deposition and thus allow on demand production of individual dosage forms with good control of drug solid state and morphology.

In this paper, we present the controlled production of amorphous dosages by the deposition of melt-based formulation using the DoD manufacturing system. Producing dosage forms in which the active ingredient is in amorphous forms can enhance the dissolution of low-solubility drugs. Amorphous drug forms are produced with alternative choice of polymers, which inhibit crystallization, along with suitable choice of operating conditions. Polymers are added to the printing material to help control drug morphology, material properties and the formulation composition [6]. Addition of crystallization inhibiting polymers, such as PVP, to the formulation, can result in the formation of dosage forms containing amorphous or semi-crystalline API. Operating conditions, i.e. crystallization temperature, are controlled using substrate temperature controller, which is used to achieve the desired product quality by influencing the dissolution behavior and therefore the bioavailability of the drug [7]. Since stability of amorphous drug forms can be challenging, design of crystallization temperature profile and precise control of substrate temperature profile are even more important than in the case of crystalline API formulations. The amorphous dosage forms are achieved by controlling the cooling profile of the deposits to solidify the drug in the amorphous form before crystallization occurs.

1. Yu LX, Amidon G, Khan MA, Hoag SW, Polli J, Raju GK, et al. Understanding pharmaceutical quality by design. AAPS J. 2014;16(4):771–83.
2. Food and Drug Administration CDER. Guidance for Industry PAT - A Framework for Innovative Pharmaceutical. 2004.
3. Simon LL, Pataki H, Marosi G, Meemken F, Hungerbu K, Baiker A, et al. Assessment of Recent Process Analytical Technology (PAT) Trends: A Multiauthor Review. Org Process Res Dev. 2015;19(1):3–62.
4. Hirshfield L, Giridhar A, Taylor LS, Harris MT, Reklaitis G V. Dropwise additive manufacturing of pharmaceutical products for solvent-based dosage forms. J Pharm Sci. 2014;103(2):496–506.
5. Içten E, Giridhar A, Taylor LS, Nagy ZK, Reklaitis G V. Dropwise Additive Manufacturing of Pharmaceutical Products for Melt-Based Dosage Forms. J Pharm Sci. 2015;104(5):1641-49.
6. Trasi NS and Taylor LS. Effect of Additives on Crystal Growth and Nucleation of Amorphous Flutamide. Crystal Growth & Design 2015; 12(6): 3221–3230.
7. Içten E, Nagy ZK, Reklaitis GV. Supervisory Control of a Drop on Demand Mini-manufacturing System for Pharmaceuticals. Comp. Aided Process Eng., 2014; 33: 535-540.

410043 Continuous Solid-Phase Crystallization

Abstract:

Continuous crystallization is one of the emerging technologies that has been widely investigated recently. Most of the advancements have happened in the pharmaceutical industry, and for the continuous crystallization of small molecules, all of the platforms have been in the liquid phase. There has not been any work in continuous solid-phase crystallization for complex matrices, when the solid domain (micron sized particles) consists of a small and large molecule mixture.

This work presents a novel technique for continuous solid-phase crystallization of dairy powders, as a model system, which contains lactose (small molecule, initially amorphous), proteins (large molecules, which have few effects on the crystallization process through different molecular interaction with lactose and water), and minerals. Experimental results and mathematical modeling of the process demonstrate the performance of the system for a wide range of processing conditions and lactose/protein ratios. The ability of the technique to manipulate the particles' micro-structure, and to design the particles' physico-chemical properties, by controlled and partial crystallization of the powders, will also be discussed.

The challenges for powder processing at highly critical conditions, crystallization within a complex matrix, and implementation of the technique for the industrial applications, will be presented. The approaches to overcome these challenges, as well as the gap in the knowledge and potentialities for further improvements will also be discussed.

The wide applications of the technique are in the Pharmaceutical, Food, and BioNanotechnology industries for solid-phase and solvent-exchange crystallization, and conditioning of the particles' structure.

423627 Think Small, Think Flexible: The Expansion Strategy of a Continuously Operated Modular Multi-Product Plant

Abstract:

As a result of product diversification and increasing volatility of markets, the capability to adapt the production output of chemical plants becomes crucial for economic success. The plant design for pharmaceutical products is determined by a shorter time-to-market and uncertain market sales. Therefore a product launch requires a production of small amounts during the market entry. Thus, pharmaceutical chemicals are conventionally produced batch-wise in a multi-product plant that is flexible regarding production rates. However, a batch-wise processing is inefficient in terms of resource consumption, batch failures and cleaning effort. Due to an uncertain market demand, for the production of increased product quantities a more efficient and flexible plant concept is required [1].

In this context, a continuously operated modular multi-product plant offers the solution of an efficient and flexible production. Continuously operated processes enable an increase of efficiency in production. The modular design enables a short time-to-market and a flexible adaptation of the production capacity. Thus, an efficient early market entry is possible due to the design of modular production line with a fixed design capacity for small product amounts. When market demand increases a stepwise numbering-up of the modular production line allows a fast and flexible adaptation of production capacities by reuse of engineering. The smaller the scale of the modular production line is, the more expansion steps are necessary to supply the increasing market demand. As a consequence, this capacity flexibility results in additional costs incurred by the loss of economy of scale. Furthermore, a flexible reallocation of the production of different products within the production capacity is possible. This planning flexibility also results in additional costs for setup changes and storage of the products. The increased flexibility provided by this modular plant concept has to be balanced with the consequent costs.

In this work the modular plant concept regarding capacity flexibility and planning flexibility is evaluated. Regarding an uncertain market development, the sequential expansion by numbering-up of a modular production line is investigated. Therefore, a long-term production planning for lot-sizing and scheduling of a multi-product production is applied. Thus, necessary expansion steps are identified when the market demand exceeds the production capacity. An economic evaluation is conducted to consider additional cost by the sequential investment and the production planning. A decision tree analysis is applied to balance the added value of flexibility with the additional costs. Therefore, the obtained flexibility of different scales of a modular production line is evaluated.

[1] T. Bieringer et al., Chemical Engineering & Technology, 36 (6), 2013, 900–910

410571 Membrane Enhanced Peptide Synthesis (MEPS) – Integrating Organic Solvent Nanofiltration (OSN) into Iterative Chemical Reactions

Abstract:

In the pharmaceutical industry, the high potency and specificity of peptides make them more attractive than traditional small-molecule drugs. The chemical synthesis of peptides involves the repetitive addition of amino acid, in between which the intermediate products have to be isolated for purification purposes. It is of paramount importance to achieve high yield for the isolation of each intermediate product or the cumulative mass loss will be detrimental to overall yield of the synthesis. The conventional peptide synthesis methods, namely Liquid Phase Peptide Synthesis (LPPS) and Solid Phase Peptide Synthesis (SPPS), have been used in the industry over the past few decades, but innovation is still required to improve process performance. Membrane Enhanced Peptide Synthesis (MEPS) integrates Organic Solvent Nanofiltration (OSN) into LPPS. A hexapeptide (Pyr-Ser(Bzl)-Ala-Phe-Asp-Leu-NH2) was synthesised successfully via MEPS at a scale of 33.65 mmol and a soluble anchor concentration of 10.4 wt% in the starting solution. Before the cleavage and global deprotection of peptide, MEPS achieved a high purity (88.1%) and a decent overall yield (71.2%), similar to LPPS (100.0% and 72.4% respectively), in which the intermediate products were precipitated and purified. Consequently, the performance of MEPS was better than LPPS in terms of material cost, process time, volumetric efficiency and E-factor, showing that OSN is indeed a more attractive option than precipitation for the isolation of intermediate products in liquid phase iterative chemical reactions.

432830 Prediction of Intracellular Nucleotide Sugar Variation Impact on Monoclonal Antibodies (mAbs) Glycan Distribution Using Glycosylation Reaction Network Models

Abstract:

Prediction of intracellular nucleotide sugar variation impact on monoclonal antibodies (mAbs) glycan distribution using glycosylation reaction network models

Sha Sha[a] and Seongkyu Yoon[a]

[a]Department of Chemical Engineering, University of Massachusetts, Lowell, MA, USA

Glycosylation is a critical post-translational process on monoclonal antibodies (mAbs) produced by mammalian cells as the glycan profile on mAbs plays significant impact on the pharmaceuticals' stability, efficacy and half-life. Glycosylation central reaction network (CRN) in cellular *Golgi* compartment is the branching paths for mAb to be attached with various intermediate or terminal glycan structures via the mediation of a series of enzymes. The recent development of reaction network models is an approach to simulate glycosylation process in *Golgi* CRN and predict the distribution of glycan on final monoclonal antibodies (including a number of intermediate or terminal glycan structures) through simulation using the model, and thus are potentially of great value in the application in the industrial bioprocess. Reaction network models are formulated by describing the kinetics of reactions in the network and involve various intracellular factors associated with glycosylation as model parameters. The capability of models being built to reflect key intercellular factor variation is one of the priorities for models to ensure proper performance.

However, the performance of model prediction for glycan distribution out from intracellular factor (e.g. nucleotide sugar) variation has not been sufficiently evaluated in the past studies. In this work, a reaction network model was developed, describing the CRN by resembling Golgi as a four-compartment continuous stirred-tank reactor (CSTR) and the glycan distribution was estimated by solving the model based on mass balance of all glycan structures in the network at a steady state. The impact of intracellular nucleotide sugar concentration variation on the glycan distribution output was then simulated using the model and further compared with nucleotide sugar- glycan variation correlation being examined by published literature. The model was then optimized to minimize the gap between model prediction and experiments-derived data, therefore becoming a more reliable tool to be utilized in the future for glycosylation prediction.

(552e) Understanding the Effect of High Gas Entrance Velocity on Chinese Hamster Ovary (CHO) Cell Culture Process

Chaudhary, G., Boehringer Ingelheim Fremont Inc.
Guillen, N., Boehringer Ingelheim Fremont Inc.
Karim, K.,
Whiteley, E.,
Lin, H., Boehringer Ingelheim
Khetan, A., Boehringer Ingelheim Fremont, Inc.
Originally presented on: 11/11/2015 13:54:00 - 14:15:00

Growing demand for biologics requires high intensity unit operations such as high cell density fed-batch processes. High yielding cell culture process requires higher aeration/oxygenation to satisfy high oxygen demand and for carbon dioxide removal in a bioreactor. Higher aeration/oxygenation through sparger can result in higher gas entrance velocity (GEV), which can lead to cellular damage and can adversely impact the cell culture performance. A study was performed to evaluate the effect of high GEV on CHO cell culture process using spargers of different number/size of holes. Our results demonstrate that high GEV imparts significant negative effect on cell growth, viability, and productivity. Similar effects were observed in large scale production bioreactors. In addition, the extent of cell damage (necrosis or apoptosis) under high GEV was examined by lactate dehydrogenase measurement and flow cytometry analysis. Insights into the cell condition under high GEV will be presented and discussed.

435020 Efficient Optimization of Tablet Coating Operation Using Hybrid CFD-DEM Simulations

Abstract:

Tablet coating is a common unit operation in pharmaceutical manufacturing. The coating process depends on a number of process variables – spray parameters, coater geometry, mass loading, drying air flow etc. Process engineers in the pharmaceutical industry are interested in optimizing coater operation from both process and product perspectives. Performing physical experiments with coaters to determine the optimal set of process parameters is, more often than not, impractical. Digital experiments using computer simulations of tablet coating offer a viable alternative to experiments. We present here an efficient hybrid CFD-DEM model framework that enables a relatively rapid and efficient optimization of the coating process. The approach described here is an effective alternative to full scale CFD/DEM co-simulation and is designed to provide useful design guidance to process engineers at a significantly reduced computational expense.

437563 Multistage Continuous MSMPR Crystallizer with Solids Recycle

Abstract:

Continuous manufacturing has been well developed in food and chemical industries, however, batch operation is still the most common method used in the pharmaceutical industry [1]. For crystallization, continuous processing has the advantages of enhanced reproducibility of results, reduced cost, improved process efficiency and flexibility in production capacity [2]. Due to these reasons, continuous crystallization has obtained great interest in both industry and academia in recent years [3].

The major challenges associated with the transition from batch to continuous crystallization process are to improve the yield of active pharmaceutical ingredients (APIS) and the purity of products. Batch process can obtain a higher yield by pushing the process to equilibrium while a continuous process is operated at steady state [4]. An appropriate recycle stream can be applied in continuous process to boost the yield. A study of cooling crystallization of cyclosporine showed a 91.8% yield by using a single-stage mixed-suspension, mixed-product removal (MSMPR) crystallizer with concentrated mother liquor recycle system [5]. However, previous research shows that for MSMPR, the mean crystal size and the purity decrease as improving the yield by recycling the mother liquor [6]. This is because mother liquor recycle will make the steady state with low supersaturation and high impurity/API ratio: the slow growth caused by low supersaturation will lead small crystals and the high impurity/API ratio causes more impurity incorporation.

A common approach of improving the yield of continuous crystallization is to lower the operating temperature and thus decrease the solubility of API. However, for some APIs, e.g. cyclosporine, whose growth kinetics are very sensitive to temperature, the growth rate is very low (~0.1 μm/min) at low temperature. Thus, low operation temperature will cause low yield or it is required a much longer process residence time. The goal of this work is to enhance the yield under low temperature and push the steady state close to equilibrium while keep the residence time short and remain the purity of the crystal. To accomplish this goal, a two-stage MSMPR crystallizer with concentrated slurry recycle stream is established for cooling crystallization of cyclosporine. By recycling the crystals back to crystallizer, the total growth surface area increases so that more crystal growth occurs even if the liner growth rate is low. Thus, the yield can be improved. As the residence time of the crystals increases by recycling, crystals can grow more and have larger crystal size. Since the residence time of the mother liquor is shorter than that of the crystals, the impurity in the mother liquor will not accumulate in the crystallizer and the crystal purity can remain.

Moreover, a population balance model in conjunction with experimentally determined crystallization kinetics and distribution coefficients is developed. The model can simulate n-stage MSMPR crystallizer with different number of slurry recycle or mother liquor recycle to any stage(s). Using this model, the optimal stage temperatures, residence time, feed concentration, and recycle ratios can be obtained. Experiments guided by the model is used to obtain higher experimental yield and further valid the model.

This work presents an alternative idea to apply recycle stream to improve the yield and purity and control the crystal size in a continuous process.

References:

[1] Parikh, D. M. Handbook of Pharmaceutical Granulation Technology; Marcel Dekker: New York, 1997

[2] Vervaet, C.; Remon, J. P. Chem. Eng. Sci. 2005, 60, 3949–3957.

[3] Poechlauer, P.; Manley, J.; Broxterman, R.; Gregertsen, B.; Ridemark, M. Org. Process Res. Dev. 2012, 16, 1586−1590.

[4] Chen, J.; Sarma, B.; Evans, J. M. B.; Myerson, A. S. Cryst. Growth Des. 2011, 4, 887−895.

[5] Wong, S. Y.; Tatusko, A.; Trout, B. L.; Myerson, A. S. Cryst. Growth Des. 2012, 12, 5701-517.

[6] Alvarez, A. J;Singh, A.; Myerson, A. S. Cryst. Growth Des. 2011, 11, 4392-4400.

422817 Characterization and Modeling of Feeders: A Critical Component in Continuous Pharmaceutical Manufacturing

Abstract:

In recent years, continuous manufacturing technologies have been widely accepted in the pharmaceutical environment given the many advantages they offer to product quality. Some of the benefits continuous processes offer over batch counterparts include: enhanced process control, reduced solids handling, small equipment footprint, reduced scale up requirements, and more uniform processing [1]. These advantages make continuous manufacturing a desirable technology for many pharmaceutical manufacturers. Nonetheless, there are many challenges to this technology: higher initial cost, extended initial development time, lack of PAT tools available for monitoring continuous processes, and difficulty of implementation for low volume and dosage products [1]. This last challenge is primarily based on the difficulty to accurately feed small amounts of materials to the process and blend them efficiently. Research has focused on understanding the behavior of both feeders and blenders with the aim of improving upon these challenges [2, 3]. During the early phases of continuous manufacturing blending was studied more extensively given the legacy importance in batch processes (i.e., in batch it is one of most important unit operations). However, feeding has the most significant effect in continuous processing as its sets the mass balance for the entire process. Lack of feeder control and understanding, leads to lack of control of any downstream unit operation [4].

Feeding accuracy is of great importance in continuous processes given the stringent requirements for product composition. To this end, Loss-in-Weight (LIW) feeders under gravimetric control are commonly used. LIW feeders deliver powders by calculating the amount of material lost (i.e., fed out of the unit) and controlling the rate of material delivery accordingly [5]. This strategy provides the feeding accuracy required in pharmaceutical processes. LIW feeders generally consist of a hopper, a flow aid and a powder conveying system, all over a load cell that measures its weight. The hopper contains the material in the feeders and is the receptacle where refills are performed. The flow aid is located at the bottom of the hopper with the purpose of breaking any powder bridges and disrupt the statics to facilitate transport of material into the conveying system. The conveying system, often a screw due to its accuracy over other systems, is used to discharge the material out of the feeder. Previous research focused on studying LIW feeders to understand how these components affect the unit's performance (i.e., the output flow rate). The conveying system was found to be the most significant component on feeder performance, followed by the refill hopper refill strategy [2]. Material properties were also found to have a significant impact on feeding performance [6].

Feeders are often not considered in the absence of other processing equipment, therefore development of models for these units is not as extensive compared to other unit operations. Multiple articles focus on the use of Discrete Element Model (DEM) simulations to characterize the effect screw and hopper configurations have on the unit's performance [7, 8]. The results from these studies are often semi-qualitative in nature and do not allow for development of models that can be used for process modeling (e.g., flowsheet models and control strategy design). In this work we developed a semi-empirical model of feeder behavior using material properties as input parameters. This work is an extension of previously developed empirical models [5], with the goal of incorporating the behavior and effects of both the flow aid and conveying screws in the model.

The semi-empirical model was developed to simulate feeder behavior, primarily focusing on modeling screw speed and flow rate as a function of time. Frequency variations (i.e., noise) of these variables were characterized using Fourier series, whose coefficients were regressed from experimental data for individual components. Screw speed and flow rate were directly correlated in the model through the use of a feed factor, whose units are mass per screw revolution. Feed factor, defined as the maximum mass fitting in a screw flight, was modeled to account for changes in hopper material properties occurring due to material changes, flow aid systems, feeder emptying and refills. Material properties were characterized before and after the feeder in order to evaluate and model their relationship to screw speed and flow rate. Lowest feeding capacity (LFC) of the feeder was evaluated based on these findings and the developed model. The effect of flow aid systems on the mixing of powders inside the hopper was experimentally characterized using residence time distribution (RTD) experiments. Models to predict mixing behavior in the unit were developed using fundamental chemical engineering models (e.g., continuous stirred tank reactors (CSTR), plug flow reactors or any combination). The study was performed for multiple feeder types with varying the flow aid systems, hopper configurations, and conveying system locations. Operating conditions (e.g., flow rate, refill level) and powder properties were also varied to investigate the effects these parameters had on the mixing. An optimization framework for refill strategy with minimal trial-and-error was developed. The ultimate goal of this work was to develop a model that can be used to predict the behavior of LIW feeders for any material *a priori*, using the results from powder characterization test. Such model can lead to a significant reduction in development time and minimize product waste.

References

1. Rogers, A., A. Hashemi, and M. Ierapetritou, *Modeling of Particulate Processes for the Continuous Manufacture of Solid-Based Pharmaceutical Dosage Forms.* Processes, 2013. **1**(2): p. 67-127.

2. Engisch, W.E. and F.J. Muzzio, *Method for characterization of loss-in-weight feeder equipment.* Powder Technology, 2012. **228**(0): p. 395-403.

3. Vanarase, A.U. and F.J. Muzzio, *Effect of operating conditions and design parameters in a continuous powder mixer.* Powder Technology, 2011. **208**(1): p. 26-36.

4. Dai, J. and J.R. Grace, *Biomass granular screw feeding: An experimental investigation.* Biomass and Bioenergy, 2011. **35**(2): p. 942-955.

5. Boukouvala, F., et al., *An integrated approach for dynamic flowsheet modeling and sensitivity analysis of a continuous tablet manufacturing process.* Computers & Chemical Engineering, 2012. **42**(0): p. 30-47.

6. Falk, J., et al., *Mass flow and variability in screw feeding of biomass powders — Relations to particle and bulk properties.* Powder Technology, 2015. **276**(0): p. 80-88.

7. Owen, P.J. and P.W. Cleary, *Screw conveyor performance: comparison of discrete element modelling with laboratory experiments.* Progress in Computational Fluid Dynamics, an International Journal, 2010. **10**(5): p. 327-333.

8. Fernandez, J.W., P.W. Cleary, and W. McBride, *Effect of screw design on hopper drawdown of spherical particles in a horizontal screw feeder.* Chemical Engineering Science, 2011. **66**(22): p. 5585-5601.

425126 An Investigation of Mixing Performance of a Continuous Powder Mixing Process

Abstract:

Continuous manufacture of drug products is receiving increased attention both from academics and industry in recent years. Pfizer's approach to continuous manufacture, utilizing a Portable, Continuous, Miniature and Modular (PCM&M) technology is the focus of this presentation. In this work, the performance of a novel, vertical in-line powder mixing technology will be described. This mixer, which offers the independent selection of mixer shear rate, powder hold up mass and system mass throughput, is studied both experimentally and computationally to first characterize the mixing performance and secondly better understand how process parameters and equipment design can be manipulated to achieve the desired performance. The links between process conditions, powder flow regime, and mixing performance are illustrated with experimental data. Discrete element method (DEM) simulations are used to further characterize the mixer operation with respect to mixing performance and the extent of powder lubrication under a variety of conditions. Additionally, the models are used to characterize the coupled impact of gravimetric feeder variability and mixer performance (residence time distribution) on the variability in the outlet stream concentration. Together, these simulations also provide grounds to assess the influence of various design considerations on performance of the PCM&M equipment train.

422062 Comparison of Continuous High Shear Granulation Processes with Twin-Screw Wet Granulation Process

Abstract:

Wet granulation (WG) is an extremely prevalent unit operation in the manufacturing of drug product due to its numerous advantages. As particle size is enlarged in the process, WG can improve the handling and flow properties of powders, content uniformity and compression characteristics; and it also can minimize dust and ingredient segregation. These advantages could realize better drug release rate and tablets physical appearance. Currently, substantial attention has been attracted by continuous manufacturing (CM) because it can provide significant technical and business advantages relative to the conventional batch processes. CM methods have been demonstrated to be more robust and controllable, which can improve product quality and process efficiency. Furthermore, scale-up challenges can be overcome in CM by running the operation for a longer time, indicating the reduction of technique transfer risk and capital-intensive equipment requirement. In recent years, several continuous WG techniques have been developed. Most commercial continuous granulators are twin-screw extruders followed by fluid bed driers. In comparison, there have been very few studies so far on continuous high shear granulation, which resembles the batch high shear granulation in terms of dry powder mixing, binder addition and wet massing. In general, the granules produced by continuous fluid bed granulators or twin-screw extruders are different than those from high shear granulation due to the different shear and compressive forces that the material experiences in these systems.

The focus of the study is to quantify performance metrics (particle size distribution (PSD), bulk density, porosity, active concentration in granule, particle shape) of granules as a function of critical process parameters (throughput, impeller rotation speed and liquid to solid (L/S) ratio) for each formulation and granulator. Residence time distribution (RTD) studies are performed to understand the mixing dynamics which can aid in further understanding the relationship between input parameters, mixing and shear and the output granule properties.

Two continuous high shear granulators, namely Lödige CoriMix CM5 and Glatt GCG70, and Thermo twin-screw granulator/extruder (EuroLab 16-mm, 25:1 L/D) were examined. Two dosages formulations were used for experimentation, i.e., a high-dose drug formulation (15% -lactose monohydrate, 15% Microcrystalline cellulose and 70% semi-fine Acetaminophen) and a low-dose drug formulation (46% -lactose monohydrate, 46% Microcrystalline cellulose and 8% semi-fine Acetaminophen). Purified water served as the binder. A Design of Experiments (DoE) had been performed on each granulator with three levels of L/S ratio, rotation speed and throughput. Eyecon (InnoPharma) offline particle size analysis were used to quantify the median particle size and span of the output granules. Bulk density was measured from calculations involving the mass of the sample and the volume of particles (solid + liquid + pores). Scanning Electron Microscope images of the samples were taken to quantify porosity, active concentration and the shape of the granule. RTD studies, conducted by injecting colored tracer active particles and implementing a Near-infrared spectrometer, were used to evaluate the concentration of dye that was further correlated to the active concentration in the sample collected. The results allow for a quantitative understanding of the design space regarding the process parameters mentioned above, which can be used to identify an optimal operation condition to a specific formulation.

427602 Experimental Investigation Linking Granulation Performance with Residence Time and Granulation Liquid Distributions in Twin-Screw Granulation

Abstract:

Twin-screw granulation is a promising technique for the continuous production of pharmaceutical solid dosage forms based on wet granulation. In a short period, solid-liquid mixing must be achieved by arrangement of transport and kneading elements to produce granules with a particle size distribution appropriate for tableting. The residence time distribution and the solid-liquid mixing governed by the field conditions (such as location and length of mixing zones) in the twin-screw granulator thus contain interesting information about mixing and different granulation rate processes such as aggregation and breakage. In this study, the impact of process (feed rate, liquid-to-solid ratio and screw speed) and equipment parameters (number of kneading discs and stagger-angle) on the residence time, the granulation liquid-powder mixing and the resulting granule size distributions during twin-screw granulation were investigated. Residence time and axial mixing information was extracted from tracer maps and the solid-liquid mixing was quantified from moisture maps, both obtained by monitoring the granules at the granulator outlet using near infra-red chemical imaging (NIR-CI). The granule size distribution was measured by sieving. The screw speed most dominantly influenced the residence time distribution based responses, i.e. mean residence time and axial mixing. Increasing the screw speed caused reduction in mean residence time and an increase in axial mixing. Also, an interaction between screw speed and material throughput dictating the fill ratio caused a reduction in residence time and an increase in axial mixing when the throughput-force increased by the high fill ratio. The frequency and amplitude of perturbation in the mean moisture profiles (derived from the moisture maps) indicated that unlike residence time, the solid-liquid mixing is dominantly driven by the material throughput and the number of kneading discs. Increasing the number of kneading discs reduced the frequency of variation indicating reduction in the heterogeneity of liquid mixing, which is desired for a good granulation yield. Correlating these results with the particle size fractions, the low residence time and high axial mixing obtained at a high screw speed resulted in a reduced oversized (>1400 μm) fraction and an increase in yield fraction (>150 μm and < 1400 μm). However, increasing the material throughput, which had a similar effect as screw speed on the residence time and the axial mixing, resulted in a reduction of the yield fraction. This was due to the inferior solid-liquid mixing at high material throughput which ultimately negatively affected the granulation performance. This improper solid-liquid mixing at high throughput resulted in more oversized particles. Thus, a balance between material throughput and screw speed should be found to achieve the required granulation time and solid-liquid mixing for high granulation yield. The results from this experimental study improved the understanding regarding the interplay between granulation time and the axial and solid-liquid mixing responsible for the granulation yield after twin-screw granulation.

421449 NANEX: A Continuous Hot-Melt Extrusion (HME) Process for Nano-Based Formulations

Abstract:

The NANEX process is a continuous one-step HME process that converts stabilized aqueous nanosuspensions into a solid oral formulation with an increased solubility of poorly soluble drugs. A nanosuspension is fed to the extruder and the solvent is removed via devolatilization. Nanoparticles are embedded in the polymer matrix and solid nano-based pellets can be produced in a single step [1,2]. However, the amount of the active pharmaceutical ingredient (API) in a nanosuspension formulation is restricted to 20-30 % (w/w) making it hard to vary the API loading in the final extrudates.

Therefore, in the current study we focus on evaluating process parameters which impact the maximum amount of water that can be added during hot melt extrusion. Moreover, additional matrix materials that are suitable for incorporation of aqueous suspensions were investigated. Kollidon® VA64, Eudragit® E PO, AQOAT® (hypromellose acetate succinate, HPMCAS; in combination with plasticizer) and poly(ethylene glycol) 20000 (PEG 20000) were used. To assess whether the polymer could be mixed with water during the NANEX process and the added water could be removed completely via devolatilization (vacuum of 200 mbar), liquid (i.e., water) was fed to the hot melt extruder (MICRO 27 GL co-rotating twin-screw extruder, Leistritz Extrusionstechnik GmbH, Nürnberg, Germany) via a micro-angular gear pump in barrel 4, while adding the matrix material in barrel 1 and degassing of the added water in barrel 8. By adding water to the extruder at fixed throughputs and screw speeds, their influence on the maximum amount of water was examined. Clogging of the degassing unit and/or inclusions of moisture within the strands indicated the end-point. Additionally, the mean residence time at fixed process parameters was evaluated and residual moisture contents of the extrudates were measured via Karl Fischer titration.

The results demonstrated that the amount of water fed to the polymer melt is strongly dependent on the filling degree of the screw, which is a function of the throughput and the screw speed. The lower the filling degree was the higher was the maximum amount of added water. Similarly, the longer the residence time, the more water could be added. However, the quantity of added and removed water was not only related to the process parameters but also to the nature of the matrix material. It was found that polymers, which are miscible with water, could be processed via NANEX. For HPMCAS/plasticizer, 110%, and for PEG 20000 and Kollidon® V64 a maximum of 88% water could be added to the molten materials. Results of the Karl Fischer titrations clearly demonstrated that the excess water was removed entirely in the case of Kollidon® VA64 and HPMCAS/plasticizer. However, PEG 20000 extrudates exhibited increased mean residual moisture contents which can be attributed to process temperatures not exceeding 70 °C hampering the devolatilization of water at a vacuum of 200 mbar. To provide a remedy, a further drying step could be performed.

In summary, it was found that via the evaluation of the most important process parameters the API loading of nano-extrudates can be controlled easily and continuously without changing the complex and sensitive formulation of the nano-suspension.

1. Baumgartner, R.; Eitzlmayr, A.; Matsko, N.; Tetyczka, C., Khinast, J. and Roblegg, E. Nano-extrusion: A promising tool for continuous manufacturing of solid nano-formulations. Int J Pharm, 477, 1-11 (2014)

2. Khinast, J.; Baumgartner, R. and Roblegg, E. Nano-extrusion: a one-step process for manufacturing of solid nanoparticle formulations directly from the liquid phase. AAPS PharmSciTech, 14, 601-604 (2013)

427040 Continuous Manufacturing of Immediate-Release Tablets from Hot-Melt Extruded Pellets

Abstract:

Tablets are the most common solid dosage forms. In order to decrease variability in tablet quality many formulations are granulated prior to tableting. Compared to common continuous granulation methods, hot-melt extrusion (HME) offers the possibility to enhance or adapt certain physicochemical characteristics of the active pharmaceutical ingredient (API) [1]. This is achieved by the formation of amorphous dispersions or solid solution of the API in a polymeric carrier. Moreover, HME has the advantage of continuous processes, such as good process stability, permanent process control, small footprint and decreased manufacturing time.

Our aim is to develop a continuous process line for the production of immediate release tablets (85% API released within 30 min) for a poorly soluble API (BCS II). The line consists of (1) a hot-melt extruder, (2) a die-face pelletizer, (3) a continuous blender and (4) a tableting machine making tablets in a continuous way. The API solubility is increased by formation of an amorphous system via HME. A die-face pelletizer is used to directly produce almost spherical pellets for subsequent tableting.

Formulation screening was performed to select the most promising carrier substances to match the desired release profile and stability. Based on literature research and experience five promising carriers were selected for the screening. Melt rheology measurements were performed to estimate processability and thermal sensitivity. Additionally, table top extrusions (D=9 mm, L/D=20) were carried out to study the solid state after exposure to shear forces during extrusion. Based on this screening the best formulation was selected for HME runs (D=18 mm L/D=40) on technical lab scale. The extrusion setup was designed based on material information gathered during formulation screening. The temperature profile was selected based on the DSC and rheology data. Finally, the formulation was characterized regarding the mechanical behavior and surface properties.

Pure pellets and mixtures of tablets with compaction aids were subsequently tableted using a compaction simulator. Then the tablets were analyzed and compared to the target properties.

References:

[1] J. Breitenbach, "Melt extrusion : from process to drug delivery technology," *Eur. J. Pharm. Biopharm.*, vol. 54, pp. 107–117, 2002.

[2] C. R. Young, C. Dietzsch, and J. W. McGinity, "Compression of Controlled Release Pellets Produced by a Hot Melt Extrusion and Spheronization Process," *Pharm. Dev. Technol.*, vol. 1, pp. 133–139, 2005.

[3] M. R. Abbaspour, F. Sadeghi, and H. Afrasiabi Garekani, "Design and study of ibuprofen disintegrating sustained-release tablets comprising coated pellets.," *Eur. J. Pharm. Biopharm.*, vol. 68, no. 3, pp. 747–59, Mar. 2008.

426723 Process Intensification through Continuous Spherical Crystallization Using a Two-Stage Mixed Suspension Mixed Product Removal System

Abstract:

Process intensification through continuous spherical crystallization using a two-stage Mixed Suspension Mixed Product Removal system

Ramon Peña and Zoltan K. Nagy School of Chemical Engineering, Purdue University, West Lafayette, IN 47907, US

Of utmost importance in the crystallization of active pharmaceutical ingredients (APIs) in the pharmaceutical industry is to produce crystals of good physical, processing, and biopharmaceutical properties. The definition of good physical properties depends on what the end goal and the drug formulation that the crystals will be a part of, but often processing and biopharmaceutical properties are competing interests. In most cases, the crystallization process is tailored to improve downstream process efficiency rather than improve drug molecule efficacy in the human body. Herein a novel concept and method to help satisfy both processing and biopharmaceutical interests is proposed, which is based on performing crystallization and spherical agglomeration in a two-stage continuous mixed suspension mixed product removal (MSMPR) system. With the suitable operating mode, the system enables to decouple the nucleation and growth from the agglomeration mechanisms, while performing efficient continuous manufacturing of particles with desired properties. Decoupling will offer more degrees of freedom for the control of each mechanisms. This in turn provides the means by which properties in the nucleation/growth stage can be tailored to those of most biopharmaceutical benefit and efficacy (e.g. bioavailability, dissolution, morphology) while allowing the agglomeration stage to be tailored to produce spherical agglomerates of the most processing efficiency (e.g. filtering, drying, friability).

Continuous spherical crystallization (CSC) of a model drug compound was first carried out by Kawashima et al.[1] after Kawashima and Capes[3] realized that resulting products from spherical agglomeration had very good flow properties. In this work, a model continuous mixed suspension, mixed product removal (MSMPR) crystallizer consisting of one stage was fed an aqueous suspension. The critical parameters were identified as agitation rate, suspension feed rate, and bridging liquid feed rate[2]. Since Kawashima et al.[1] very little work has been done on CSC and few references make use of process analytical technology (PAT) tools. An abundance of work exists on spherical crystallization in batch operation. The first batch spherical crystallization (BSC) was carried out by Kawashima et al.[2] who crystallized salicylic acid in ethanol by pouring the solution into a water-chloroform mixture and found the critical parameters to be similar to that in a continuous system: agitation rate, temperature, bridging liquid content, and residence time. In this study a ternary solvent system was used. This consisted of good solvent to dissolve the drug, poor solvent to precipitate it, and bridging liquid to promote agglomeration of the precipitated crystals.

In this work, a novel two-stage continuous spherical crystallization system in which the nucleation and growth processes are separated (in the first stage) from the agglomeration mechanism (in second stage), enabling precise and independent control of the internal crystal size distribution (CSD) and the agglomerate size distribution (ASD). This can enable the simultaneous control of product properties (e.g. dissolution profile) and processing requirements (flowability, compressibility, etc.)

1. Kawashima, Y.; Kurachi, Y.; Takenaka, H. Preparation of spherical wax matrices of sulfamethoxazole by wet spherical agglomeration technique using a CMSMPR agglomerator. *Powder Technol.* **1982**, 32, 155–161.

2. Kawashima, Y.; Okumura, M.; Takenaka, H. Spherical crystallization: direct spherical agglomeration of salicylic acid crystals during crystallization. *Science.* **1982**, 20–21

3. Kawashima, Y.; Capes, C. An experimental study of the kinetics of spherical agglomeration in a stirred vessel. *Powder Technol.* **1974**, 10, 85–92.

431635 Development of a Novel PAT for in-Line Measurement of Powder Dynamic Density inside Feed Frame of Rotary Tablet Presses

Abstract:

Consistent filling of the tablet press die with a uniform weight of powder has always been an empirical approach with little understanding of the powder flow pattern within the feed frame, especially at the last moment before powder is fed into the die and tablet potency is set. Compression is a complex unit operation comprising different regimes of powder flow (depending on the shear rate and packing fraction) which need to be fundamentally understood in order for this to be a predictable process.

In this work we studied powder flow (with application to die filling) in the feed frame of various rotary tablet presses by developing a novel in-line process analytical technology (PAT), the dielectric probe. This tool enabled us to capture powder dynamic density regime maps in different sections of a feed frame. The basic measurement principle of the probe is to estimate the physical properties of bulk solids (e.g. solid fraction or density) by standard correlation with the measured electrical properties (e.g. dielectric permittivity) through proper calibration for each formulation.

We investigated effects of press speed, feed frame speed, paddle design, and formulation properties on the die filling process by looking at dynamic density of powder and its fluctuations inside the feed frame. We discuss how these measurements can be used for faster and more effective set-up of the tablet press and to find and maintain the optimum operating condition throughout the entire compression. Furthermore, we show how these data are helpful in scale-up of the tablet press, especially for formulations that are sensitive to over-lubrication where it is important to maintain similar level of shear energy among different scales.

420074 Development of a Control Strategy for Continuous Manufacturing of Solid Dosage Forms

Abstract:

During the last years the advantages of continuous manufacturing have been increasingly demonstrated (and publicized) by academia, industry and regulators. Slowly, this is leading to a paradigm shift in pharmaceutical manufacturing, away from batch towards advanced continuous processing based on real-time quality assurance. As a result, different equipment suppliers are providing turn-key solutions for continuous manufacturing systems, ranging from powder feeding, to blending, granulation, tableting and/or coating. While several systems have been reported and studied in the literature, open questions still remain, especially with respect to system robustness, online measurement and process control.

In this work a control strategy for a novel industrial-scale production line for oral-dosage forms was developed. The objective of the control strategy is to consider all possible variability in raw material properties, process-settings, equipment design and environmental conditions. Constant product quality is maintained by proper equipment design and integration of unit operations, utilization of process understanding, a risk-based PAT strategy and appropriate control actions.

The presentation will highlight a successful implementation of an in-line monitoring system for critical quality attributes. In order to enable Real Time Release, PAT tools were selected with appropriate data acquisition systems to support control actions. This includes chemometric models for start-up and steady state. Out-of-spec materials are detected and eliminated at different stages of the process. All sensors are embedded in a PAT data management system to treat and register process data. Additionally, a concept for 100 % quality control of final tablets will be presented.

420103 Investigation of Segregation Effects in a Direct Compaction Process Via Multipoint NIR Monitoring and NIR Chemical Imaging

Abstract:

During the last years the application of near-infrared spectroscopy (NIRS) has become an established technology for in-line process monitoring of the API-content, as it is a fast and non-destructive method. Common methodologies have been further developed to multipoint applications allowing the timely aligned assessment of real time process data from different positions or even processes. Additionally, hyperspectral imaging (HSI) even allows measuring not only parts of powder streams or single tablets, but a large number of tablets. Even mapping of each single tablet becomes feasible and enables detection of API agglomerations.

In this study the evolution of API concentration within a direct compaction process is monitored in-line by means of a multipoint spectrometer using separate fibres at different locations within the press. In addition, a HSI unit assesses the quality of the final tablets. This approach enables the determination of content uniformity along the powder path and therefore, investigation of segregation effects along the processing route. Goal of the work is the evaluation, if those effects influence the final product quality and if different operating conditions affect the extent of segregation.

421775 Characterization of Tablet Core and Coating Roughness By Means of Optical Coherence Tomography

Abstract:

Although tablet coating is a well-established unit operation in pharmaceutical industry, the achievable quality of coating is still limited by the fact that it is a highly complex process depending on many parameters. Specifically for functional coatings slight changes of the coating equipment and process parameters may impact the physicochemical properties of the film and may consequently affect the coating quality. There are significant gaps remaining in the full scientific understanding of the process, which makes it a difficult task to produce tablets with an exact thickness and density and with only small tablet-to-tablet variations between batches or even within one single batch. The use of advanced quality control tools facilitates the understanding of physico-chemical changes of the film.

Such a quality control tool is optical coherence tomography (OCT) showing not only high potential for off-line quality control of film-coatings but also for in-line monitoring of coating processes. OCT is a high resolution imaging methodology to produce cross-sectional and three-dimensional (3D) images of film coatings in a non-destructive and contactless manner. This modality allows the direct measurement of the coating thickness under consideration of the refractive index of the coating. Beside the analysis of the coating thickness, it additionally facilitates the characterization of the tablet core and coating roughness.

An automated data evaluation algorithm extracts information about coating thickness as well as tablet core and coating roughness. This analysis can be carried out either on the basis of cross-sectional images or by using 3D data of samples. The focus in this study is on the analysis of 3D images of samples removed periodically from a pan coating process. A thickness map and profile maps of the tablet core and coating is computed from about 480,000 depth measurements per sample. This data enables among others the calculation of the arithmetical mean deviation, the root mean square deviation, the skewness and the kurtosis of the assessed profiles. Moreover, the impact of the tablet core roughness on the coating thickness is analyzed by correlating the tablet core profile and the coating thickness map. This analysis is further transferred to in-line OCT measurements allowing the investigation of core and coating roughness during the production. Such detailed information specifically emphasizes the high capability of the OCT technology to improve process understanding.

421912 In-Line Monitoring of a Pharmaceutical Fluid-Bed and Pan Coating Process By Optical Coherence Tomography

Abstract:

This study demonstrates a new in-line measurement technique for monitoring coating processes. Process monitoring is performed by an optical coherence tomography (OCT) sensor providing cross-sectional images of film-coatings in a non-destructive and contactless manner. Typically tablets or pellets are coated. Tablets are mostly coated in (perforated) pans or drums, whereas pellet coating is typically carried out in a fluid-bed. However, the monitoring of the coating growth as a function of process time is a key issue for both processes. This study shows results of coating thickness measurements from in-line OCT data for pellets and tablets coated in a fluid-bed and pan coater, respectively.

The coating thickness can be determined directly and automatically from OCT images and without the need of chemometric calibration models. Results of the in-line OCT measurements were validated by analyzing samples that were periodically removed from the process during the production. These samples were characterized off-line by OCT and dynamic image analysis in case of pellets or dimension measurements in case of tablets. Moreover, this methodology enables a multiple direct measurement of the coating thickness on individual pellets/tablets rather than providing the temporal and spatial average of a large number of pellets/tablets. This gives substantially more information about the coating quality, i.e., intra- and inter-particle coating variability, than standard quality control methods. The information about those parameters specifically emphasizes the high capability of the OCT technology to improve process understanding and to assure a high product quality. Specifically, the additional information as intra- and inter-particle coating variability as captured with the OCT sensor could be a valuable critical quality attribute (CQA). An in-line OCT system could act as a key tool for the development and control of coating processes within the process analytical technology (PAT) and quality by design (QbD) framework.

428833 Development and Demonstration of a Suite of Tools Aimed at Process Development Intensification and Efficient Experimentation Practices

Abstract:

Since its inception in early 2014, the Process Development Intensification Laboratory (iLab) within Chemical Process Development and Commercialization at Merck has been focused on the development of tools and techniques to enable acquisition of deeper, more fundamental process understanding in a resource sparing manner. This is predominantly centered around the concept of data intensification, specifically, around enabling more efficient data capture, data reduction and data analysis.

In this submission, we highlight the development and demonstration of novel, enabling technologies such as process modeling, bench-top automation, parallel processing, process analytical technologies and statistically designed experimentation. Furthermore, this submission uses a saponification deprotection reaction as a case study, highlighting the synergistic effects which can be realized through the application of many of the developed tools.

Specifically, we highlight the use of many of the tools the iLab has been employing, beginning with traditional DOE for selection of preliminary experimental conditions. We go on to discuss some of its limitations when the number of factors for investigation is high, prompting application of kinetic modeling for acquisition of deeper, more mechanistic process understanding. From there, it becomes apparent that alternative experimental techniques, specifically use of the Amigochem (a bench top automation tool combining a liquid handling robot for LC sampling and dilution with a parallel reaction station) offers an opportunity for building information-rich data sets rapidly. This enables kinetic profiling of the aforementioned DOE with minimal user intervention. We also demonstrated the use of probe-based sampling and dilution for unattended sampling of automated lab reactors. When coupled with online FTIR spectroscopy and chemometric techniques such as partial least squares regression, it becomes possible to gather compositional information and very detailed kinetic profiles rapidly in a wide variety of equipment types.

Simply collecting significant quantities of data is not always fruitful, however, unless that data can be transformed in some way to intimate process knowledge. We have also been developing means by which collected data can be used to build process models, and more importantly, how those process models can be used to drive future experimental design decisions. We have demonstrated two separate mathematical algorithms, one geared towards selection of optimal experiments to aid in model discrimination, and the other geared at selection of optimal experiments for kinetic parameter refinement.

Separately, many of the technologies developed by the iLab represent significant achievements in data collection, analysis and application in the development space. When combined as a holistic experimental design paradigm, there is significant synergy realized, and the complete toolkit stands to be truly transformative in the way in which process development is conducted in our area moving forward.

417087 Spectroscopy Assisted Process Design for Complex Multi-Phase Organic Reactions

Abstract:

Chemical reaction impurity control is an essential component of quality risk management in pharmaceutical drug substance manufacturing. Transformations involving reactive/unstable intermediates or multi-phase reaction media often present a challenge for conventional analytical techniques (e.g. HPLC, GC, NMR). During process development and plant execution of such processes, utilization of in-situ spectroscopic techniques offers a significant advantage from both sensitivity and time-of-analysis perspectives. In this work, two industrial case studies implementing spectroscopy will be discussed in the context of small molecule pharmaceutical intermediate laboratory process design and pilot plant execution.

In the first case study, process development for preparation of a triazole-based intermediate was performed utilizing Raman and Near IR spectroscopy. Such in-situ measurements were necessary as the reagents were both non-chromophoric and unstable using conventional off-line analytical methods. The data allowed for rapid analysis of kinetic information as well as mechanism elucidation. The approach was ultimately leveraged to define robust processing parameters (e.g. temperature, concentration, order of addition) for optimal impurity control. In the second case study, Near IR spectroscopy was utilized to monitor ammonia concentration in a telescope reductive amination process. Control of residual ammonia proved critical to limit formation of a des-halogenation impurity during the hydrogenation step. Application of Near IR to trigger additional vessel purges was successfully demonstrated in both the laboratory and pilot plant scale.

414998 Investigation of a Suitable Lubricant for Increasing Slip in a Continuous, Co-Rotating Twin Screw Dry Granulation Process

Abstract:

Investigation of a Suitable Lubricant for Increasing Slip in a Continuous, Co-Rotating Twin Screw Dry Granulation Process

Granulating technologies are generally described in two categories: wet and dry. Wet methods apply solution on top of an agitated powder bed. Dry methods compact powders into agglomerates; for example feeding powders through two counter rotating rolls under pressure such that the powders press together to form a thin wafer that is milled into the granulation.

Recently, the industry and universities interested in pharmaceutical processing shifted focus towards continuous granulating technology and the flexibility this technology offers. Many institutions are taking advantage of equipment readily available within their network and for this reason co-rotating twin screw extruders are currently being investigated more intensively. While often considered for making amorphous composites, because of their versatility, they can be efficient continuous granulators.

The preference when granulating with a co-rotating twin screw extruder is to employ a wet granulation method when possible. In this technique, powder is fed into the extruder and conveyed towards a mixing block via two co-rotating screws while being sprayed with a binder solution. The binder solution facilitates granulation upon compaction in the kneading zone. The granules are subsequently dried either down the barrel of the extruder, or externally in by a fluidized bed. The granule attributes are strongly dependant on the binder addition rate, screw speed, and mixing block design.

In the case for dry twin screw granulation a binder solution is not used. Instead the material is softened using the barrel heating mechanism of the extruder and the energy created by mixing. Agglomerate creation is facilitated by compaction of the soften binder. The granule attributes are strongly dependant on barrel heating/cooling temperature, screw speed, feed rate, and the screw design.

Experiments have shown that the density of the granulation is proportional to the energy applied to the granule. The applied energy can be sufficient to induce degradation in the active pharmaceutical ingredient (API). Furthermore, for highly cohesive, low bulk density pre-blends, fast screw speeds are required to convey the powder to the granulating zone while maintaining the same feed rate from batch to batch. The faster screw speeds result in additional energy input with the potential for excessive temperature generation attributed to plastic and frictional energy dissipation (PED and FED).

Evidence of these phenomena is shown by significant shifts in the steady state of the system parameters (e.g. torque, specific energy, and product temperature). Physical observations indicate adhesion of product on the screws in the vicinity of the mixing element with the potential for longer residence times and exposure to heat. Recent experiments have shown that it is possible to reduce these adverse effects of intensive mixing and shear through cooling at the mixing zone.

In addition to cooling at the mixing zone, the shear can further be reduced by adding a suitable level of lubricant, in the pre-blend, that is compatible with the active pharmaceutical ingredient (API). The lubricant allows a desirable amount of slip between the powder and the equipment surfaces. The lubrication results in lower torque readings an indication of lower energy absorbed by the product. The level of lubricant should not compromise the structural integrity of the granule or any downstream granule process. Levels of up to 1.0% magnesium stearate have been shown to reduce the torque in small scale twin-screw granulation trials.

416112 The Development of a Suitable Screw Configuration for the Use in a Continuous, Co-Rotating Twin Screw Dry Granulation Process

Abstract:

The Development of a Suitable Screw Configuration for the use in a Continuous, Co-Rotating Twin Screw Dry Granulation Process.

Granulation processes are unit operations used in the development, manufacture, and characterization of oral solid dosage forms in industry and academia. The goal with every granulation is to transform fine, poor flowing powders into larger, free flowing agglomerates with the aid of a binder. Granulating technologies are generally described in two categories: wet and dry. Wet methods apply solution on top of an agitated powder bed. Dry methods compact powders into agglomerates; for example feeding powders through two counter rotating rolls under pressure such that the powders press together to form a thin wafer that is milled into the granulation.

Recently, the industry and universities interested in pharmaceutical processing shifted focus towards continuous granulating technology and the flexibility this technology offers. Many institutions are taking advantage of equipment readily available within their network and for this reason co-rotating twin screw extruders are currently being investigated more intensively. While often considered for making amorphous composites, because of their versatility, they can be efficient continuous granulators.

The preference when granulating with a co-rotating twin screw extruder is to employ a wet granulation method when possible. In this technique, powder is fed into the extruder and conveyed towards a mixing block via two co-rotating screws while being sprayed with a binder solution. The binder solution facilitates granulation upon compaction in the kneading zone. The granules are subsequently dried either down the barrel of the extruder, or externally in a fluidized bed. The granule attributes are strongly dependant on the binder addition rate, screw speed, and mixing block design.

In the case for dry twin screw granulation, a binder solution is not used. Instead, the material is softened using the barrel heating mechanism of the extruder and the energy created by mixing. Agglomerate creation is facilitated by compaction of the soften binder. The granule attributes are strongly dependant on barrel heating/cooling temperature, screw speed, chilling feed rate, and the screw design.

Experiments have shown that the density of the granulation is proportional to the energy applied to the granule. The applied energy can be sufficient to induce degradation in the active pharmaceutical ingredient (API). Furthermore, for highly cohesive, low bulk density, pre-blends, fast screw speeds are required to convey the powder to the granulating zone while maintaining the same feed rate from batch to batch. The faster screw speeds result in additional energy input with the potential for excessive temperature generation attributed to plastic and frictional energy dissipation (PED and FED). The design of the mixing elements, along with cooling at the mixing zone has been shown to reduce the PED and FED. Results for granulations made with highly cohesive, low bulk density, pre-blends are presented.

417793 A DEM Model of Roller Compaction Including Powder Feeding

Abstract:

A DEM Model of Roller Compaction Including Powder Feeding

In our work we studied roller compaction by means of Discrete Elements Method (DEM) modeling. Roller compaction is an industrial process in the field of dry granulation, aiming to obtain granules from fine powder mixtures. Our interest is to understand the impact of powder feeding from hopper via a screw on the compaction process and the product quality .

In the presentation, we will compare our results with particle tracking results and ribbon analysis from experiments to qualitatively validate the model. The Coarse Graining framework is employed to compute the mass density, velocity and pressure fields in both, the feeding and the compaction zone. As well known, the screw feeder induces a period flow of particles, inducing a periodic flow structure of the particle bed close to the compator rolls. This effect and its impact on the sheet quality will be discussed.

This work is our first approach to DEM modeling of this process. The ability of DEM to better understand roller compaction will be highlighted as well in the talk. Finally, we will define the framework for studying roller compaction by means of DEM in the future.

Acknowledgements: this work was supported by the IPROCOM Marie Curie initial training network, funded through the People Programme (Marie Curie Actions) of the European Union's Seventh Framework Programme FP7/2007-2013/ under REA grant agreement No. 316555.

418505 Phytochemical Profile and DPPH Free Radical Scavenging of Amazonian Plant Extracts

Abstract:

1. Introduction: The medicinal plants use is based in ethnopharmacology information, and in development countries is the available form to many disease treatments. The Amazonia Rain Florets have the biggest biodiversity in the world, which leads to use of biodiversity, especially plants, for the treatment of diseases in folk medicine. Nevertheless, the popular use of medicinal plants has no security or efficacy proves. Thus, the medicinal plants studies are important. One of the activities most studied is antioxidant potential, due the free radicals (FR) or oxygen reactivity species (ROS) are involved in cell aging, inflammation process, cardiovascular diseases and cancer [1]. In Amapá State, The Pharmacy of Scientific and Technological Institute Research of Amapá (IEPA) sells plant extract use in folk medicine. Hence, the present work aimed to evaluate the phytochemical profile and DPPH free radical scavenging of commercial extracts from *Dalbergia subcymosa* Ducke (common name, *Verônica*), used from uterine inflammation and anemia treatments; *Croton cajucara* L. (common name, *Sacaca*), used as hepatoprotector and hipocholesterolemic; and *Licania macrophylla* Benth (common name, *Anauerá*), used in infection processes.

2. Material and methods: The commercial extracts were purchased in September 2014 from Scientific and Technological Institute Research of Amapá (IEPA). The phytochemical profiles were determined by colorimetric methods [2]. The compounds analyzed were phenolic compounds, flavonoids, tannins and saponins. The free radical scavenging was evaluated by the DPPH assay [3], using rutin and silymarin (*Silybum marianum* standardized extract) as positive controls. The DPPH methanol solution was prepared at 60μM, and the spectrophotometric reads were performed at 515nm, after 30 min. The results expressed in EC_{50}.

3. Results and Discussion – The radical scavenger activity of the ethanolic extracts was found that the reduction of DPPH occurred in a concentration-dependent manner, as observed with the decrease in the absorbance of DPPH. The EC_{50} of radical scavenger activity of the *Dalbergia subcymosa* Ducke was 36.77±3.55 μg/mL and the phytochemical profile shown the presence of phenolic compounds, flavonoids, tannins and saponins. The *Croton cajucara* EC_{50} of radical scavenger activity value obtained was 35.21±9.13 μg/mL, shown antioxidant activity, and were found in this extract the presence of phenolic compounds, flavonoids and saponins. However, according to Tieppo (2006) [4] the *Croton cajucara* chloroform extract no show the antioxidant activity, the difference between the results may be due different extraction forms. The popular use of *Croton cajucara* as anti-inflammatory was proved [5]. The *Licania macrophylla* Benth ethanolic extract shown the presence of phenolic compounds, flavonoids and tannins, and in revised literature were related the presence of steroids and triterpenoids [6]. The *Licania macrophylla* Benth EC_{50} of radical scavenger activity value obtained was 28.50±7.45 μg/mL, result was similar the founded in other specie of the same genre [7]. The present work is the first report of the DPPH scavenger activity from these commercial extracts.

4. Conclusions – The screening of plant extracts using DPPH free radical method proved to be effective for the selection of those, which could have an antioxidant activity. These extracts are rich in radical scavengers, such as flavonoids and tannins, known as antioxidants. Further, more studies are necessary to characterize the biological antioxidants.

5. References

[1] J.F.S. Daniel, et. al. J. Brazil. Chem. Soc., 16, (2004) p.634-638

[2] M. Kalegari, et.al. Nutrition., 30, (2014) p. 713-718

[3] W. Brand-Williams, M. E. Cuvelier and C. Berset. LWT-Food Sci. Technol., 28, (1995) p. 25-30.

[5] Perazzo, F. F. et al. Rev. bras. farmacogn. 17 (4), (2007), p.521-528

[6] Gomes, M. L. et al. Rev. Bras. Farm., 87 (1), (2006), p. 26-29

[7] Silva, J.B.N.F. et al. Arch. Biol. Sci., 64 (2), (2012) p. 459-464

425976 Analysis of Post-Translational Regulation of Engineered CHO Using Proteomics

Abstract:

Cell engineering has been applied to improve the post-translational modification (PTM) by manipulating one or multiple enzymes in CHO. However, the effects of the interaction between intracellular proteins, heterologous protein expression and the regulation of host cells on PTM have not been fully investigated at systems biology level. In this study, the comparative and quantitative proteomics was performed to investigate seven host cells, including parental cells (CHO K1 and CHO S) and engineered cells ($Pro^{-}5$ CHO, $Pro^{-}5Lec20$, $Gat^{-}2CHO$, $Gat^{-}2Lec20$, and Lec13) expressing different β4-galactosyltransferase (β4GalT) or down-regulating α-1,6-fucosyltransferase, and their anti-cancer monoclonal antibody (mAb) production cell lines. The intracellular proteomic maps of these host cells and production lines were created. The expression of the proteins associated with PTM of mAb, the metabolism of glucose, glutamine and amino acids, and the transcription, translation and folding of mAb were investigated and compared. The PTM regulation of mAb and the interaction of intracellular proteins caused by the genetic engineering were analyzed. This study demonstrated that proteomics is a powerful tool to understand the mechanism of protein expression and protein quality regulation. The Omics knowledge developed could be used to optimize the cell engineering strategy to achieve high quality of biopharmaceuticals.

427252 Vacuum Compression Molding As Screening Approach for Hot-Melt Extruded Dosage Forms

Abstract:

Hot melt extrusion (HME) is a continuous process with growing importance for the pharmaceutical industry [1]. It is a robust production process for solid dispersions. Solid dispersions can increase the solubility of the active pharmaceutical ingredient (API) in case of poor crystalline solubility or they can help to tailor a formulation to a certain API release behavior [2].

Often, HME supplies intermediates in form of pellets, which are processed via conventional tablet compaction or capsule filling to the final dosage form. Recently, injection molded dosage forms gained the interest since they offer several advantages [3]. They can have high hardness or high elasticity which are requested for example for abuse resistant formulations or implants. However, the effect of interactions between formulation, process parameters and product properties is not well understood. Thus, in general, HME formulation development usually requires a host of experimental attempts. In many cases, formulation screening is performed with small-scale extruders, for example with conical twin-screw extruders, which typically require 5 gram of material. The extruder delivers molten material which is typically collected as strand and cut into small pieces, which are then analyzed with respect to their suitability for the indented dosage form and release profile. However, many of the analyzed properties are impacted by several parameters concurrently. For example, the drug release profile is strongly impacted by the type of solid dispersion and the available exchange surface. Thus, for example defined geometries can simplify formulation screening since less parameters superimpose.

This work presents an alternative screening approach that requires less than 1 gram of material for the production of one sample tablet. A novel vacuum compression molding (VCM) tool has been developed. A powdery mixture of raw materials is filled into the sampling chamber. Subsequently, the tool is evacuated and heated above the melting point forming a homogeneous bubble-free tablet. Stickiness issues are prevented by the used PTFE coated release foils. Formulations consisting of matrix polymers and an API have been produced via the proposed method. The obtained tablets are analyzed by standard screening analysis tools (DSC, dissolution testing, etc.). The results are compared to results obtained from extruded material analyzed by the same methods. The comparison shows that similar results can be obtained with less effort. Thus, it has been demonstrated that VCM can be a useful tool in formulation development as it reduces workload and the required material amount.

References:

[1] M. A. Repka, N. Langley, and J. DiNunzio, *Melt Extrusion - Materials, Technology and Drug Product Design*. New York, Heidelberg, Dordrecht, London: Springer, 2013.

[2] E. Roblegg, E. Jäger, A. Hodzic, G. Koscher, S. Mohr, A. Zimmer, and J. Khinast, "Development of sustained-release lipophilic calcium stearate pellets via hot melt extrusion.," *Eur. J. Pharm. Biopharm.*, vol. 79, no. 3, pp. 635–45, Nov. 2011.

[3] L. Zema, G. Loreti, A. Melocchi, A. Maroni, and A. Gazzaniga, "Injection Molding and its application to drug delivery.," *J. Control. Release*, vol. 159, no. 3, pp. 324–31, May 2012.

DEVELOPMENT OF RETINOID ST1926 NANOPARTICLE FORMULATION VIA FLASH NANOPRECIPITATION

Zaynab Jaber[1], Melody Saikali[1], Claudio Pisano[2], Nadine Darwiche[1] and ***Walid Saad****[3], (1)Department of Biochemistry and Molecular Genetics, American University of Beirut, Beirut, Lebanon, (2)Biogem, Research Institute, Ariano Irpino (AV), Italy, (3)Chemical Engineering, American University of Beirut, Beirut, Lebanon*

Abstract

Natural retinoid compounds such as all-*trans* retinoic acid constitute a promising class of anti-cancer compounds that have been used against various types of cancers. However, their clinical usage is hindered by undesirable side effects and acquired resistance. As a result, synthetic retinoids exhibiting increased specificity and reduced toxicity were developed. In particular, ST1926 is a hydrophobic synthetic retinoid that has shown promise in several solid and liquid tumors, and has reached Phase I clinical trials for ovarian cancer treatment. Further development of ST1926 is hindered by its reduced bioavailability and therapeutic efficacy due to glucuroconjugation in cancer patients, leading to its rapid excretion in urine.

In this research project, a nanoparticle-based formulation of ST1926 is developed to overcome limitations in formulation, efficacy, and bioavailability. Formulating ST1926 into controlled-size nanoparticle suspensions offers several advantages, including (1) overcoming the formulation challenge due to the drug's low water solubility, (2) enhancing the drug's efficacy through preferential tumor accumulation due to the Enhanced Permeability and Retention (EPR) effect observed with long-circulating nanoparticles, and (3) potentially reducing drug glucuronidation by providing nanoparticle surface protection during *in vivo* circulation using amphiphilic block copolymers.

ST1926 nanoparticles (STNP) were prepared using a recently developed particle formation technology termed Flash Nanoprecipitation (FNP) which uses fast mixing in a confined volume to afford co-precipitation of the solute and a stabilizing amphiphilic diblock copolymer, yielding stable, controlled size nanoparticles.

STNP with an average diameter size of 230 nm as determined by dynamic light scattering were prepared using poly(styrene-b-ethylene oxide) stabilizing polymer at a drug/polymer weight ratio or 1/5, respectively. *In vitro* cell growth assays were conducted to compare the anti-cancer activity of ST1926 and its corresponding STNP formulation. Cells were seeded in 96-well plates, treated with ST1926 or STNP, and monitored over 72 hours. Preliminary results show that STNP exhibits potent anticancer activities similar to the native ST1926 in colorectal cancer cells, at pharmacologically relevant concentrations.

Further studies are planned to characterize, optimize, and investigate the efficacy of STNP using *in vitro* and *in vivo* tumor models.

Introduction

Cancer is one of the major causes of death worldwide; in particular, colorectal cancer is the third amongst men and women, which highlights the need for developing novel, more efficient, and safe therapies.

Natural retinoid compounds such as all-*trans* retinoic acid constitute a promising class of anti-cancer compounds that have been used against various types of cancers.[1] However, their clinical usage is hindered by undesirable side effects and acquired resistance. As a result, synthetic retinoids exhibiting increased specificity and reduced toxicity were developed.[2] In particular, ST1926 ((2E)-3-[30-(1-adamantyl)-40-hydroxy[1,10-biphenyl]-4-yl]-2-propenoic acid), is a hydrophobic synthetic retinoid that has shown promise in several solid and liquid tumors, and has reached Phase I clinical trials for ovarian cancer treatment.[3-7] Further development of ST1926 is hindered by its reduced bioavailability and therapeutic efficacy due to glucuroconjugation in cancer patients, leading to its rapid excretion in urine.[8] The molecular structure for ST1926 is shown in Figure 1.

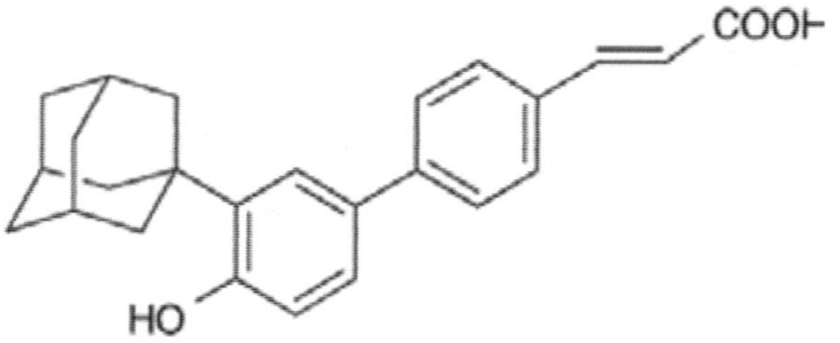

Figure 1: Synthetic Retinoid ST1926

In this research project, a nanoparticle-based formulation of ST1926 is developed to overcome limitations in formulation, efficacy, and bioavailability. Formulating ST1926 into controlled-size nanoparticle suspensions offers several advantages, including (1) overcoming the formulation challenge due to the drug's low water solubility, (2) enhancing the drug's efficacy through preferential tumor accumulation due to the Enhanced Permeability and Retention (EPR) effect observed with long-circulating nanoparticles,[9] and (3) potentially reducing drug glucuronidation by providing nanoparticle surface protection during *in vivo* circulation using amphiphilic block copolymers.

Nanoparticle suspensions are produced using a recently developed particle formation technology termed Flash Nanoprecipitation (FNP) which uses fast mixing in a confined volume to afford co-precipitation of the solute and a stabilizing amphiphilic diblock copolymer, yielding stable, controlled size nanoparticles.[10-11] The fast mixing is achieved using a tangential flow mixing cell where one stream containing the drug and an amphiphilic diblock copolymer dissolved in a water-

n Figure 2.

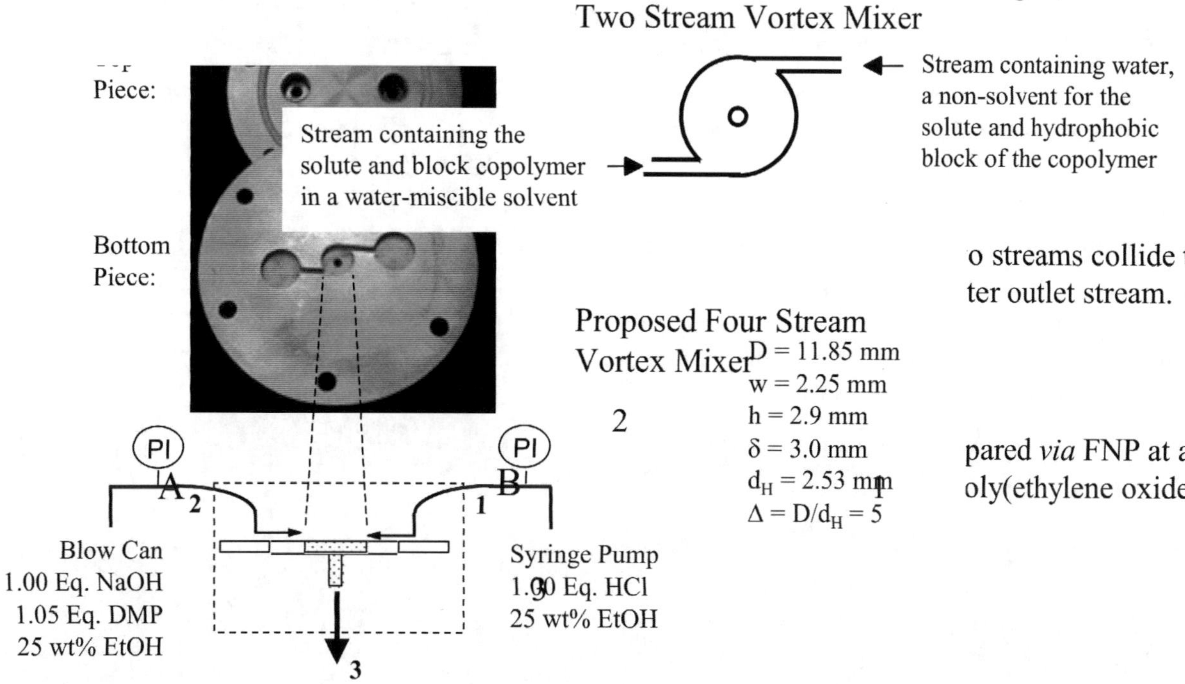

o streams collide tangentially
ter outlet stream.

pared *via* FNP at a drug to
oly(ethylene oxide) block

copolymer (PS-PEO) with a polystyrene hydrophobic block size of 1,500 g/mole, and a poly(ethylene oxide) block size of 2,400 g/mole were dissolved in 3 mL of tetrahydrofuran (THF). The resulting solution was mixed at 12 ml/min with water at 108 ml/min using the tangential flow mixer. STNP were collected at the mixer outlet and the particle size determined using dynamic light scattering (DLS).

In vitro cell growth assays: The efficacy of STNP was assessed in *in vitro* human colorectal cancer models: HCT-116 (wild-type p53) and HCT-116 p53$^{-/-}$ cell lines. Cells were seeded in 96-well plates, treated with ST1926 or STNP, and monitored over 72 hours. Cell growth was assayed in triplicate wells using the cell proliferation MTT CellTiter 96 Assay. The results are expressed as percentage of control (0.1% DMSO) and are the average ± SE of three independent experiments.

Cell Death Assay: Cells were seeded in 100 mm dishes and treated with 1 μM ST1926 or STNP. Apoptosis was assessed by TUNEL assay 48 hours post-treatment.

Results and Discussion

STNP were produced *via* FNP with average hydrodynamic diameter of 230 nm as determined by dynamic light scattering. The number average particle diameter distribution is shown in Figure 3. The particle diameters remain relatively unchanged over 24 hours at room temperature.

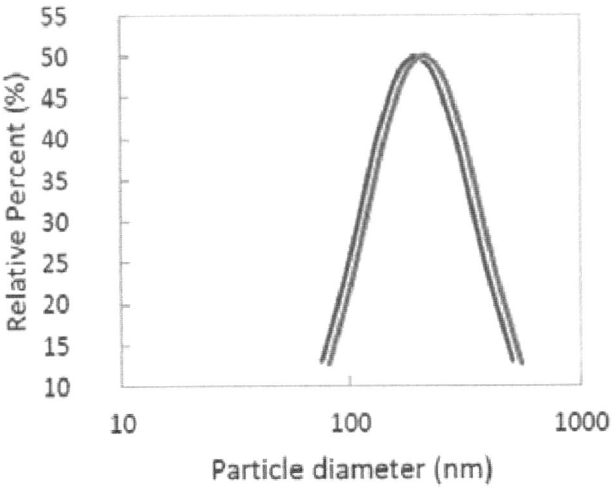

Figure 3: Number-averaged particle diameter distribution for STNP at a drug to polymer weight ratio of 1/5 at the initial (blue) and 24 hour time points (grey). Particle diameters remain relatively unchanged over a 24 hour period.

In vitro cell growth assays were conducted to compare the anti-cancer activity of ST1926 and its corresponding STNP formulation. Cells were seeded in 96-well plates, treated with ST1926 or STNP, and monitored over 72 hours. Our results show that STNP exhibits potent anticancer activities similar to the native ST1926 in colorectal cancer cells, at pharmacologically relevant concentrations. The results are summarized in Figure 4 for HCT-116 WT, and Figure 5 for HCT-116 p53$^{-/-}$ cell lines.

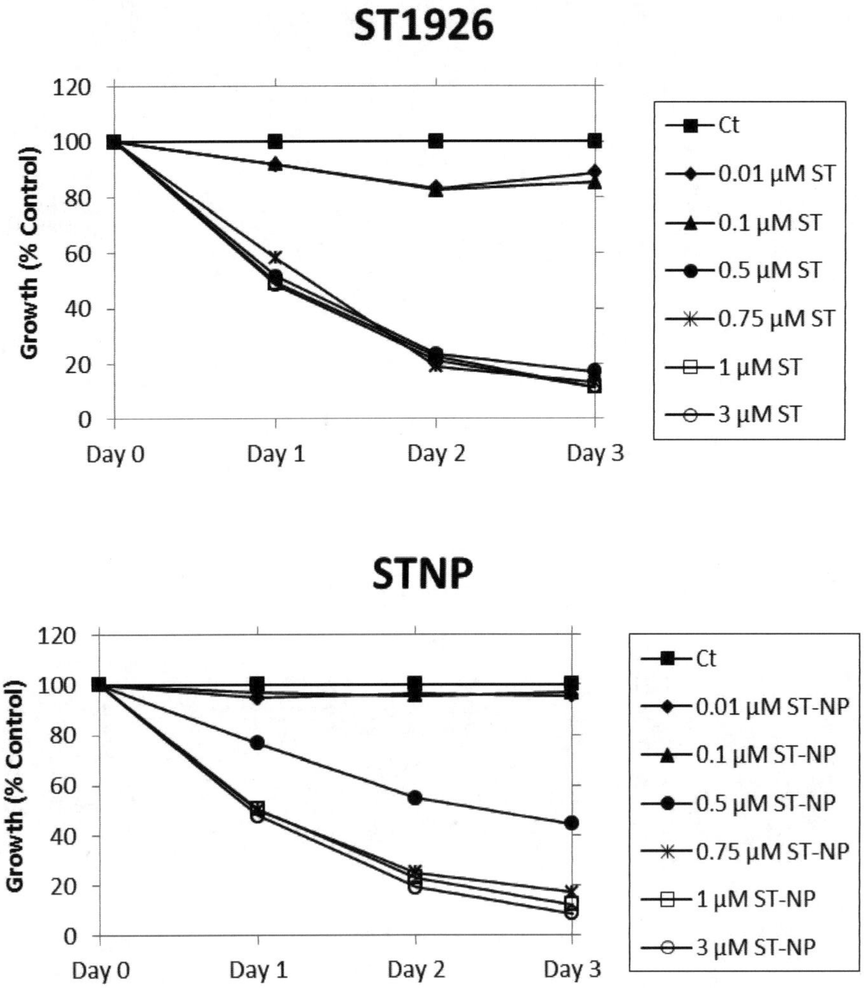

Figure 4: HCT-116 cell growth assay results. ST1926 (top) and its corresponding NP formulation (STNP, bottom) inhibit proliferation of colorectal cancer cell lines at submicromolar concentrations.

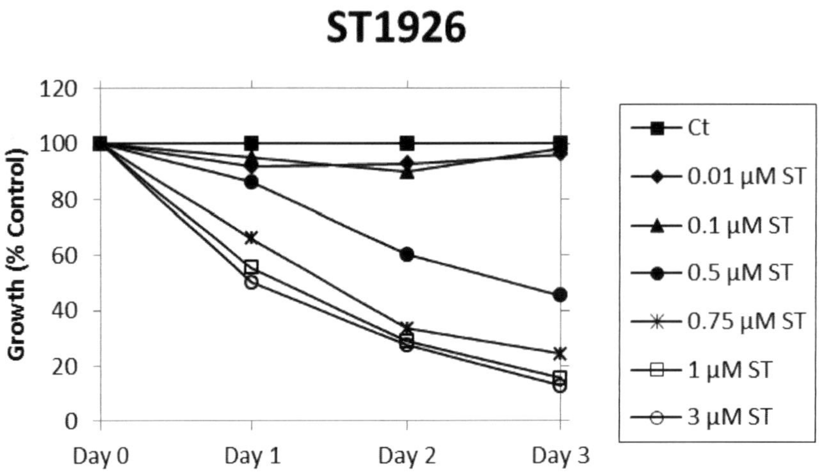

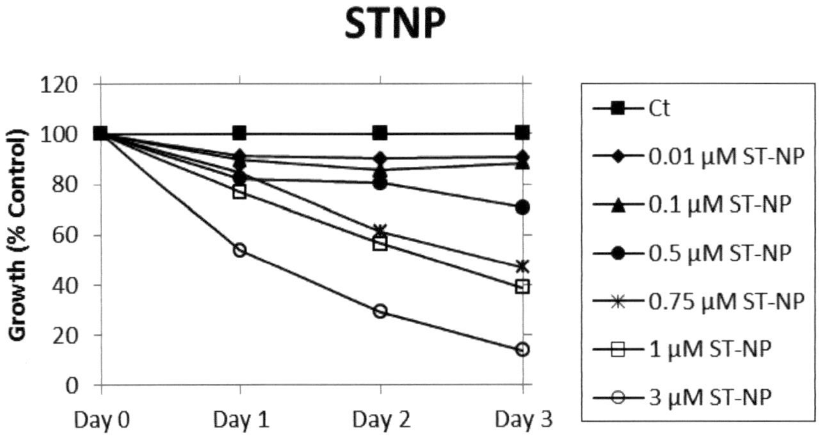

Figure 5: HCT-116 $p53^{-/-}$ cell growth assay results. ST1926 (top) and its corresponding NP formulation (STNP, bottom) inhibit proliferation of colorectal cancer cell lines at submicromolar concentrations.

Additionally, apoptotic assay results suggest that STNP induced apoptosis in colorectal and breast cancer cells similarly to native ST1926, as demonstrated by TUNEL assay (data not shown).

Conclusion

ST1926 was formulated into stable nanoparticle suspensions via FNP. The anticancer activity profile exhibited by STNP is similar to the native ST1926 in colorectal cancer cells, at pharmacologically relevant concentrations.

Further studies are planned to characterize, optimize, and further investigate the efficacy of STNP both *in vitro* and *in vivo*. Ultimately, *in vivo* testing will provide valuable insight into the efficacy of the STNP formulation and any improved performance compared with ST1926.

References

1. Schenk, T., Stengel, S., Zelent, A. (2014) "Unlocking the potential of retinoic acid in anticancer therapy," *British Journal of Cancer*. 111(11):p. 2039-2045.
2. Dallavalle, S. and F. Zunino (2005), "Synthetic retinoids as potential antitumour agents," *Expert Opinion on Therapeutic Patents*. 15(11): p. 1625-1635.
3. Cincinelli, R., Dallavalle, S., Merlini, L., Penco, S., Pisano, C., Carminati, P., Giannini, G., Vesci, L., Gaetano, C., Illy, B., et al. (2003), "A novel atypical retinoid endowed with proapoptotic and antitumor activity," *Journal of Medicinal Chemistry*, 46:p. 909-912.
4. Pisano, C., Merlini, L., Penco, S., Carminati, P., Zunino, F. (204), "Cellular and pharmacological bases of the antitumor activity of a novel adamantyl retinoid, ST1926, " *Journal of Chemotherapy*, 16 Suppl 4:p. 74-76.
5. Garattini, E., Parrella, E., Diomede, L., Giannì, M., Kalac, Y., Merlini, L., Simoni, D., Zanier, R., Ferrara, F., Chiarucci, I., et al. (2004), "ST1926, a novel and orally active retinoid-related molecule inducing apoptosis in myeloid leukemia cells: modulation of intracellular calcium homeostasis," *Blood* 103:p. 194-207.
6. El Hajj, H., et al. (2014), "Preclinical efficacy of the synthetic retinoid ST1926 for treating adult T-cell leukemia/lymphoma," *Blood*, 124(13): p. 2072-2080.
7. Nasr, R.R., et al. (2015), "ST1926, an orally active synthetic retinoid, induces apoptosis in chronic myeloid leukemia cells and prolongs survival in a murine model," *International Journal of Cancer*, 137(3): p. 698-709.
8. Sala, F., et al. (2009), "Development and validation of a liquid chromatography-tandem mass spectrometry method for the determination of ST1926, a novel oral antitumor agent, adamantyl retinoid derivative, in plasma of patients in a Phase I study," *Journal of Chromatography B-Analytical Technologies in the Biomedical and Life Sciences*, 877(27): p. 3118-3126.
9. Maeda, H. (2015), "Toward a full understanding of the EPR effect in primary and metastatic tumors as well as issues related to its heterogeneity," *Advanced drug delivery reviews*, 91: p. 3-6.
10. Shen, H., et al. (2011), "Self-assembling process of flash nanoprecipitation in a multi-inlet vortex mixer to produce drug-loaded polymeric nanoparticles," *Journal of Nanoparticle Research*, 13(9): p. 4109-4120.
11. D'Addio, S.M. and R.K. Prud'homme (2011), "Controlling drug nanoparticle formation by rapid precipitation," *Advanced Drug Delivery Reviews*, 63(6): p. 417-426.

434115 Simplifying Tablet Film-Coating Process-Scaleup and Transfer through Droplet-Size Measurement

Abstract:

Objective

Describe a simple method of using laser diffraction for droplet-size measurement to characterize coating gun performance and simplify scaling and transfer of pharmaceutical film-coating processes.

Introduction/Background

Scale-up of pharmaceutical tablet film-coating often involves using a variety of two-fluid coating guns which have very different coating-suspension atomization characteristics and which may rely on different control parameters (atomization pressure vs. atomization air flow). Still, the film-coating suspension droplet size may deemed low risk and receive little attention during process development until coating problems are detected at the pilot and manufacturing scales. Vendor data on coating gun performance may be incomplete or too generalized for direct use in specific applications. Laser diffraction can readily be used to determine the spray characteristics produced by different guns over a range of process conditions. This approach provides data that allow matching of droplet size across scales and elimination of droplets size as a variable in scaleup considerations.

Methods

Several coating guns (spray nozzles) from commercially available coating pans were compared using film-coating suspensions with varied composition and concentration. The viscosity and refractive index of each suspension was measured and recorded. The guns were mounted above a Malvern Spraytec open-frame laser-diffraction system. The guns were positioned at a range of heights above the laser beam to simulate typical gun-to-bed distance in coating applications. A spray was produced over a range of liquid spray rates, atomization air pressures, atomization air flow rates, pattern air pressures, and pattern air flows which mimicked actual process conditions. The spray was measured over a 5 second period using a capture rate of 50 Hz. The resulting 250 measurements per run were averaged to provide a droplet-size distribution histogram and D10,D50, and D90 data for each condition. The droplet-size results were compared between guns and with respect to available vendor nozzle data. The data allowed determination of parameter setpoints for each coating gun which could produce a given target droplet size.

Results

Matching droplet size between guns was possible. Data was reproducible with a tolerance of approximately 2 to 3 microns. D_{50} appeared to be the most reliable and repeatable characteristic of the distribution. Suspension viscosity played a major role in droplet size. Pattern air had some influence on the droplet size for given conditions. Nozzle height had no effect to a minor effect on droplet size, though a slight increase in droplet size was observed at increasing heights.

Droplet size was measured by mimicking atomization for a tablet-coating process which previously was restricted to low spray rates and had produced a visually rough film coating (Process A). The measurements indicated the droplet sizes being produced were larger than those produced on different coating equipment where the process had been run successfully and robustly (Process B). The atomization parameters for Process A were adjusted until the coating gun produced droplets similar in size to those produced in Process B. Process A was then successfully run at a higher spray rate using the new atomization parameters.

Conclusions

Laser diffraction can be used to provide simple performance data that allows comparison of droplets produced by coating guns across scales and manufacturers.

Implications

This approach can be used to provide droplet-size data that lowers the risk associated with scaleup and transfer of film-coating sprays.

434812 Computer Aided Search for Drug Candidates: Cathepsin-L

Abstract:

Drug candidates make up a small portion of known compounds. To find the candidates among the dead ends, extensive compound libraries are tested against a given target in high-throughput screens, with the hopes that at least one will be active. To increase the efficiency of the high-throughput screens, researchers are looking to computational resources for help. If researchers can build accurate models that can examine compounds and predict activities, they can cut the amount of compounds they would have to screen to only those with the most potential, saving a great deal of time and resources.

To this end, we developed a new pipeline. Given some previously obtained data, the pipeline will develop quantitiave structure-activity relationship (QSAR) models capable of predicting a compound's activity. Implemented in virtual high-throughput screens, the models can single out compounds which have high predicted activity, focusing the experimental library of compounds to those that have high predicted activity, reducing the amount of compounds tested and the reagents needed to test them.

To test our pipeline, we took NCBI PubChem Bioassay AID 825 as a test case. AID 825 looks at Cathepsin-L, a receptor implicated in several disease pathways, including Ebola. Using classification and activity data from the assay, our pipeline trained our models and used them to search NCBI's PubChem Compound database, looking for potential active compounds. From the pipeline results, we isolated a select group of compounds for which we have high confidence in our prediction. Of those compounds, 15 were commercially available and obtained for evaluation, verifying the value and accuracy of our pipeline and as new data for further model improvement.

436254 Development of a Challenging Salt Formation Step for a Phosphate API

Abstract:

The development of a final API step involving salt formation is described. Challenging properties of this system include: (1) the free base API has low solubility in most organic solvents; (2) residual solvent is difficult to reduce below desired limits; (3) salt disproportionaton occurs at mild temperatures. A process that addresses these challenges was developed, generating API with desired physical properties.

Improving Purity and Size Distribution in Biopharmaceutical Crystallization Using PAT-Based Feedback Control Strategies

Elena Simone[1]*, Wei Zhang[2,3], Zoltan K. Nagy[1,3]

[1]Department of Chemical Engineering, Loughborough University, UK; [2]North China Pharmaceutical Group Corporation, Shijiazhuang, China; [3]School of Chemical Engineering, Purdue University, USA

*e.simone@lboro.ac.uk

Introduction and motivation

- ❑ **Vitamin B12** is a water-soluble vitamin with a key role in the formation of blood often used in the treatment of pernicious anaemia

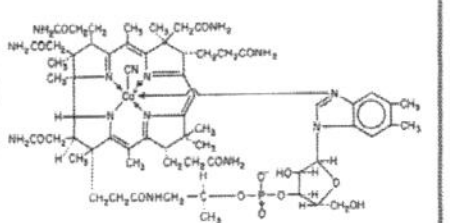

- ❑ A common semi-synthetic form of the vitamin is produced industrially by **bacterial fermentation** and used in many pharmaceuticals and supplements, as well as a food additive
- ❑ The crystallization step of the synthetic vitamin B12 is usually problematic because of the presence of growth inhibiting **impurities**
- ❑ Different **feedback control strategies** will be tested on an **industrial sample** to improve the quality of the final product

Crystals of synthetic vitamin B12

Methodology

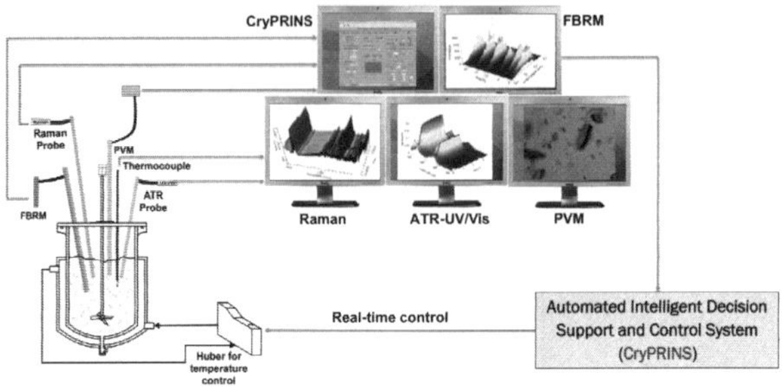

- ❑ Direct nucleation control (DNC) and calibration-free supersaturation control (CF-SSC) were both tested with the compound
- ❑ FBRM and ATR-UV/Vis were actively used for the feedback control strategies
- ❑ PVM and Raman were used as monitoring tools

Results and discussion

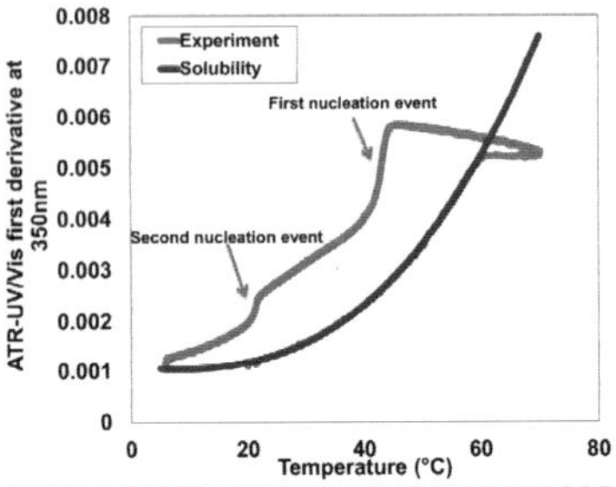

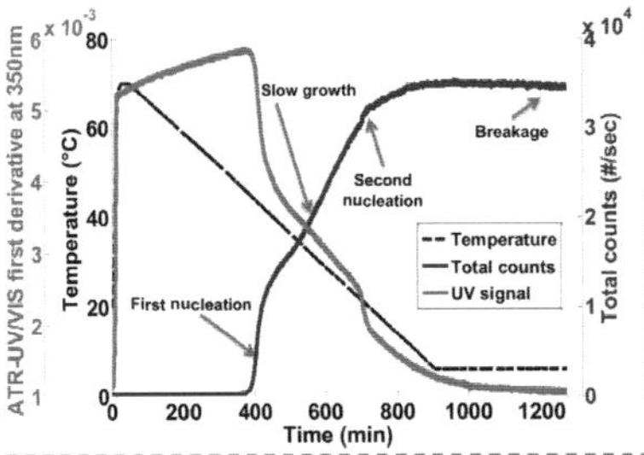

Simple slow cooling (-0.075 °C/min)

- ❑ The presence of impurities inhibits growth after the first nucleation event
- ❑ Supersaturation increase during cooling

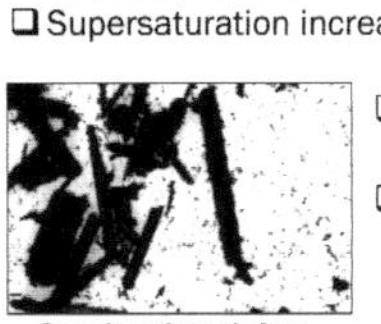

Crystals at the end of run

- ❑ A **second nucleation event** due to high supersaturation happens
- ❑ The slight decrease in counts at the end of the run is caused by breakage of the bigger crystals

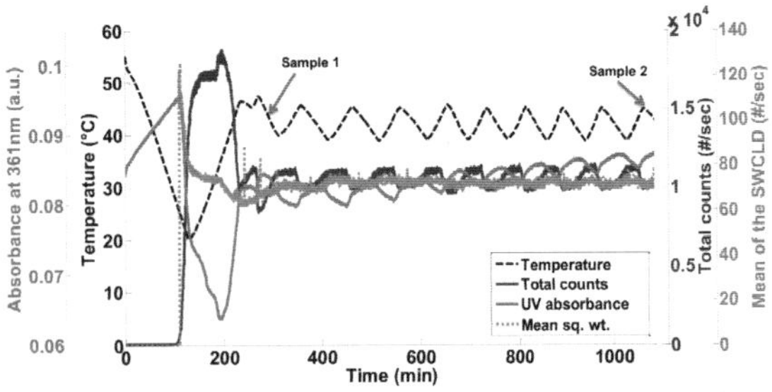

Direct nucleation control

- ❑ Crystal are nucleated and the number of counts is kept constant at 10000 #/sec using the direct nucleation control strategy implemented in CryPRINS
- ❑ The DNC is **not suitable** for these crystals since they grow as long **needles**: it prevents nucleation but also growth and the temperature keeps oscillating without converging

Crystals at the end of the first temperature cycle

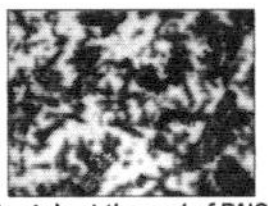

Crystals at the end of DNC

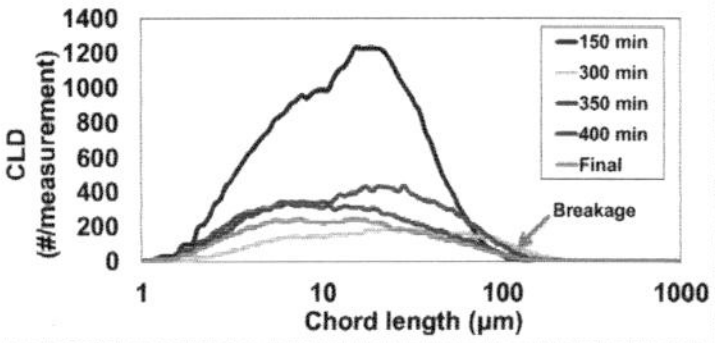

Supersaturation control (calibration free)

- ❑ High purity crystal are seeded (2.9% of total solute) and supersaturation is kept constant through the SSC strategy implemented in CryPRINS

Crystal at the end of run

- ❑ A high supersaturation is required to induce growth
- ❑ At the end of the run abundant **nucleation** still happens

PVM: after seeding

PVM: end of run

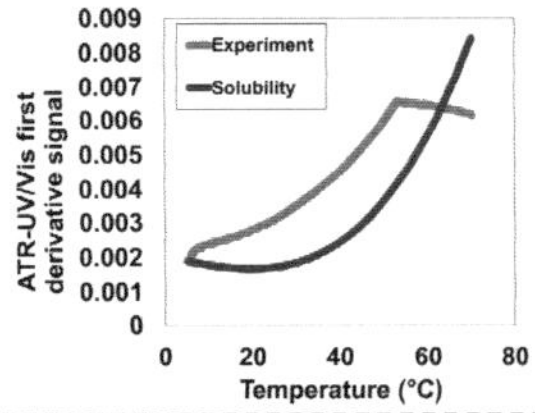

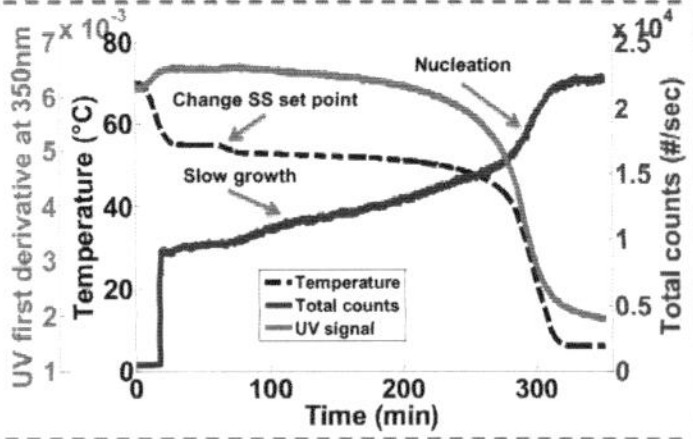

Supersaturation control in purified solution (calibration free)

- ❑ A clear solution is prepared using purified vitamin B12
- ❑ SSC is applied after seeding high purity crystals (2.8% of total solute)

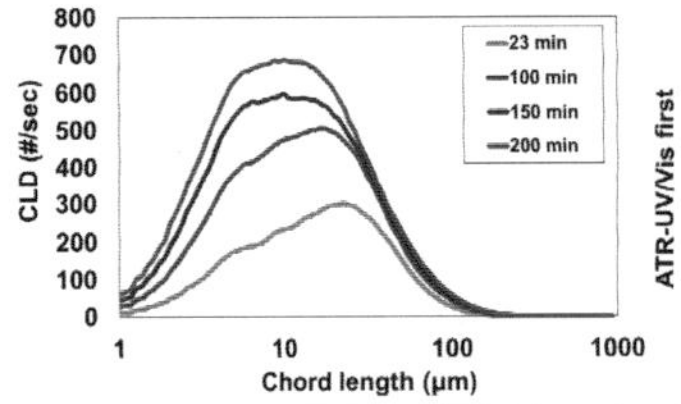

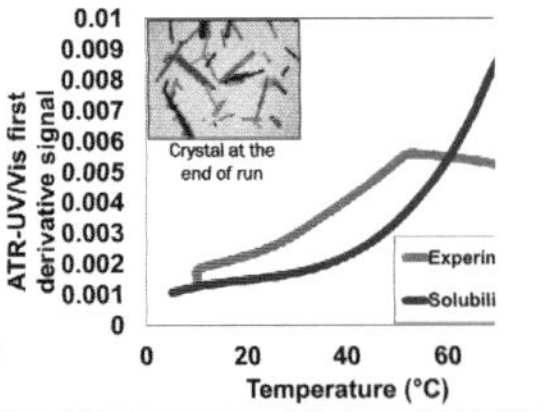

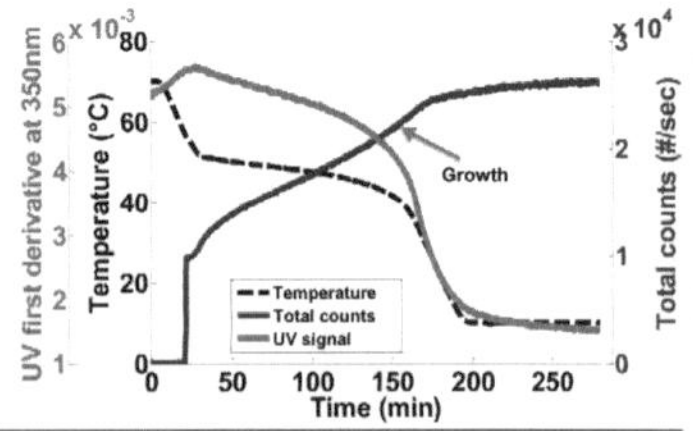

Conclusions

Different feedback control strategies were implemented for the crystallization of vitamin B12:

- ❑ DNC was found to be **not suitable** because of the **needle shape** of the crystals
- ❑ **Calibration free SSC was successfully implemented** and found to be more effective on material which has already been crystallized once (less impurities)

Acknowledgements

Financial support provided by the European Research Council grant no. [280106-CrySys]. Travel grant awarded from the Royal Society of Chemistry.

ROYAL SOCIETY OF CHEMISTRY erc European Research Council Loughborough University PURDUE UNIVERSITY

417236 The Effect of Hot Melt Extrusion Process, Polymer Type, and Molecular Weight on Polymer-Water Interactions of Pharmaceutical Excipients for Extrusion

Abstract:

The Effect of Hot Melt Extrusion Process, Polymer Type, and Molecular Weight on Polymer-Water Interactions of Pharmaceutical Excipients for Extrusion

Kevin O'Donnell[1], Jin Zhao[1], Florin Dan[2], Uma Shrestha[1], William Porter[1]

(1) Pharmaceutical Excipients R&D, Dow Pharma & Food Solutions, Midland, MI 48674

(2) Analytical Sciences, Midland, MI 48674

The presence of water in an amorphous pharmaceutical solid dispersion can greatly impact the final performance of the solid dosage form(1). Thus, the objective of this study was first to understand the interactions of pharmaceutical polymers often used in hot melt extrusion (HME) with water based on the polymer type and molecular weight. The second objective was to determine the changes in polymer-water interactions after being processed by hot melt extrusion.

Hot melt extrusion of the pure polymers was performed on a Leistritz Nano16 HME. Extruded polymers were pelletized and milled in an Alpine impact mill. Extruded and non-extruded polymers were analyzed by Dynamic Vapor Sorption at 25 °C over a relative humidity range of 0-90%. The glass transition temperature as a function of relative humidity was determined for extruded and non-extruded samples by modulated differential scanning calorimetry.

Hypromellose, also commonly named as hydroxypropyl methylcellulose (HPMC), is well recognized as a crystallization inhibitor in amorphous solid dispersions in both the solid state and during dissolution. AFFINISOL™ HPMC HME polymers demonstrated reduced moisture uptake compared to polyvinyl pyrrolidone-vinyl acetate copolymer prior to extrusion. Polyethylene glycol – polyvinyl caprolactam – polyvinyl acetate copolymer had similar moisture uptake to HPMC type 2910. AFFINISOL™ HPMC HME moisture sorption was significantly lower than that of HPMC type 2208 or 2910, as shown in Figure 1. Variation on molecular weight of AFFINISOL™ HPMC HME did not impact moisture absorption except for a slight change at RH=90% (25 °C). Various models were used to fit sorption and desorption data. It was found that Henderson(*2*) and Flory-Huggins/Vrentas (FHV)(*3*) Models fit well at moisture absorption levels less than 80% and 50% respectively, but not as well at high moisture levels (R^2: 0.97 to 0.99). The Guggenheim-Anderson-de Boer (GAB)(*4*) model gave the best overall fit (R^2>0.99). The fitting parameters of these three models are sensitive to the polymer type and processing conditions, and therefore can provide insights of polymer-water interaction. The FHV model fit the best for PVPVA 64 which indicated that the mutual miscibility and plasticizing power of the water dominated the isotherm process. The pseudo c' parameter of PVPVA derived from isotherm is about 1/3 of the c' parameter derived from AFFINISOL™ HPMC HME polymers. This indicates that PVPVA has a much stronger interaction with water than that of AFFINISOL™ HPMC HME polymers which is consistent with the physical parameters derived from the other models. The water content to saturate the monolayer for PVPVA is about three times as large as that of AFFINISOL™ HPMC HME polymers. Hot melt extrusion process did not impact AFFINISOL™ HPMC HME isotherm when RH<70% (25 °C) and further reduced moisture absorption when RH>70% as shown in Figure 1.

The findings from this research can help pharmaceutical scientists design stable amorphous drug dispersions using amorphous polymers for hot melt extrusion. The evaluation of various predictive models can provide an insight into the mechanisms of water-polymer interaction, inspire further development of universal model that can predict the isotherms of polymers with various hydrophobicity and help to identify the key factors involving residual water in amorphous polymers that are critical in drug formulation and processing.

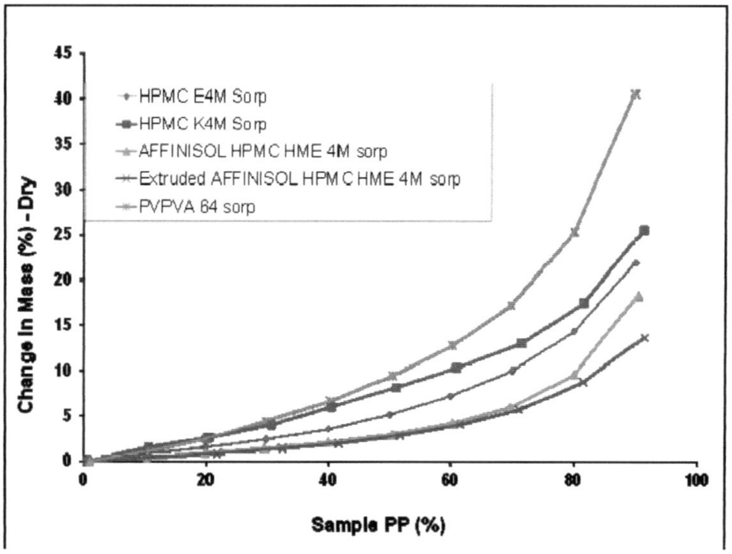

Figure 1. Comparison of water vapor sorption isotherms for AFFINISOL HPMC HME polymer, HPMC, and PVPVA 64.

References

1. D. S. Jones *et al.*, Thermodynamically stable amorphous drug dispersions in amorphous hydrophilic polymers engineered by hot melt extrusion. *Chemical Engineering Research and Design* **92**, 3046-3054 (2014).

2. S. M. Henderson, A hasic concept of equilibrium moisture. . *Agric. Eng.* **33**, 29-31 (1952).

3. B. C. Hancock, G. Zografi, The use of solution theories for predicting water vapor absorption by amorphous pharmaceutical solids: A test of the Flory-Huggins and Vrentas models. *Pharm. Res.* **10**, 1262-1267 (1993).

4. R. B. Anderson, Modifications of the Brunauer, Emmett and Teller Equation. *J. Am. Chem. Soc.* **68**, 686 (1946).

438620 Fast Screening and Designing Drug-Loaded Polymeric Nanoparticles via Thermodynamic Prediction

Abstract:

Active pharmaceutical ingredients (API) -polymer miscibility is considered to be an essential prerequisite for the successful formulation of a physically stable solid dispersion. The objective of the development of an API-polymer miscible binary system is to provide an environment in which the crystallinity of API is so altered as to manipulate its solubility and nucleation rate. In an API-polymer miscible system, it is expected that the local environment of API and the physical stability of API are altered due to molecular level mixing of API with polymers. Owing to the potential for the successful formulation of a poorly water-soluble API, the study of miscibility of API with polymers has become the important topic in both academic and industrial research. To establish a method for predicting API and polymer compatibility as a means to guide formulation development, we predict thermodynamics of binary system through free energy calculations based on Flory-Huggins interaction parameter and group contribution method (GCM). The degree of compatibility between API and polymers has been predictable by GCM which is of importance in the design of a wide range of delivery systems including block copolymer micelles.

In experiments, ibuprofen is the model drug to validate our prediction models. Three kinds of block copolymers (BCPs) were used to encapsulate ibuprofen. The results reveal that the prediction of interaction between API and polymers by GCM is efficiently for fast screening candidates. After validation, this method can be developed for the other common pharmaceutical polymers for their ability to yield a stable formulation system in pharmaceutical industry and as an initial tool for fast screening of immiscible combination of a polymer and API.

424701 The Role of Modeling, Simulation and Optimization Approaches in the Design of Pharmaceutical Production Processes

Abstract:

Achieving cost-effectiveness while maintaining good manufacturing practice (GMP) is an emerging topic in pharmaceutical manufacturing. In this presentation, the role of modeling, simulation and optimization (SMO) approaches is explored by showcasing two recent research examples on active pharmaceutical ingredients (APIs) and excipients, respectively.

The first example is on optimizing the purification process of thermally unstable APIs. The target here was an industrial process for sterilizing API as the raw material of a Parenterals (or injectable) product. The process consists of dissolution of crude API with water by heat exchanger, sterile filtration, crystallization and drying, and as a product sterile API crystal is obtained. In the heat exchanger, the crude API needs to be dissolved completely while degradation reaction of the API occurs in the liquid phase. This dissolution process of thermally unstable API was modeled with the overall yield as the objective function, and the flow rate and the temperature in the heat exchanger as design variables. The process model also considered a couple of characteristics such as particle size distribution, non-linear temperature profile in the heat exchanger, and sterility requirements by GMP. The simulation result indicated an optimal set of the design variables as well as certain ranges where the yield decreases dramatically due to the degradation or non-dissolution.

The second example is on the multiobjective optimization of the use process of pharmaceutical excipients, i.e., reception, storage, testing, and application of the material. As an example, benzyl alcohol, an oxidation-sensitive material, was investigated, and a model was created, which defines the amount of oxidized impurities, and the costs associated with testing as the quality and economic objectives, respectively. The process was modelled and simulated, considering container volume and purchase frequency as the design variables, and also the use of Raman spectroscopy as an alternative to conventional manual testing. The multiobjective evaluation indicated options on the Pareto frontier of the design variables as well as the effectiveness of the use of Raman spectroscopy.

In these two examples, the SMO approaches showed its supportive role for shifting the design concept of pharmaceutical production processes from "by-testing" to "by-design". The goal of the design can cover not only a single aspect but also to multiple ones, such as quality, economy, supply stability or risks regarding environment, health and safety.

431772 Modeling of Drug Substance Manufacturing in a Multi-Purpose Plant

Abstract:

This work presents different modeling strategies to address the manufacturing of drug substances in a multi-purpose plant. In multi-purpose plants the process equipment needs to have flexibility to tackle a wide range of chemical and physical operations, often with multi-phasic systems, and a broad range of solvents, temperature and pH. It is not uncommon that such process equipment in multi-purpose plants are not fully adequate to specific highly demanding chemical processes (eg a given impeller may be adequate for intense local mixing, but not fit for axial mixing of solids). Hence, it is crucial to use adequate modeling tools and methodologies to optimize the process performance as much as possible considering the equipment constraints; such strategies are aligned with the Quality by Design initiave being promoted by the regulators.

The case-study to be shown comprises the scale-up of the manufacturing of two distinct drug substances from the laboratory to the same full scale reactor. The kinetics of both processes is assessed in detail in the laboratory scale through the fitting of the transient differential equations to the analytical data (using Dynochem); the intrinsic kinetics of both processes is determined through decoupling of the fluid dynamics portion of the mass transfer resistance in the kinetic constants using both Dynochem and CFD (Computational Fluid Dynamics) - this is made possible since the laboratory reactor geometry, impeller and baffle system are relatively simple. In constrast, the geometry, impeller and baffle system of the full scale reactor are not simple, so this is addressed differently. The fluid dynamics of both processes in the full scale reactor is assessed solely with CFD, which enables the determination of the local mass transfer coefficient distribution. It is then shown how the use of the local mass transfer distribution along with the intrinsic kinetics of each process enables an improved understanding of the scale-up to the full scale.

426939 Systematic Retrofitting Methodology for Pharmaceutical Drug Purification Processes

Abstract:

Pharmaceutical production processes are gaining research attentions in the field of CAPE/PSE. In this work, we present a systematic method for retrofitting pharmaceutical manufacturing plants, together with an industrial case study.

The method consists of five tasks: (I) collection of process understanding/data, (II) creation of process model, (III) adaptation of the model for optimization, (IV) optimization and (V) task outcome interpretation. In Task I, an overview of the process is obtained by performing Path Flow Decomposition (PFD), which determines the sources and sinks of the all substance in a batch, and thus facilitates the creation of a model. In Task II descriptive mathematical models are created for every unit, which will then be evaluated for feasibility in Task V. The aim of Task III is to transform the descriptive model into an analytical model, which relates the optimization variables to the objective function. In parallel to identifying variables to be optimized, analysis is performed on constraints, especially on the pharma-specific, GMP and product quality aspects. All these information are considered in Task IV, i.e., optimization, where MINLP plays a central role. Task V runs simultaneously to all the other tasks, evaluating the feasibility of the outcomes and the necessity to iterate corresponding Task(s), which resembles iterative nature of the design.

The methodology was applied to a manufacturing plant of a solid drug, where crude crystal is purified through the process of Dissolution-Filtration-Crystallization (DFC). In Task I, a trade-off was identified: the higher the process temperature, T_H, the more degradation of the crude crystal; the lower the T_H, the more amount of undissolved crystal. The process flow rate, F, affects also the overall yield. In Task II, process model was created with considering measured data, e.g. particle size distribution (PSD) during the dissolution for the heat exchanger. The descriptive model was adapted for optimization by incorporating accurate information, e.g., liquid-wall-liquid heat transfer. We were able to model reactive-dissolution of crystals with PSD inside the heat exchanger, which is the secondary novelty of the work. As Task IV, three alternative process layouts were considered. The insertion of coolers before and after the filtration unit enabled a substantial improvement, lowering the degradation reaction and thus increasing the yield and the net present value. The feasibility of the model was supported by the result of the performance of lab scale experiments on a plant with the same new layout.

The methodology is a practical decision-supporting tool for the retrofitting of pharmaceutical purification processes, and its application is not only limited to the DFC processes. The novelty lays in its integrative nature, indeed the combination of different technics, e.g., information gathering in GMP environment, mathematical modelling and multiobjective optimization, and decision making, in the field of pharmaceutics. The methodology can show its weakness in case the data measurement is difficult, however, elucidation of trends and behaviours inside the plant can still serve as a basis for process retrofitting.

Model based approach to develop a ring closure reaction with high product selectivity

Marimuthu Andiappan, Adam McFarland, Jared Fennell

Small Molecule Design and Development (SMDD)

Eli Lilly and Company, Indianapolis, USA

AICHE Annual Meeting 2015

1

Outline

- Goal – Ring closure reaction with high product selectivity
- Characterization of reaction: Flow NMR, Hypothesized mechanism
- High Selectivity (high yield, low impurity) criteria
- Kinetic Model prediction
- Proof of concept experiment

2

Ring closure reaction - Example

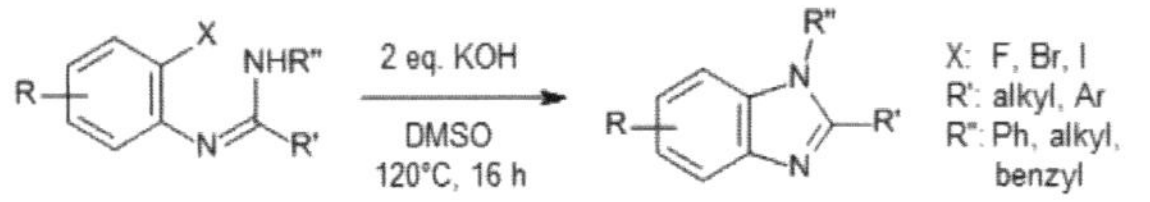

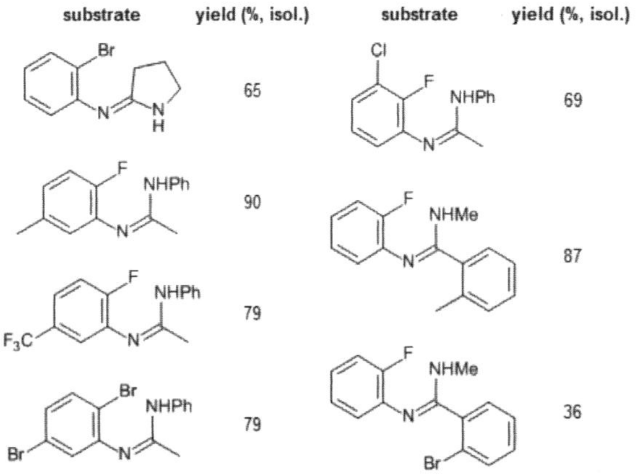

- In-situ product selectivity, impurity levels and isolated product yield depends on the nature of substrate
- How to design reaction with high product selectivity?

H. Baars, A. Beyer, S. V. Kohlhepp, C. Bolm, *Org. Lett.*, **2014**, *16*, 536-539

3

Our study: Ring closure reaction in toluene

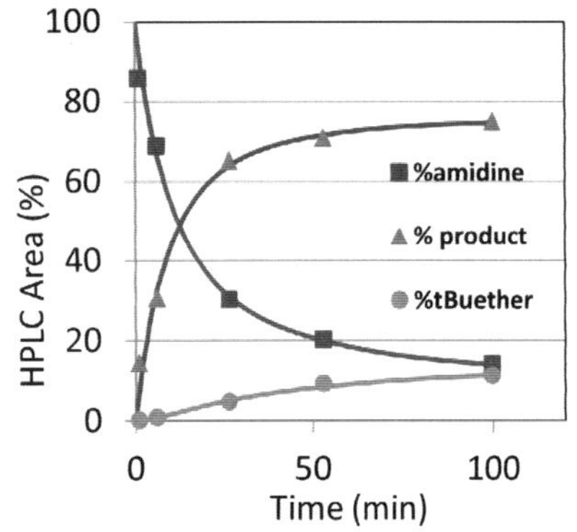

70 °C

- Desired reaction

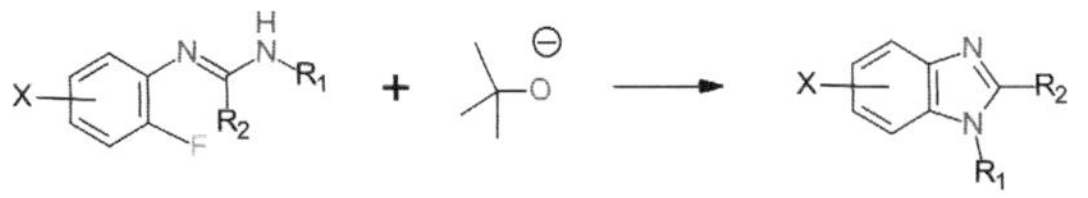

Amidine t-butoxide Product

4

Our study: Ring closure reaction in toluene

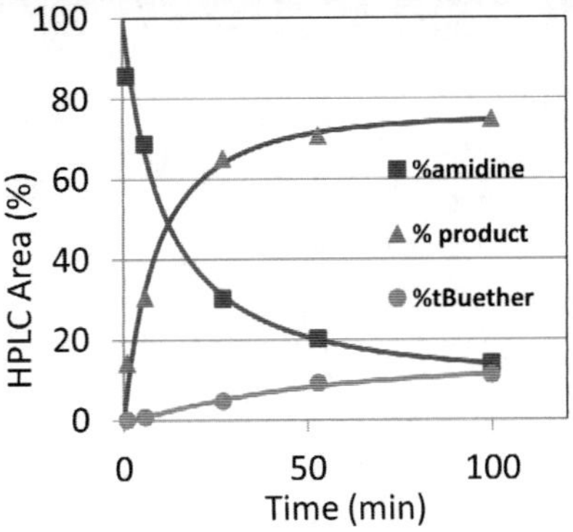

70 °C

- Problem: t-butyl ether impurity formation due to competing reaction

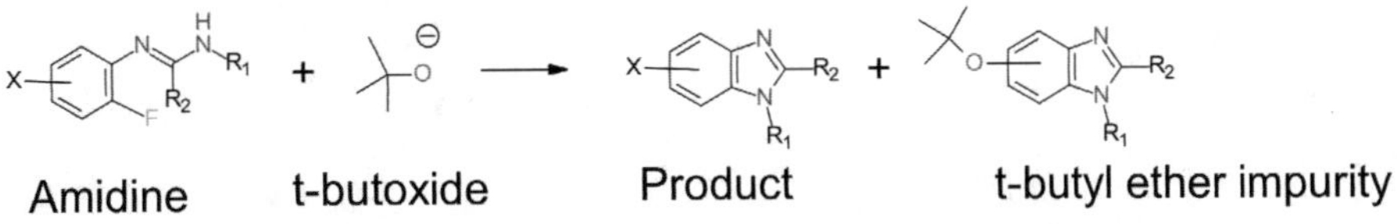

5

Our study: Ring closure reaction in toluene

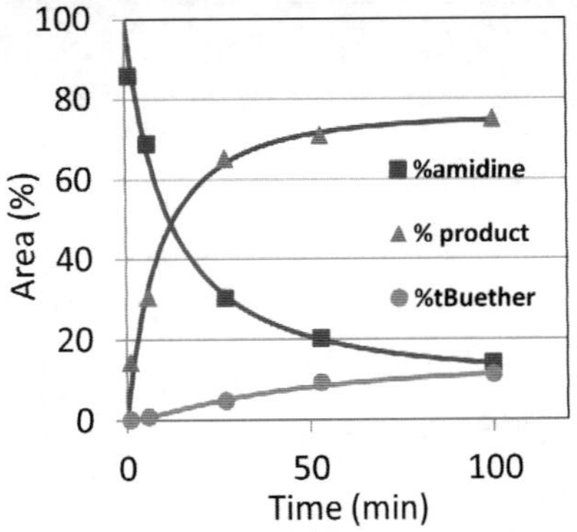

70 °C

- Acceptable level of impurity in isolated product: Not more than 0.5%

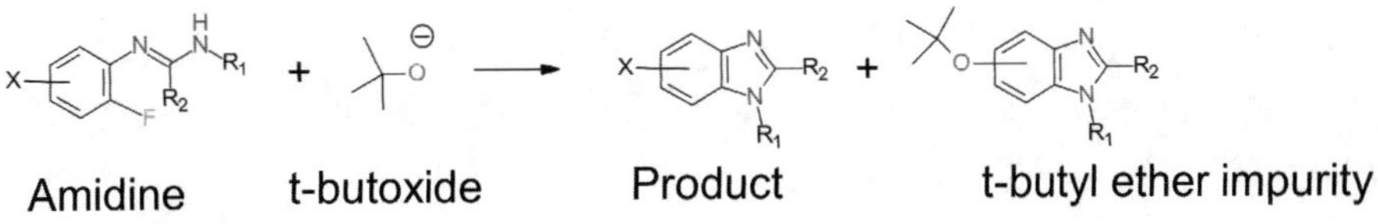

6

Our study: Ring closure reaction in toluene

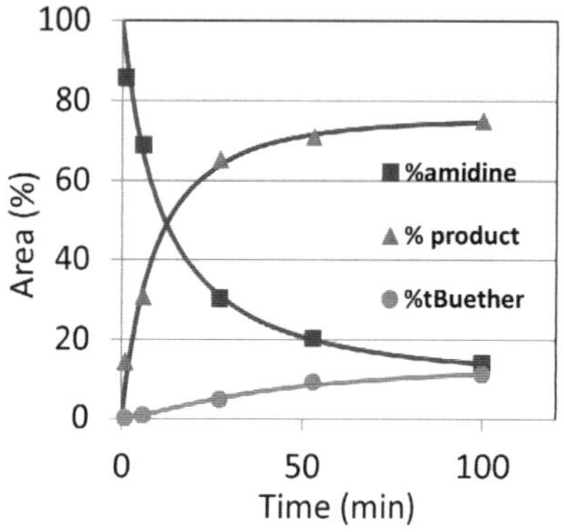

70 °C

- Can we design reaction with impurity level less than 0.5% at EOR using Mechanistic approach?

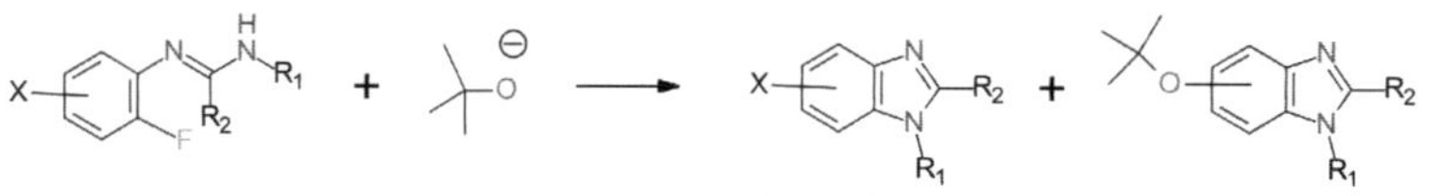

7

Flow NMR - Ring closure reaction at 25 °C

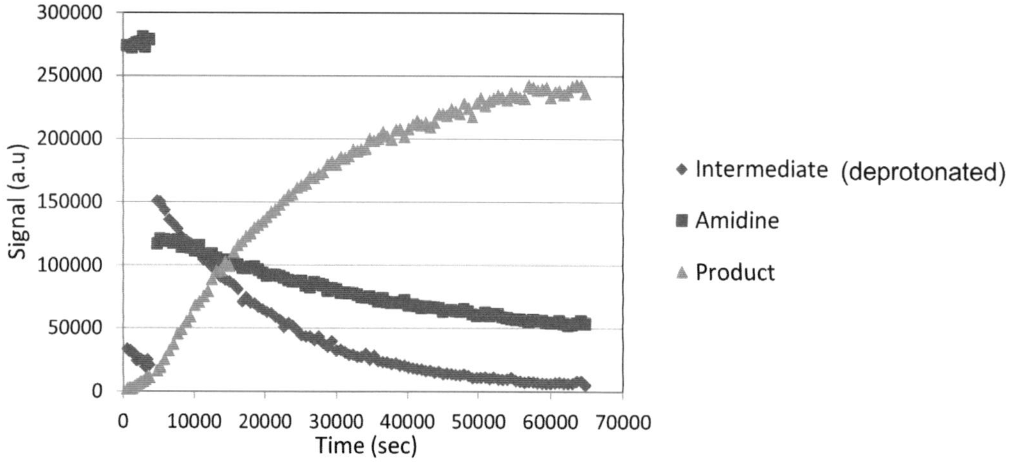

Signal from Proton NMR

- Equilibrium between amidine and deprotonated intermediate (equilibrium depends on t-butoxide concentration)
- Product forms from deprotonated intermediate through irreversible reaction

8

Mechanism – Production formation

Overall Reaction

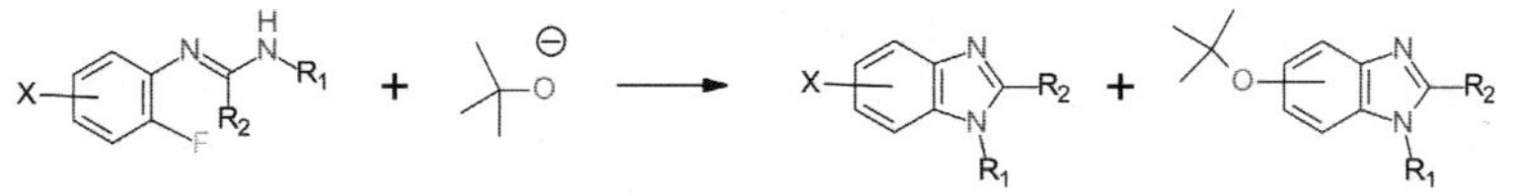

Amidine t-butoxide Product t-butyl ether impurity

Step 1

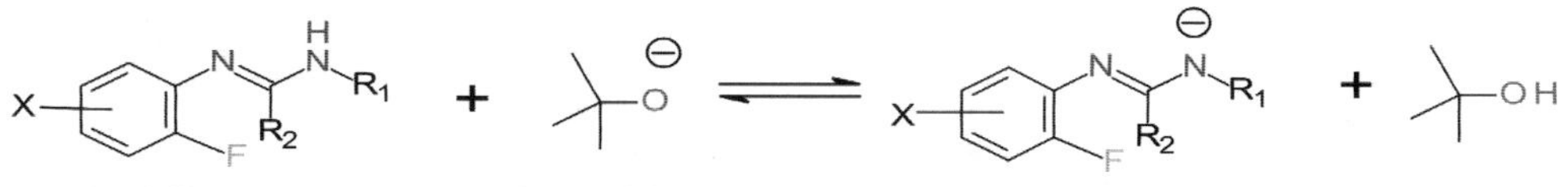

Amidine t-butoxide intermediate t-butanol

Step 2

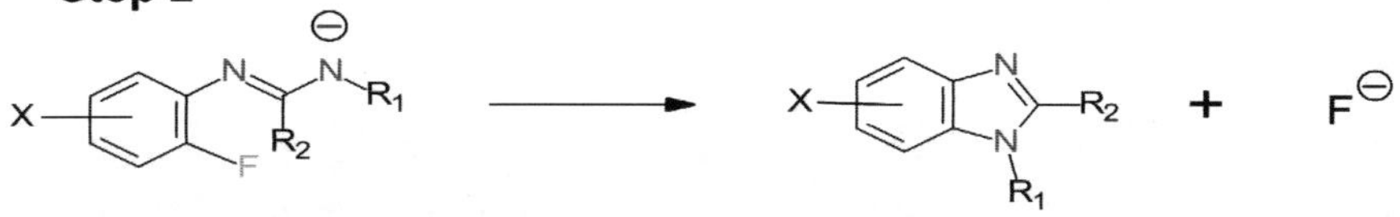

Intermediate Product

9

Possible mechanisms for t-butyl impurity formation

I. t-butyl impurity formation from product

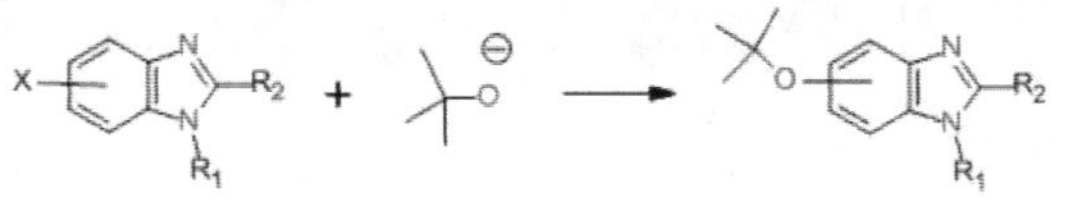

Product t-butoxide t-butyl ether impurity

II. t-butyl impurity formation from amidine substrate

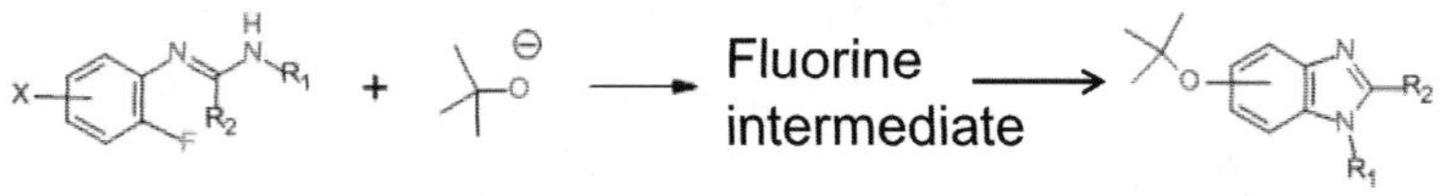

Amidine t-butoxide t-butyl ether impurity

III. t-butyl impurity formation from deprotonated amidine intermediate

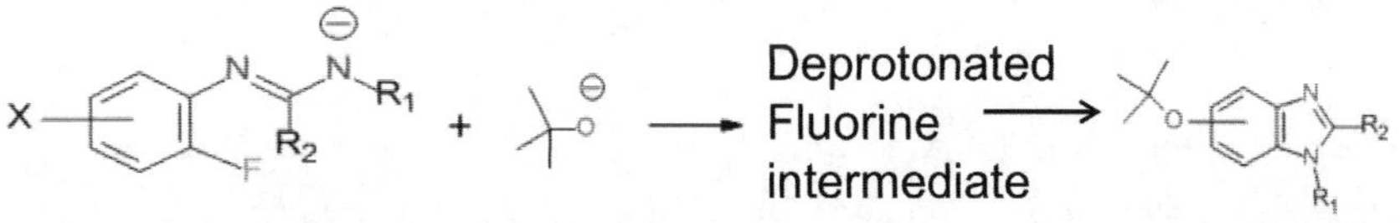

Deprotonated intermediate t-butoxide t-butyl ether impurity

10

Flow NMR: Reaction at 50 °C, Impurity formation from product or amidine?

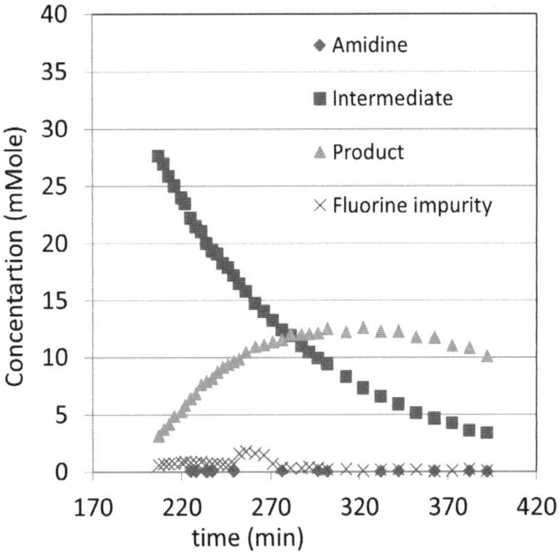

Signal from Fluorine NMR

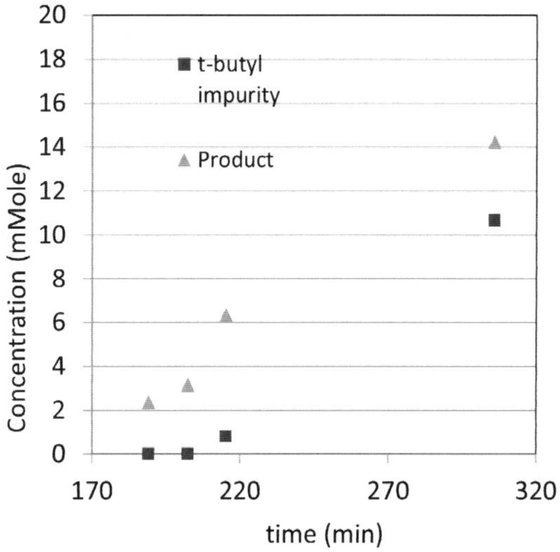

Signal from Proton NMR

- ~10 equiv. of t-butoxide: Transient Fluorine impurity (t-butyl ether impurity could form from amidine) – Disappears to t-butyl impurity

11

Flow NMR: Reaction at 50 °C, Impurity formation from product or amidine?

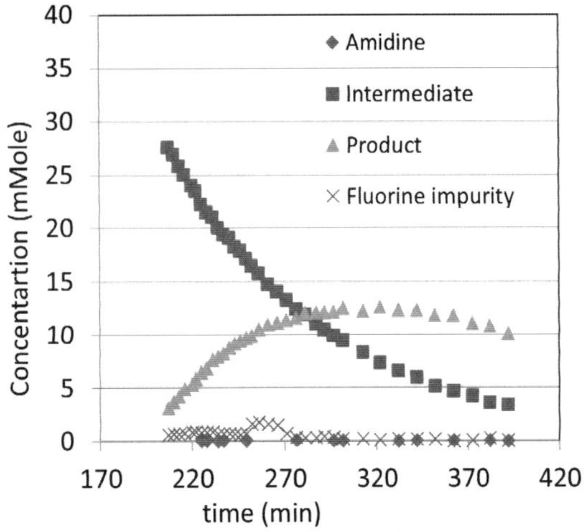

~10 equiv. t-butoxide

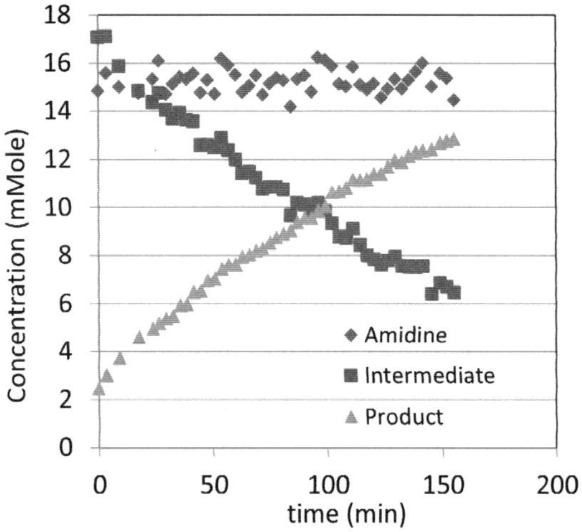

~2 equiv. t-butoxide

- ~10 equiv. of t-butoxide: Transient Fluorine impurity (t-butyl ether impurity could form from amidine) – Disappears to t-butyl impurity
- ~2 equiv. of t-butoxide: No significant transient Fluorine impurity (t-butyl ether impurity formation from amidine may not be significant)

12

Mechanism (Hypothesized)

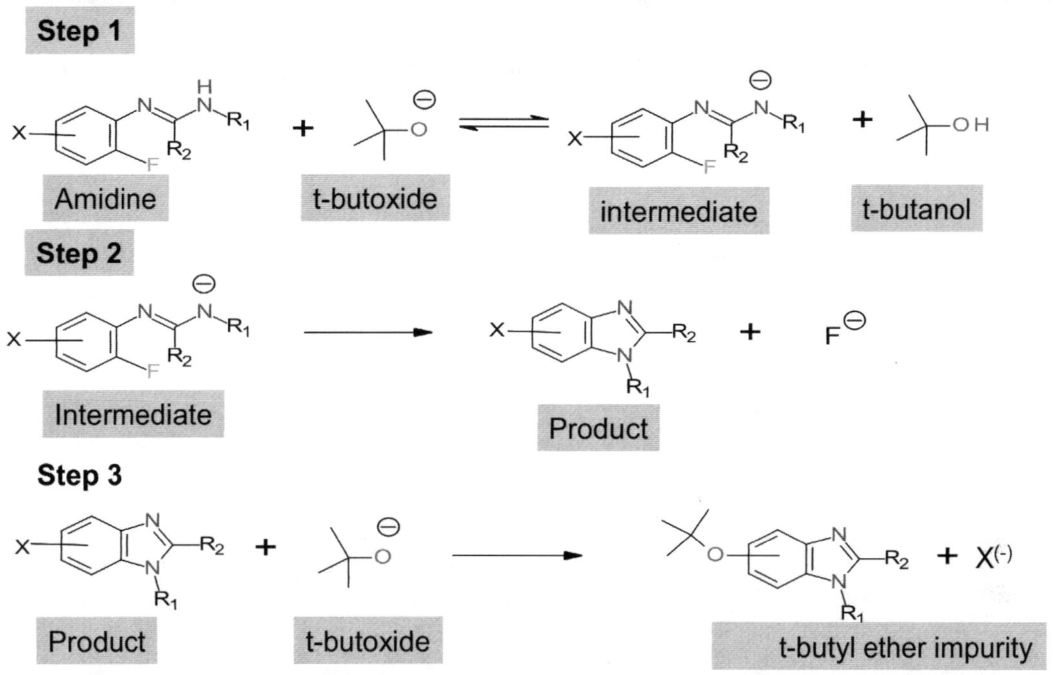

13

High Product Selectivity Criteria

- Reaction 1
  - Amidine + t-butoxide ⇌ Intermediate + t-BuOH **(Keq)**
- Reaction 2
  - Intermediate ⟶ product + byproduct2 **(k_2)**
- Reaction 3
  - product + t-butoxide ⟶ impurity + byproduct2 **(k_3)**

- High product selectivity criteria, Selectivity=$r_p/(r_p+r_{imp})$

$k_2 K_{eq}$ [Amidine] [t-butoxide]/[t-butyl alcohol] >> k_3 [Product] [t-butoxide]

- Controlling selectivity
  - I. Metering addition of t-butoxide may not be help
  - II. Dilution may help
  - III. Removing t-BuOH from solution may help

14

Model build in DynoChem

- K_{eq} is calculated using Marvin software

- Reaction 1
  - Amidine + t-butoxide ⇌ Intermediate + t-BuOH (**K_{eq} = 0.001**)
- Reaction 2
  - Intermediate ⟶ product + byproduct2 (**k_2**)
- Reaction 3
  - product + t-butoxide ⟶ impurity + byproduct2 (**k_3**)

15

Exp. vs. Model (fitted in DynoChem)

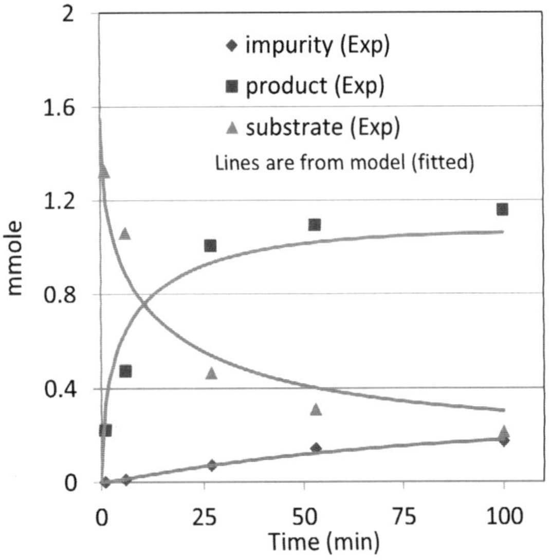

70 °C

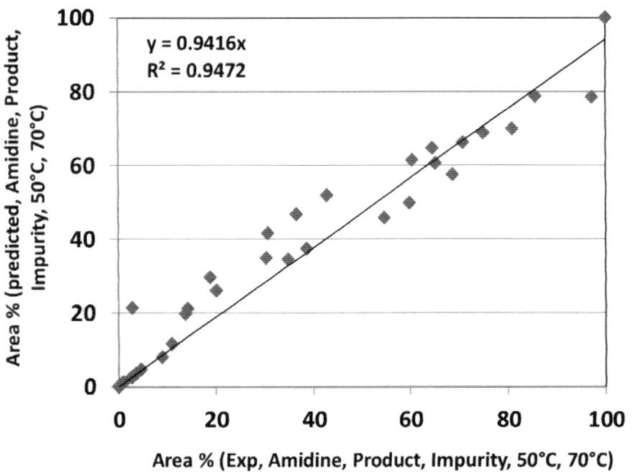

Overall fit

- Mechanistic kinetic model is reasonably well fitted to experimental data

16

Exp. vs. Model (fitted in DynoChem)

	Final Value	Units	Confident Interval (%)
Rxn2.Ea>	129.682	kJ/mol	17.724
Rxn3.Ea>	106.257	kJ/mol	11.785
Rxn3.k>	7.14E-4	L/mol.s	18.023
Rxn2.k>	5.001	1/s	33.343

- Reaction 1
  - Amidine + t-butoxide ⇌ intermediate + t-BuOH (**Keq=0.001**)
- Reaction 2
  - Intermediate → product + byproduct2 (**k_2**)
- Reaction 3
  - product + t-butoxide → impurity + byproduct2 (**k_3**)

17

Prediction from Model : I. Effect of metering addition of t-butoxide

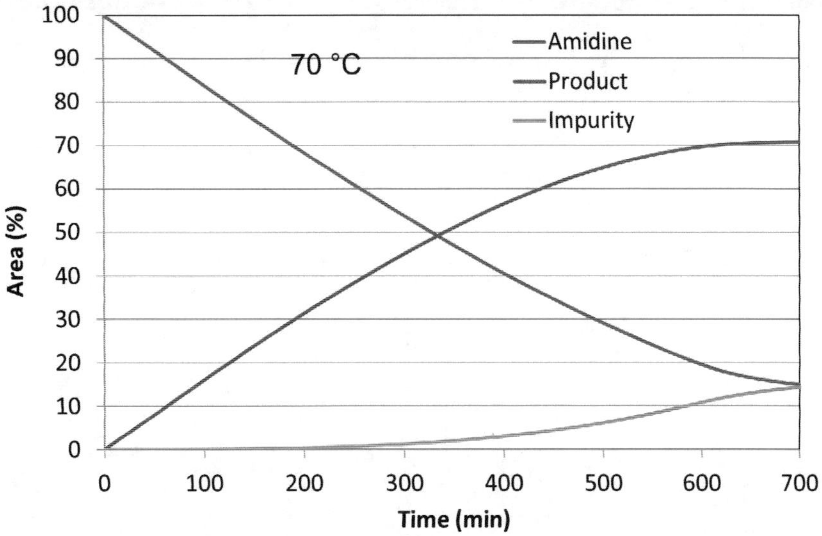

- Metering addition of t-butoxide through 600 mins – Impurity level is similar to batch reactor

18

Prediction from Model : II. Dilution effect

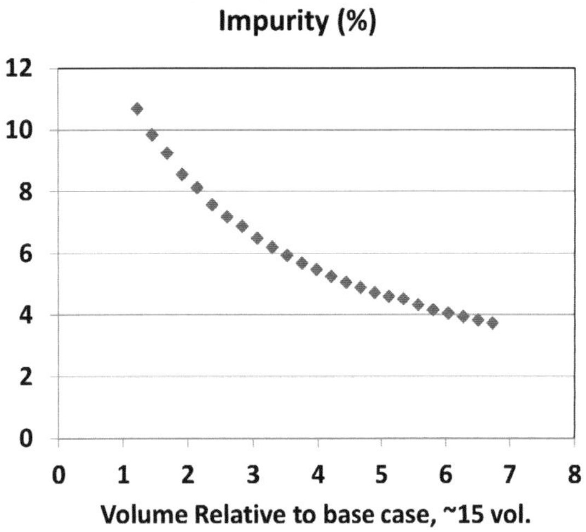

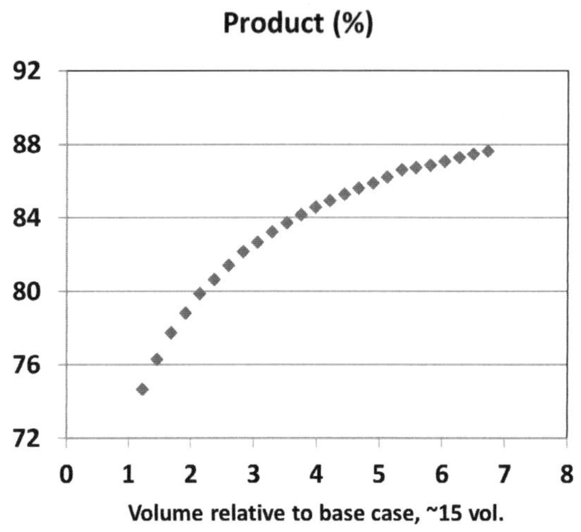

$k_2 K_{eq}$ [Amidine] [t-butoxide]/[t-butyl alcohol] >> k_3 [Product] x [t-butoxide]

- Super-dilution may be required – is not practically possible in the plant
  - ~100 vols. to minimize impurity less than 4%
  - ~ 300 vols. to minimize impurity about ~ 1%

19

Prediction from Model: III. Effect of t-BuOH removal (removal rate= 0.05 mole$L^{-1}Sec^{-1}$)

$k_2 K_{eq}$ [Amidine] [t-butoxide]/[t-butyl alcohol] >> k_3 [Product] [t-butoxide]

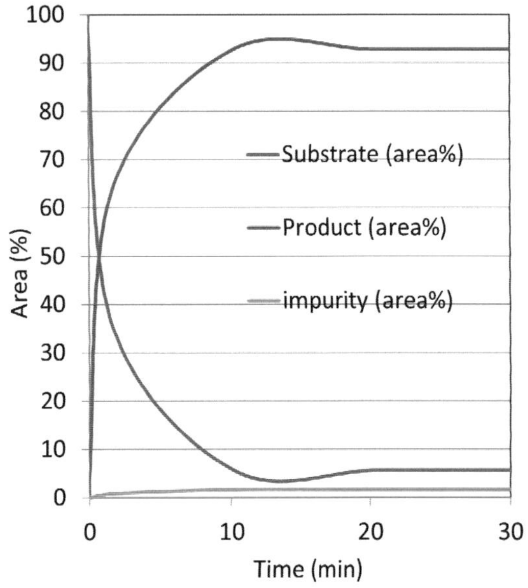

With removal of t-butyl alcohol

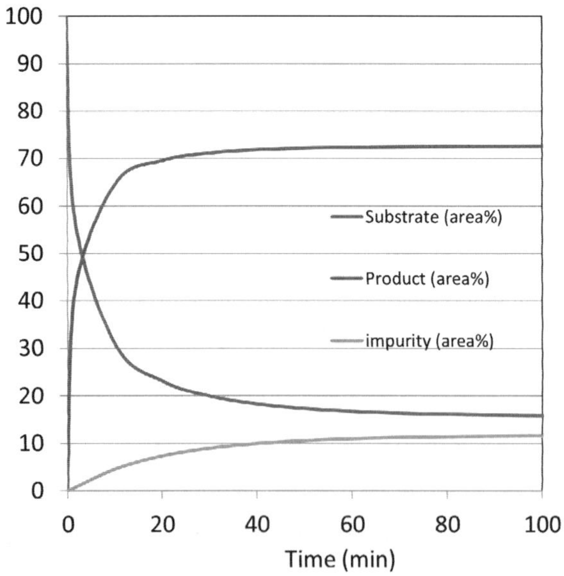

Without removal

20

Binary VLE of t-BuOH and toluene at 1 atm (DynoChem NRTL)

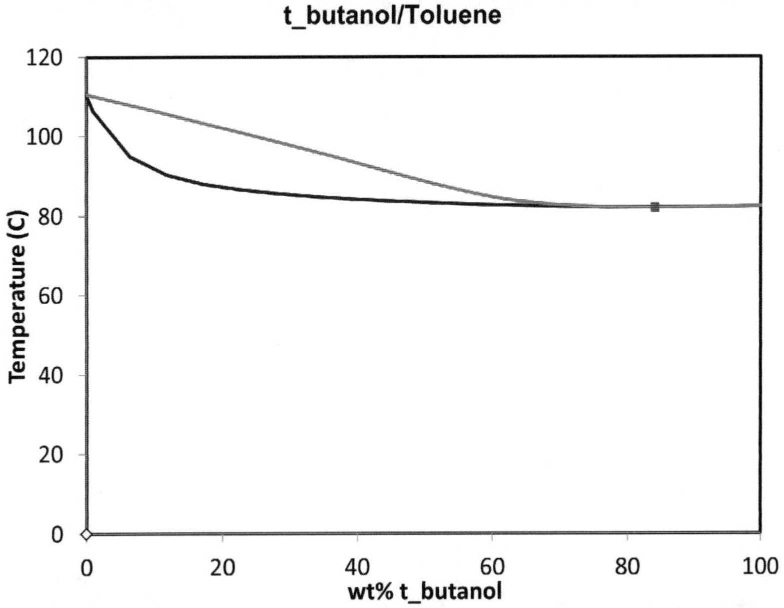

- Distillation can be an option to remove t-BuOH

21

Reactive distillation

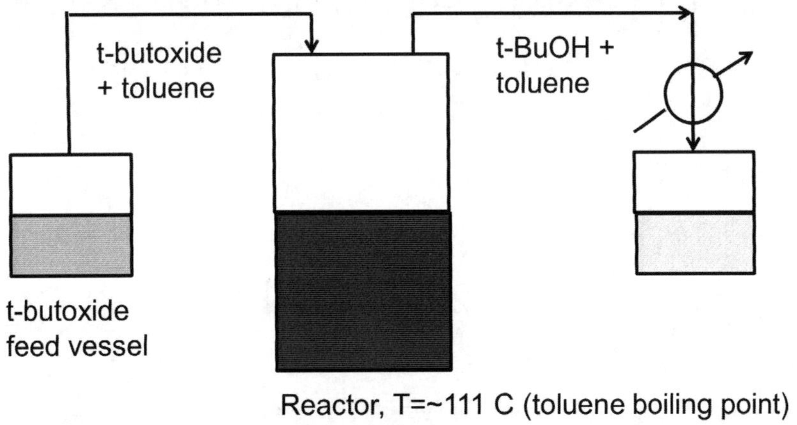

- Fed-batch reactive distillation
  - t-BuOH removal rate ≈ t-BuOH formation rate
  - Control rate of reaction: t-butoxide (in toluene) addition rate
  - Constant volume reactive distillation, distillation rate = t-butoxide (in toluene) addition rate

22

Proof of concept experiment - Reactive distillation

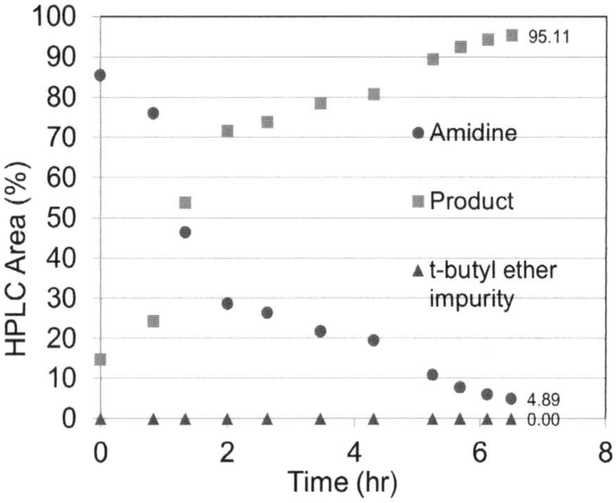

Reactive distillation at ~111 °C

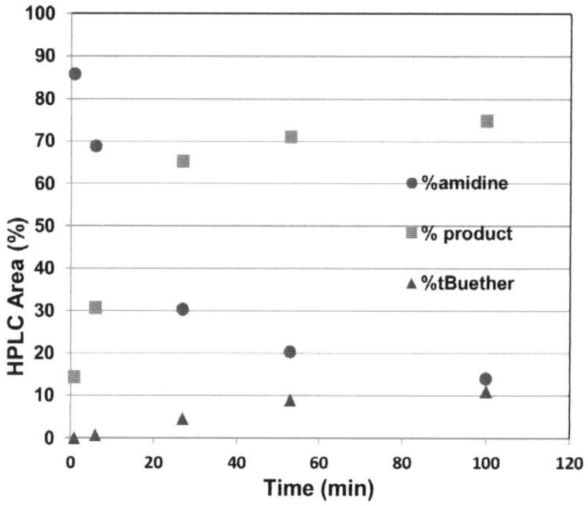

Reaction at 70 °C (No distillation)

- Reaction rate - controlled by metering addition of t-butoxide in toluene (+ little bit THF)
- No significant t-BuOH in the reaction samples
- No significant t-butyl ether impurity : t-BuOH removal effect

Summary

- Mechanistic model based approach led to
  - Ring closure reaction with extremely high selectivity (in-situ, EOR) for the possible use in the future campaign
- The approach is applicable for other reactions to minimize impurity formation from similar competing reaction pathways.

Acknowledgement

- Matt Embry
- Stan Kolis
- Doug Kjell
- Jeremy Merritt
- Shankar Vaidyaraman
- Shekhar Viswanath

25

Thanks

Questions ?

26

Proof of concept experiment - Reactive distillation (continuation)

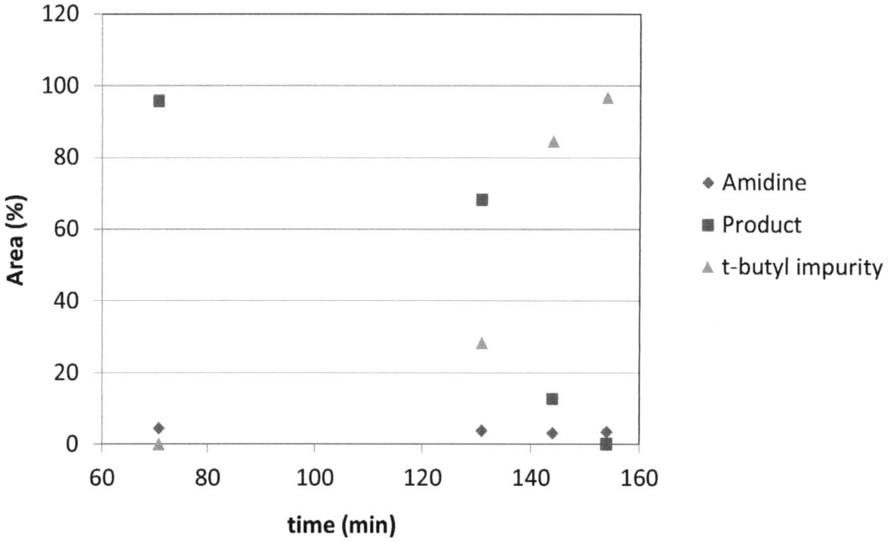

Reactive distillation at ~111 °C (continuation)

- Confirms t-butyl ether impurity formation from product and t-butoxide

27

Proof of concept experiment - Reactive distillation, reason for the trend

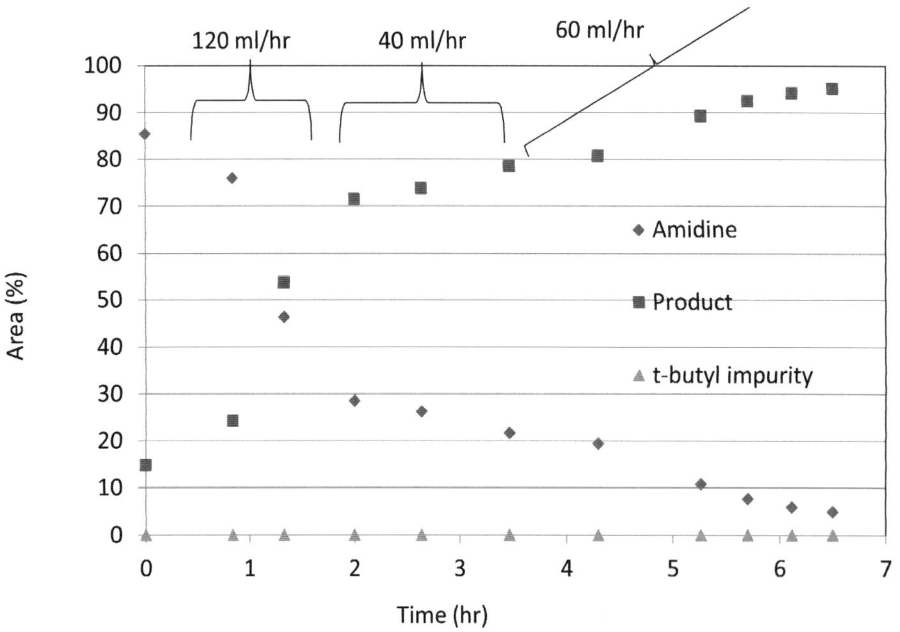

28

Proof of concept experiment 2

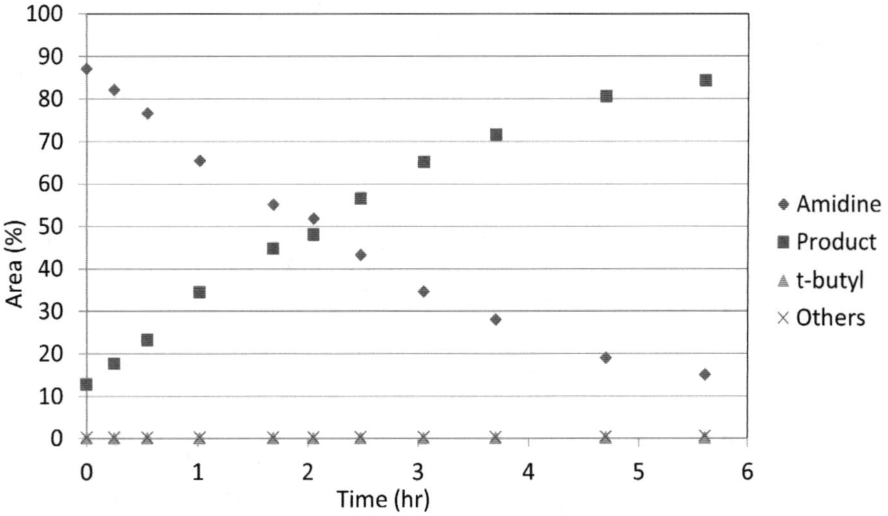

Reactive distillation at ~111 °C

- ~1 equiv. of t-butoxide (in toluene+THF): ~constant addition rate

29

Proof of concept experiment 3

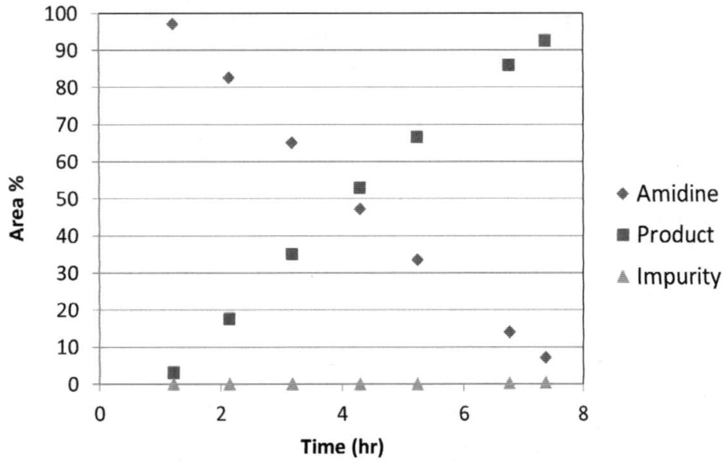

Reactive distillation at ~111 °C, Tj=115 C

- Initial Amidine: 1 g, Distillation working volume: ~20 ml, Tj=115 C
- K t-butoxide (in toluene+THF): 1 ml/hr addition rate (total: 6-7 ml)

30

430528 Simulation of Biopharmaceutical Bioreactor Using an Integrated CFD-Population Balance Modeling Approach

Abstract:

Since mid-1980's the specific productivity and density of cells in animal cell cultures have been increased several folds. These have been achieved via improving vectors, host cell engineering, selection, gene amplification and cell line screening [1] and optimum design of medium and feeding strategy and process engineering [2]. The necessity of integration of these goals has been established and partially addressed but a comprehensive bioreactor model in which extracellular and intracellular phenomena are coupled is still missing.

To represent the effect of agitation, aeration and feeding strategy on productivity, growth and viability of cells, the model of a large-scale bioreactor has multiple compartments which capture fluid dynamics, cell source variability, interphase mass transfer, gas-liquid interaction, and cell metabolism. The turbulent multiphase flow inside the bioreactor has been successfully modeled using *k-e* models of eddy viscosity provided in commercial CFD software packages [3]. The output of the CFD model is local velocity fields, shear and energy dissipation rates. The most important effect of shear on cells is lysis. To take into account the effect of this phenomenon on the viable cell density a population balance model can be implemented and integrated in which the death kernel includes lysis in addition to other causes of death. The merit of this type of modeling is in the prediction of the average behavior of a group of individuals showing randomness. Therefore it has been a trusted approach to model population of cells and to compute the distribution of cells over cellular states [4]. Due to low solubility of oxygen in water, high densities of cells quickly consume all of it in a saturated culture and produce enough carbon dioxide that it can have inhibitory effects. As a result addition and removal of gases are inevitable parts of operation of a large-scale fermenter. Dissolved oxygen concentration is calculated through estimation of gas-liquid mass transfer coefficients under the operating condition. The gas phase volume fraction can be calculated using gas hold-up correlations available in the literature or from the CFD with an additional Eulerian or discrete phase. The energy dissipated due to bubble rupture is usually higher than that in the bulk of the tank. So gas-liquid interaction directly affects the number of viable cells. Population balance modeling has the capability to capture the stochasticity involved in breakage and coalescence of bubbles and therefore is a good candidate for modeling the gas phase. Another advantage of this approach to capture the effects of bubbles is the more precise calculation of the gas-liquid interphase area; hence a better estimation of cell entrapment rate in the surface of bubbles. Besides lysis, bioreactor hydrodynamics also affects distribution and number of surface receptors, production of specific proteins, certain metabolic processes and can induce apoptosis [5]. Furthermore, nutrients and oxygen accessibilities to cells as well as the toxic effects of the material discharged from cells should also be taken into account. So a connection to cell metabolism seems necessary which can be built via a single cell model. Animal cells contain a large number of molecules so the main decision is the level of the complexity of the single cell model. Metabolic engineering and flux analysis can provide a framework to measure the importance of certain biochemical reactions and their effects on the systemic behavior of cells.

In summary, the model is composed of three main parts which represent bioreactor hydrodynamics, cells population and cell metabolism. The first component consists of a CFD coupled with a population balance model of bubbles. The result of solving this sub-model is nutrients and dissolved oxygen distributions, velocity fields and shear. The cells population model needs this information for calling the single cell model and calculation of its kernels; birth, death, growth and transition. The three components together connect protein production and viable cell density to operational parameters such as aeration and agitation rates and feeding schedule.

References

1. Wurm, F.M., *Production of recombinant protein therapeutics in cultivated mammalian cells.* Nature Biotechnology, 2004. **22**(11): p. 1393-1398.

2. Xie, L.Z. and D.I.C. Wang, *High cell density and high monoclonal antibody production through medium design and rational control in a bioreactor.* Biotechnology and Bioengineering, 1996. **51**(6): p. 725-729.

3. Schmalzriedt, S., et al., *Integration of physiology and fluid dynamics*, in *Process Integration in Biochemical Engineering*. 2003. p. 19-68.

4. Sidoli, F.R., S.P. Asprey, and A. Mantalaris, *A Coupled Single Cell-Population-Balance Model for Mammalian Cell Cultures.* Industrial and Engineering Chemistry Research, 2006. **45**(16): p. 5801 - 5811.

5. Shuler, M.L. and F. Kargi, *Bioprocess Engineering*. 2 ed. 2002: Prentice Hall.

436072 Systematic Design and Optimization of Fines Removal in Mixed-Product Mixed-Suspension Crystallizer

Abstract:

Fines removal in mixed-product mixed suspension crystallizer (MSMPR) via controlled growth and dissolution is systematically studied in this work. The problem is formulated as a multi-objective optimization, with the fitness functions defined as (1) the ratio of fines particle to the total number of particles (or the "fine index"), (2) the particle mean size, and (3) yield. The decision variables are the temperature set-points of each MSMPR stage, in which the optimization is carried out under different operating constraints, including seeded vs. unseeded operations, size-dependent vs. size-independent crystallization systems, different number of stages and different residence times. The results demonstrate that there exist clear trade-offs between the different objectives and that the various operating constraints significantly dictate if the obtained Pareto solutions are suitable for experimental implementations.

416933 Accounting for Interfacial Solvent Effects within a Mechanistic Crystal Growth Model

Abstract:

The choice of solvent has a significant impact on crystal growth habit. It is therefore one of the most important design decisions in crystallization and its effect is one of the most desirable for a crystal growth model to predict.

Our group has developed a mechanistic crystal growth model, accounting for both spiral and two-dimensional nucleation & growth mechanisms [1,2]. Critical quantities in the model are kink and edge energies on the crystal surface, which affect the predicted relative face growth rates under these layered mechanisms. These energies are modified by solvation of the crystal surface, which must be effectively accounted for. However, the determination of this solvent-induced energetic modification must not compromise the ability to obtain fast growth shape predictions, in order for our model to remain a practical engineering tool.

Interfacial energy models based on the bulk interface approximation are suitable and their application to our mechanistic model is the focus of this work. The energy of an interface is expressed through a penalty term of each phase's cohesive energy and an adhesive reward. For the work of adhesion, the van Oss, Chaudhury & Good (vOCG) model [3] provides a functional form that appropriately separates interactions according to how they match across an interface. In applying this to our model, solid-state crystal interactions are determined using a force field and solvent interactions are determined using solubility parameters and appropriate correlations to interfacial energies. These cohesive energies can be used to determine the required parameters for the vOCG model and although there is some uncertainty in the required partitioning, shape predictions are often not sensitive to this. Heuristic rules are developed for the correct application of this interfacial energy model to different types of systems.

To test the success of this within our mechanistic growth model, predictions for vapor and solution grown crystal shapes of various systems have been compared to experimental results, which isolates the effect of the solvent and has indicated that this approach is successful.

References:

1. Snyder, R. C. and Doherty, M. F. Proceedings of the Royal Society A: Mathematical, Physical and Engineering Science 465(2104), 1145–1171 (2009).
2. Lovette, M. A. and Doherty, M. F. Crystal Growth & Design 12(2), 656–669 (2012).
3. Van Oss, C., Chaudhury, M., and Good, R. Advances in colloid and interface science 28, 35–64 (1987).

(659b) Influence of Diacylglycerol on the Physicochemical Properties and Crystallization Behavior of Palm Oil

Huang, X.,
Hao, H., Tianjin university, Tianjin, China
Wang, J., National Engineering Research Center for Industrial Crystallization Technology, School of Chemical Engineering and Technology, Tianjin University

Originally presented on: 11/12/2015 8:55:00 - 9:15:00

Abstract: In this study, diacylglycerol (DAG) oils were produced through enzymatic glycerolysis and chromatographic respectively and they were added to purified tri-glycerides of palm oil at different concentrations (2, 5, 10, 30 and, 50% w/w). The physical and chemical characteristics, i.e., iodine value, fatty acid composition and triglycerides composition were evaluated and only slight differences were observed. Then an exhaustive analysis of the crystallization behavior of the samples were per-formed by using Differential scanning calorimetry (DSC), Raman Spectroscopy, po-larized light microscopy, low-resolution magnetic pulsed nuclear resonance and X-ray diffraction over different ranges of crystallization temperatures. The aim of this study was to elucidate the mechanism changes of palm oil crystallization by the addition of two types DAG. The experiment results show that the influence of these DAGs on the crystallization behavior (e.g. nucleation rate, crystal growth rate, polymorph transform) of palm oil was depended on the DAG type, supercooling and diglycerides' concentration, e.g. the presence of a low concentration of enzymatic-DAG was found to significantly delay nucleation and crystal growth velocity of palm oil and induce the formation of β form at low degree of supercooling. Finally, we use Avrami model to interpret these influences and the results match the experiment successfully. This study will provide essential information for the palm oil industry for food application purposes.

427862 Investigation of the Formation of Nanoscale 2CL-20 · HMX Cocrystal

Abstract:

Energetic nanoscale 2CL-20·HMX, a cocrystal of CL-20 (2,4,6,8,10,12-hexanitro-2,4,6,8,10,12-hexaazaisowurtzitane) and HMX (1,3,5,7-tetranitro-1,3,5,7-tetrazocine) at a 2:1 molar ratio, was prepared by a novel method of bead milling an aqueous suspension of ε-CL-20 and β-HMX. Conversion of the coformers to the cocrystal form was monitored by powder X-ray diffraction (XRD) and scanning electron microscopy (SEM) analysis of specimens sampled at various milling times. Complete conversion to the cocrystal form was achieved. Rounded 2CL-20·HMX cocrystal particles with a mean size below 200 nm were produced. The conversion was further studied by varying the processing variables. A mechanism for the conversion of the coformers to the cocrystalline form is postulated based on the experimental results. As an inherently safe manufacturing method, the aqueous bead milling process has a great potential in advancing the cocrystal research and applications in the field of energetic materials.

427940 Kidney Stone Growth Modification: Insights from First Principles Calculations

Abstract:

Nearly 1 in 11 Americans will develop kidney stones during their lives[1]. Kidney stones are largely composed of calcium oxalate monohydrate (COM) crystals and develop as a result of heightened concentrations of oxalate in urine [2]. Previous work [2] has identified citric acid (CA) as an effective COM crystal growth modifier. Recently, Dr. Rimer's group at the University of Houston has shown that hydroxycitric acid (HCA) is a highly effective COM crystal growth modifier with notably different growth modification properties than CA. HCA is a molecule with the same structure as CA except for an additional hydroxyl functional group. In this work we use first-principles calculations to elucidate the growth modification mechanisms of CA and HCA at the molecular level. Ab-initio pKa calculations[3]demonstrate that HCA exhibits lower deprotonation constants than CA due to increased intramolecular hydrogen bonding. Accelerated Density Functional Theory calculations (accounting for solvent effects) revealed that HCA exhibits increased folding capability and binding affinity to both calcium ions and oxalate surfaces than CA. Most importantly this work introduces concepts that may be used in the rational design of effective kidney stone growth modifiers.

References

1. Scales, C. D., Smith, A. C., Hanley, J. M., and Saigal, C. S. European Urology 62, 160–165 (2012).
2. Qiu, S. R., Wierzbicki, a., Orme, C. a., Cody, a. M., Hoyer, J. R., Nancollas, G. H., Zepeda, S., and De Yoreo, J. J. Proceedings of the National Academy of Sciences of the United States of America 101(7), 1811–1815 (2004).
3. Feng, S., Bagia, C., and Mpourmpakis, G. Journal of Physical Chemistry A 117, 5211–5219 (2013).

429995 Inhibition of Calcium Oxalate Monohydrate Crystallization Using Organic Acids

Abstract:

Inhibition of Calcium Oxalate Monohydrate Crystallization using Organic Acids

Jihae Chung[1], Michael G. Taylor[2], Isadora Carnaval[2], Giannis Mpourmpakis[2], John R. Asplin[3], and Jeffrey D. Rimer[1]

[1]University of Houston, Department of Chemical and Biomolecular Engineering, Houston, TX 77204

[2]University of Pittsburgh, Department of Chemical and Petroleum Engineering, Pittsburgh, PA 15261

[3]Litholink Corporation and Renal Section, School of Medicine, University of Chicago, Chicago, IL 60637

Kidney stone disease is one of the most common disorders where the number of people afflicted is on the rise; however, there have been no significant advancements in therapeutic treatments for this disease over the past three decades. More than 80% of human kidney stones are calcium oxalate monohydrate (COM) polycrystalline aggregates. The current therapeutic for COM stones is citrate, an over-the-counter supplement with moderate inhibitory effect on COM crystal growth and relatively low compatibility with patients when consumed for an extended period of time. We have explored the effects of hydroxycitrate (HCA), which is a molecular analogue of citrate that is a more potent inhibitor of COM crystal growth, and is thus a possible replacement of citrate as a therapy. HCA differs from citrate by only one hydroxyl group, yet interestingly, this subtle structural difference leads to completely different modes of interaction with COM crystal surfaces. Here, we will compare the effects of citrate and hydroxycitrate on COM crystallization. Our studies focus on the effects of growth inhibitors at both macroscopic and microscopic length scales. In bulk crystallization assays, we quantified the effect of inhibitors on the bulk crystal habit as well as their impact on the kinetics of crystallization. We employ *in situ* atomic force microscopy (AFM) to probe the interfacial interactions between inhibitors and COM crystal surfaces in real time. Working in conjunction with collaborators, we are exploring the effects of HCA and other modifiers *in vivo* and *in silico* to establish a synergistic platform to rationally design crystal growth inhibitors, which can then be used to develop new drugs for pathological diseases involving mineralization.

432022 Continuous Production of Polymer Coated Drug Crystals, Particles and Nanoparticles

Abstract:

Polymeric coating can provide protection for fragile drugs from hydrolysis and degradation. With an appropriate polymer coating, nanoparticles are known to traverse the physiological human mucous barrier. Conventional approaches to such fabrication are quite cumbersome and often employ batch techniques. We have adapted the technique of solid hollow fiber cooling crystallization (SHFCC) as well as the technique of porous hollow fiber based anti-solvent crystallization (PHFAC) to continuously produce polymer-coated drug crystals and particles. Both submicron particles and nanoparticles have been coated. We have also continuously synthesized polymer-coated drug crystals by anti-solvent crystallization from a solution of the drug and the polymer. The synthesized polymer-coated drug crystals involve crystals of the drug Griseofulvin (GF) coated by a thin layer of the polymer Eudragit RL100. In the case of particles, submicron (550 nm) and nano-sized (12 nm) silica particles acted as the host particles; Eudragit RL 100 and poly (lactide-co-glycolide) were the coating polymers. For the polymer-coated drug crystals, the surface morphology, particle size distribution, and the polymer coating thickness were characterized by SEM, STEM, laser diffraction spectroscopy (LDS) and thermogravimetric analysis (TGA). To study the properties of the coated drug crystals, x-ray diffraction (XRD), Raman spectroscopy, and dissolution test were implemented. Coated silica particles were analyzed after vacuum drying using SEM, STEM, energy dispersive spectrometry, LDS and TGA. Scaleup has been demonstrated using hollow fiber modules of different dimensions. Extended duration runs have been conducted for two hours to demonstrate no variation in product quality.

432455 Mechanism of Hematin Crystallization and Inhibition in Biomimetic Solutions

Abstract:

Crystallization is governed by thermodynamic and kinetic events that result in molecule (solute) incorporation into crystal surfaces. These fundamental processes mediate both natural and synthetic crystallization pathways. Our aim is to understand the molecular pathways for the crystallization of hematin, which is a byproduct of heme detoxification in malaria parasites. Upon entering its human host, the parasite catabolizes hemoglobin in red blood cells and releases free heme (Fe^{2+}), which is rapidly oxidized to toxic hematin (Fe^{3+}) within its digestive vacuole (DV). The parasite utilizes this process to sequester the toxin into innocuous crystals within the complex DV environment. Hematin crystal growth ensures parasite survival within the human host and is thought to be suppressed by quinolone-class antimalarial drugs. These crystals are referred to as hemozoin (*in vivo*) or β-hematin (*in vitro*). Identification of antimalarial drug action could provide a foundation for drug design to overcome the continued resurgence of antimalarial drug resistance. Here, we will discuss the molecular mechanism of hematin crystallization and inhibition by quinolone-class antimalarial drugs in a biomimetic growth environment.

In situ atomic force microscopy (AFM) revealed that hematin crystallization occurs through a classical layer-by-layer mechanism involving the addition of individual molecules to the growth sites present on the crystal surface.[1] Our studies employed a two-phase medium which we designed to mimic the complex environment of parasite DVs, which is comprised of an acidic aqueous phase and lipids. We used citric buffer-saturated octanol (CBSO) as a biomimetic solution to grow hematin crystals.[2] During *in situ* AFM experiments, we directly observe the effects of hematin surface diffusion and molecular level fluctuations in the rates of attachment/detachment from step edges. Towards understanding the relationship of solvent interactions on molecule addition, we quantified the rate of layer generation and the velocities of anisotropic step advancement as a function of supersaturation in the absence and presence of antimalarial modifiers. These quantifications provide a foundation for determining the modes of action of antimalarial drugs (i.e. inhibition of layer formation and/or step propagation). We identify that current antimalarials act by unique modes of inhibition. Here, we will summarize these findings and discuss our current studies to elucidate the role of functional moieties in current antimalarials and how the organic solvent affects drug-crystal interaction. To this end, we use a combination of *in situ* AFM and chemical force microscopy (CFM) measurements wherein AFM tips are functionalized with moieties that mimic specific regions of antimalarial drug molecules. Our findings reveal interesting trends of molecular recognition between antimalarial drugs and crystal surfaces that when coupled with bulk crystallization assays is a potential platform for the design and screening of new drug compounds for malaria.

References

1. Olafson, K.N., M.A. Ketchum, J.D. Rimer, and P.G. Vekilov, *Mechanisms of Hematin Crystallization and Inhibition by the Antimalarial Drug Chloroquine*. Proceedings of the National Academy of Sciences U S A, 2015. **112**: p. 4946-4951.

2. Olafson, K.N., J.D. Rimer, and P.G. Vekilov, *Growth of Large Hematin Crystals in Biomimetic Solutions*. Crystal Growth & Design, 2014. **14**(5): p. 2123-2127.

432317 Characterization of Lab and Plant Reactors Using Fast Competitive Reactions and a Predictive Model

Abstract:

Scale-up of reaction and crystallization conditions often involves assumptions about equivalence of mixing between lab and plant. Scale-down of process for lab studies requires similar assumptions. A question that often arises is how small can the lab equipment be, so that the results translate to scale. With modern lab synthesis workstations providing good control of addition rates and temperature in reactors that resemble their larger scale equivalents, scale-up and scale-down work sometimes takes place in lab systems with volumes as low as 100 mL. It is possible using mixing-sensitive test reactions to quantify the mixing environment in vessels at any scale and to thereby identify lab and plant conditions that are more quantitatively equivalent.

The fast competitive reaction system involving neutralization and ester hydrolysis (a so-called 'Bourne Reaction', ref 1) has been used to study mixing characteristics in several common lab reactors used for routine development of chemical processes. The final product distribution after running the reaction in fed batch mode reflects the history of relative mixing and reaction rates in the plume or reaction zone that forms near the feed location.

Most previous applications of this technique have been in larger vessels where a critical feed time can be identified and longer feed times represent micromixing-controlled conditions. It is possible using a predictive model containing both kinetics and mixing parameters to estimate the product distribution as a function of feed time and mixing energy dissipation rate. This latter parameter, often referred to as epsilon, with units of W/kg, is of interest for a wide range of mixing phenomena. Using the predictive model together with measured product distribution, it is possible to estimate the corresponding epsilon. This represents a useful level of vessel characterization, in addition to the more widely known heat and mass transfer characteristics (UA and kLa).

We will present and discuss results for different reactors and operating conditions and also how best to use predictive models to design and analyse the experiments, which require a good dynamic range of selectivity to be informative. We will compare mixing performance at operating conditions that are commonly used in the lab and plant. Finally, we will deduce some guidelines towards obtaining equivalent mixing between lab and plant reactors.

1. Investigation of micromixing in stirred tank reactors using parallel reactions, John R. Bourne , Shengyao Yu, Ind. Eng. Chem. Res., 1994, 33 (1), pp 41–55

420196 Case-Studies of Enhancing Manufacturing Flexibility for a Commercial Biologic

Abstract:

During the manufacturing process of a commercial recombinant molecule, CHO cells are cultured through the inoculum expansion in shake-flasks and seed bioreactors followed by inoculation of the production bioreactor. During primary recovery, cell culture is centrifuged and the centrate is passed through a series of depth filters and sterile filters prior to transfer to downstream operations. Three case-studies will be presented to illustrate typical study workflows carried out to support the manufacturing of a commercial biologic moiety. The first case-study entails the validation of three optional, additional passages during inoculum expansion. Evaluation of process parameters against acceptance criteria derived from the in-process control strategy for inoculum expansion will be described. The second case-study details how a selective pressure agent used in inoculum expansion, and originating from an alternate vendor, was evaluated by comparison to the same compound originating from the primary approved vendor. The third case-study discusses a process qualification study for changing out a depth filter and a membrane filter during primary recovery. These case-studies successfully improved manufacturing flexibility allowing full utilization of capacity while handling several multi-product campaigns.

432326 The Confined Impeller Stirred Tank (CIST): A New Mixing Device Designed to Properly Scale-Down Local Mixing Effects in Industrial Applications

Abstract:

The confined impeller stirred tank (CIST): a new mixing device designed to properly scale-down local mixing effects in industrial applications

Marcio B. Machado, Suzanne M. Kresta

University of Alberta, Edmonton AB T6G 2V4

Scaling-down and -up an industrial application remains challenging. The effectiveness of these industrial processes is improved if efficient dispersion and dissolution of the additives is achieved. It is known that current bench scale tests do not always represent the industrial scale behavior. One of the most critical steps for a process scale-down or scale-up is to have a fully turbulent flow regime in the bench scale mixing vessel, because if the flow falls into transitional flow, the results are not reproducible. This work presents the confined impeller stirred tank, as shown in Figure 1, which is a mixing device designed to provide uniform intensity of mixing over the entire volume of the tank. The CIST is a 1 L tank filled with five impellers. Such configuration confines the flow. The tank has a diameter of T = 7.6 cm and a height H = 3T. Mean velocity profiles were measured at different axial positions of the tank using a Laser Doppler Velocimeter. Several impeller geometries, impeller diameters and fluids were used in order to check their effect in the flow. This tank is able to sustain fully turbulent flow throught the entire vessel at the Reynolds numbers less than 3000 while conventional stirred tanks require Reynolds numbers up to 300 000 to achieve fully turbulent flow at the surface. Having fully-developed turbulent flow at the surface, where the chemicals are usually added, is ideal for decreasing the dissolution time, avoiding mesomixing effects and providing reproducible bench scale experimental data which can be used to specify local mixing conditions in the industrial plant. These results have implications both for scale-up and for many industrial applications with surface feed or with dip pipes in the top third of the tank. A methodology to scale-down or –up an industrial process based on the total mixing energy provided to the system is also presented.

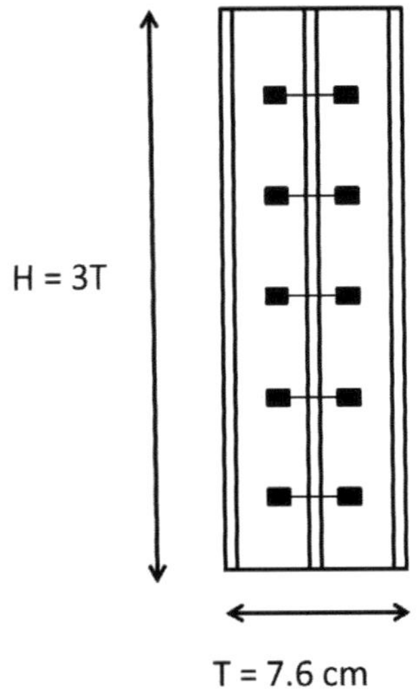

Figure 1: The confined impeller stirred tank (CIST) operating with a set of five Rushton turbines

429039 A Hybrid Experimental-Computational Method to Scale-up Powder Dissolution in Viscous Solvent: An Industrial Approach

Abstract:

Dissolution of solid particulates in a liquid solvent is an important process in several pharmaceutical operations. Our application relates to topical ointment manufacture where a particulate ingredient must be dissolved in a viscous liquid. From a lab-scale process, we seek to scale-up this operation to pilot scale, and finally to commercial scale.

In this work, we developed a hybrid experimental-computational approach to simulate and scale-up the dissolution process. A population balance model (PBM), implemented in-house, served as the framework to track the size and distribution of particles as they evolve with time. Experimental dissolution kinetics data generated from lab-scale reactor vessels (Mettler-Toledo OptiMax-500) was used to calibrate the mass-transfer rate law. Our PBM framework also incorporates the effect of fluid-particle flow on the dissolution rate—the flow behavior for varying agitation conditions at different scales was characterized by performing a series of computational fluid dynamics flow simulations. The calibrated PBM framework was then used to execute "virtual experiments" in lieu of expensive, actual experiments at the larger scales.

The initial particle size distribution is an inherent source of uncertainty in our process since size distributions of the solute may vary lot-to-lot. The solute powder also is prone to *in-situ* agglomeration when added to the viscous solvent—the extent of agglomeration is random and unpredictable. Therefore, we quantitatively captured this size-distribution uncertainty in our modeling framework and developed confidence intervals, instead of point estimates, for the scaled-up dissolution time predictions.

Our work showcases a successful example where confidence-based model predictions have guided process development and reliably addressed scale-up of an industrial pharmaceutical operation.

419741 Scale-up of Agitated Drying: Using Agitator Motor Power Draw to Control Changes in Powder Properties

Abstract:

Scale-up of agitated drying operations to produce Active Pharmaceutical Ingredients (APIs) with the desired powder properties can be challenging. Traditional laboratory drying equipment has not successfully reproduced the degree of agglomeration or attrition observed at scale. The development of new lab-scale tools that can reproduce the magnitude of forces observed during agitation of bulk powders has provided insight into the factors that influence changes in powder properties during API manufacturing. Several recent reports have shown that agitator torque and the total work per unit mass imparted on the drying cake through agitation can be correlated to changes in API particle size. Factors such as API characteristics, batch size, dryer geometry, residual solvent content, and the defined agitation protocol impact the agitator torque, work input, and extent of changes in powder properties during drying. This work builds off these previous reports by using agitator motor power draw, which can be easily measured at both laboratory and pilot plant scale, to compare agitated drying studies conducted at different scales. The application of power measurement to define and implement control strategies for agitated drying of two APIs will be described. In the first case, the power measurement was used to identify an upper limit on agitation and avoid breaking up agglomerates during the drying process. For the second API, the agitator power draw was used to define a lower limit for agitation time to ensure sufficient breakage of needles occurred to provide material with the target bulk density.

424868 Rational Scale-up of Hot Melt Extruders: 1D Predictive Modeling Based on 3D Smoothed Particle Hydrodynamics

Abstract:

Rational Scale-Up of Hot Melt Extruders: 1D Predictive Modeling based on 3D Smoothed Particle Hydrodynamics

Josip Matić[*,**], Andreas Eitzlmayr[*], Johannes Khinast[*,**]

*Institute for Process and Particle Engineering, Graz University of Technology, Inffeldgasse 13/III, 8010 Graz, Austria

**Research Center Pharmaceutical Engineering, Inffeldgasse 13/III, 8010 Graz, Austria

Email for correspondence: khinast@tugraz.at

Introduction

Hot melt extrusion (HME) is a continuous manufacturing process primarily using co-rotating twin-screw extruders (TSE). The potential to increase the bioavailability of poorly soluble drugs, the steady product quality of a continuous process, the increased efficiency as well as a reduction of operational costs led to the specific interest of pharmaceutical industries in HME. In order to effectively predict the impact of screw configuration and process parameters on the HME process, an improved understanding of flow and mixing in individual screw elements as well as in the entire extrusion process is of high interest.

In our previous work[1], we showed that the smoothed particle hydrodynamics [2] (SPH) simulation method provides significant benefits for extruder screw simulations. In contrast to computational fluid dynamics (CFD), SPH inherently accounts for convective mixing and free-surface flows, allowing a detailed investigation of flow and mixing in fully filled and partially filled extruder screw elements. In order to provide correct boundary conditions for the complex screw geometries, a novel wall interaction method was developed [3].

Methodology

Due to the high computational expense of spatially resolved flow modeling, simplified modeling approaches are still required for the modeling of entire extrusion processes. One-dimensional (1D) models consider the spatial dependencies only along the screw axis and thus, provide the benefit of comparably low computational costs, suitable for the industrial use in design, optimization and scale-up of extruders. Following the idea of a continuous stirred tank reactor cascade with back-mixing, as shown by Choulak et al. [4], a 1D extruder model was developed [5], [6] in order to predict profiles of filling ratio, pressure, specific energy input, material temperature and mixedness along the screw axis. Due to its strong simplifications, the 1D approach requires pre-computed data about flow and mixing in individual screw elements, which were obtained from the spatially resolved SPH model. This combination of the three-dimensional (3D) SPH approach for the detailed investigation of individual extruder screw elements and the 1D approach for the description of process variables along entire screw configurations results in a comprehensive tool, which increases the process understanding, supports design, optimization and scale-up, and reduces the need for experiments.

Results

An extruder scale-up scenario was analyzed based on proposed scaling laws for process parameters as throughput and screw speed which result in similar conditions at different length scales. Our results show the impact of the extruder dimensions (scale) on filling ratio, pressure, specific energy input, temperature and mixedness.

References

[1] A. Eitzlmayr, G. Koscher, and J. Khinast, "Co-Rotating Twin-Screw Extruders: Detailed Analysis of Conveying Elements Based on Smoothed Particle Hydrodynamics. Part 1: Hydrodynamics." Chemical engineering science, Oxford, p. In Press, 2015.

[2] J. J. Monaghan, "Smoothed particle hydrodynamics," *Reports Prog. Phys.*, vol. 68, no. 8, pp. 1703–1759, Aug. 2005.

[3] A. Eitzlmayr, G. Koscher, and J. Khinast, "A novel method for modeling of complex wall geometries in smoothed particle hydrodynamics," *Comput. Phys. Commun.*, vol. 185, no. 10, pp. 2436–2448, May 2014.

[4] S. Choulak, F. Couenne, Y. Le Gorrec, C. Jallut, P. Cassagnau, and a. Michel, "Generic Dynamic Model for Simulation and Control of Reactive Extrusion," *Ind. Eng. Chem. Res.*, vol. 43, no. 23, pp. 7373–7382, Nov. 2004.

[5] A. Eitzlmayr, J. Khinast, G. Hörl, G. Koscher, G. Reynolds, Z. Huang, J. Booth, and P. Shering, "Experimental Characterization and Modeling of Twin-Screw Extruder Elements for Pharmaceutical Hot Melt Extrusion," *AIChE J.*, vol. 59, no. 11, pp. 4405–4450, 2013.

[6] A. Eitzlmayr, G. Koscher, G. Reynolds, Z. Huang, J. Booth, P. Shering, and J. Khinast, "Mechanistic Modeling of Modular Co-Rotating Twin-Screw Extruders," *Int. J. Pharm.*, vol. 474, no. 1–2, pp. 157–176, 2014.

(661g) Designing for Compaction in Tablets with Spray Dried Intermediates

Ploeger, K., Merck & Co, Inc.
Temtem, M., Hovione FarmaCiência SA
Cordeiro, P., Hovione
Aghajanian, A.,
Originally presented on: 11/12/2015 10:30:00 - 10:50:00

Title: Designing for Compaction in Tablets with Spray Dried Intermediates

Authors: Kristin Ploeger (kristin.ploeger@merck.com), Marcio Temtem (mtemtem@hovione.com), Paula Cordeiro (pcordeiro@hovione.com), Angela Aghajanian (angela.deese@merck.com)

Abstract:

In an environment with increasing numbers of poorly-soluble drug substances, spray drying is used to enhance the solubilization of some of these molecules. Inclusion of this unit operation in the manufacture of oral solid dosage forms results in a new challenge, where the compactability for the formulation is frequently impacted in three separate unit operations (spray drying, roller compaction, and tabletting). The interconnection between the unit operations requires a Quality by Design approach to development, where impact of process parameters across multiple unit operations is considered since the design space in one unit operation has the potential to restrict the design space in the others.

Approaches for designing the spray dried particle attributes will be discussed. The combination of modeling and simulation tools with process understanding and prior knowledge were crucial to deliver a commercially viable process. Moreover, the description of the process and its design space through meaningful and scale-independent parameters (some of them, such as relative saturation of the drying gas and mean droplet size, accurately estimated by modeling tools) were used in the benefit of the scale-up (Figure 1). Corrections between process parameters and material attributes will be presented.

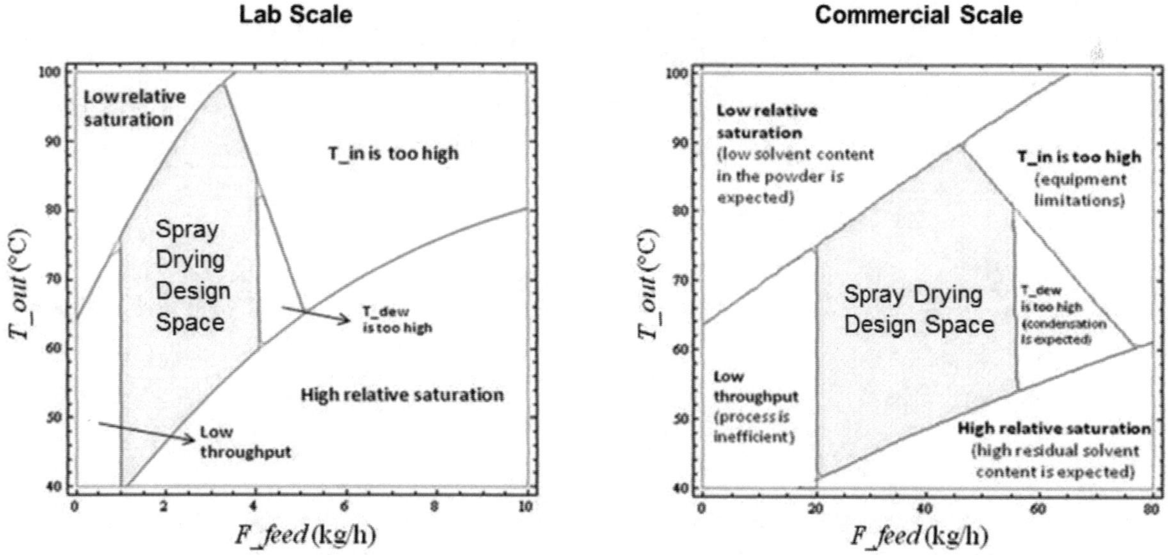

Figure 1 ? Schematic representation of the lab and commercial thermodynamic design space

Spray dried particle attributes were evaluated to determine which, if any, could be a surrogate measure of particle compactability. A multivariate study of compactability across the three unit operations was performed (Figure 2), to evaluate if enhancement of compactability in the spray dried particles could permit more aggressive roller compaction conditions, without impact to the tabletting process or final attributes (such as disintegration, dissolution, and friability). While traditional studies of roller compaction typically include independent variables like roll pressure, this work will discuss the potential benefits (and challenges) of designing the DoE with a parameter like tensile strength (analogous to the practice of designing tabletting DoEs with tablet hardness as a parameter).

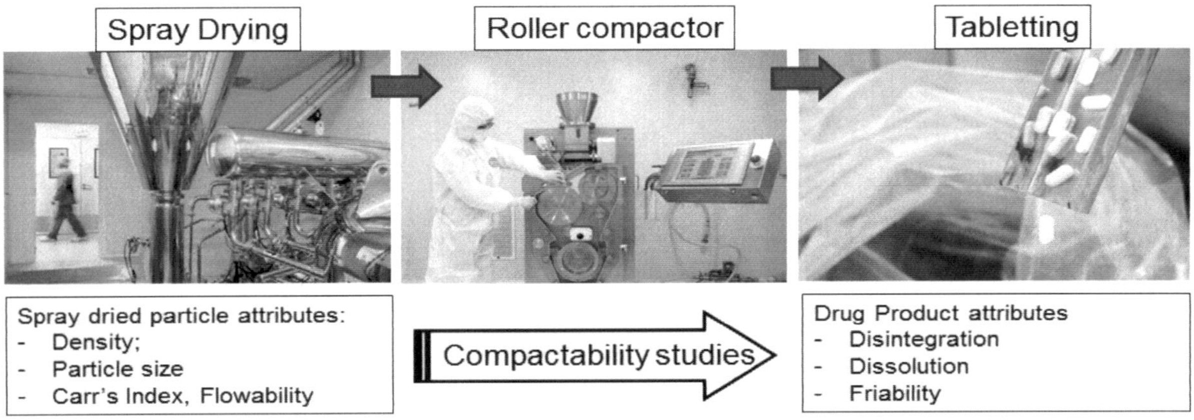

Figure 2 ? Schematic of the proposed experimental protocol (Images ? Hovione copyright)

408475 Molecular Mobility As a Tool for Understanding the Impact of Polyvinylpyrrolidone (polymer) and Tpgs (surfactant) in Crystallization Kinetics of Amorphous Celecoxib

Abstract:

Formulating drugs (active pharmaceutical ingredients) as amorphous solid dispersions is an alternative for improving the solubility of poorly water soluble drugs. An amorphous solid dispersion ideally consists of the amorphous drug homogeneously distributed and thermodynamically stable within a polymer matrix. In addition to the polymer, the formulation may contain other materials such as surfactants to improve processability, dissolution and flow behavior, among others. Understanding the crystallization tendency of such compounds is crucial when assessing it's feasibility for a solid dispersion formulation, as re-crystallization of the drug is an undesired effect and may lead to potential market withdrawal of a commercial drug product. Celecoxib is a poorly water soluble (BCS class II) non-steroidal anti-inflammatory drug. When compared to other BCS class II compounds, it has relatively low molecular mobility (high viscosity) and crystal growth rate upon cooling from the undercooled melt state [1,2]. The presence of surfactants or polymers has demonstrated to significantly increase or inhibit the radial crystal growth rate of Celecoxib, respectively. Interestingly but not necessarily advantageous, the combination of both, results in a complex behavior where the polymer is able to reduce the growth rate acceleration caused by the surfactant resulting in an increase or decrease of the overall crystal growth rate depending on the miscibility of Celecoxib and the surfactant [3]. Even though, there is a large number of surfactants and polymers available for amorphous solid dispersions, few studies elucidate the impact of all of these in the crystallization kinetics of drugs. While the impact of surfactants and polymers in the crystallization growth rates of pure amorphous Celecoxib has been studied and reported, this impact analyzed in terms of the molecular diffusivity is a knowledge still not available. Crystallization of a drug is a diffusive process where the nucleation rate is dictated by the fluidity of the liquid (viscosity). Fundamental knowledge may be obtained when the crystallization tendency (or lack thereof) is analyzed in terms of the "glass forming ability" of the liquid, *i.e.*kinetic fragility. Here, the liquid's viscosity with temperature may follow Arrhenius or Vogel-Tamman-Fulcher (VTF) dependence, depending on the relaxation of the liquid molecules. It is commonly known that glassy liquids at high temperature follow Arrhenius dependence (linear), however upon lowering the temperature deviates from Arrhenius and follows VTF law (non-linear). Interestingly, some speculate that the VTF behavior transitions back to an Arrhenius dependence at low temperatures [4]. The viscosity is basically the resistance of a fluid to flow under an external applied stress, it is then related to the molecular relaxation and hence mobility of a material. We have characterized the rheological properties of Celecoxib with small amounts of PVP (polymer) and/or TPGS (surfactant) with the purpose of understanding the impact on the molecular diffusivity of the Celecoxib during crystallization from the undercooled melt state. We expect these results to complement the knowledge available of molecular mobility of Celecoxib with regards to thermodynamic quantities for solubility prediction [5]. Additionally, in general for when assessing a drug as a candidate for a solid dispersion formulation during developmental stages of a drug product. Amorphous samples where prepared using a cryogenic mill. Binary and ternary mixtures consisted of 10 wt.% TPGS or 10 wt.% PVP dispersed in a 90wt.% Celecoxib, and ternary mixtures of 10wt.% PVP, 10 wt.% TPGS and 80 wt.% Celecoxib. Samples were characterized in an Anton Paar MCR 301 series rheometer, using 20mm diameter parallel Peltier Plates. Temperature dependent Dynamic Viscosity was measured while cooling the sample from melting point at a cooling rate of 10°C/min. We measured the dynamic rheological properties of the Celecoxib samples. The dynamic viscosity data was analyzed in terms of molecular mobility, Activation Energy (E_a), and fragility (Strength Parameter) (D) of the drug near the glass transition (T_g) and melt temperature (T_m), this was possible by analyzing the viscosity data with Arrhenius and Vogel-Tamman-Fulcher (VTF) equations. The temperature dependent dynamic viscosity curves resulted in overall viscosity increase upon cooling from the melting temperature, which was expected due to the nature of the measurement. The rate at which the viscosity increased with temperature varied with sample. Generally speaking, pure Celecoxib and Celecoxib/TPGS resulted in a small degree of undercooling compared to Celecoxib/PVP and Celecoxib/PVP/TPGS, this due to sample solidification during measurement. Pure Celecoxib resulted in an increase in viscosity of approximately one decade at roughly 10°C degrees of undercooling; at this point the viscosity suddenly increased dramatically almost three orders in magnitude. This sample could not be measured more than 30°C degrees of undercooling. Celecoxib/TPGS resulted in a similar behavior as pure Celecoxib, with a reduced degree of undercooling, as the sample showed a marked increase in viscosity near the melt temperature. Celecoxib/PVP resulted in a significant increase in the degree of undercooling (~95°C) as the dynamic viscosity could be measured up to the glass transition temperature of the sample with an increase of roughly one decade of viscosity every 10°C in temperature. Interestingly, almost 10°C before the glass transition, the viscosity increase rate with temperature decreased significantly. Near the glass transition temperature, this sample showed roughly a viscosity increment of seven orders of magnitude. Similar behavior was observed for Celecoxib/PVP/TPGS, having similar viscosity values as the Celecoxib/PVP sample near melt temperature, as the glass transition approached during cooling, the Celecoxib/PVP/TPGS sample showed a decrease of one decade in viscosity compared to Celecoxib/PVP. Celecoxib/PVP/TPGS viscosity increase rate also was reduced near the glass transition. Analyzing the viscosity data in terms of diffusivity indicates that the molecular mobility of the Celecoxib is impeded by the presence of the PVP, this occurs by the hindering of local arrangements of the Celecoxib molecules within a crystal lattice (crystallization inhibition). This was evidenced by the significant increase in viscosity and degree of undercooling compared to pure Celecoxib. This induced hindering of Celecoxib molecules allowed for viscosity measurements near the glass transition temperature. On the other hand, the TPGS acted as a plasticizer where miscibility between TPGS and Celecoxib aided in the molecular mobility of Celecoxib molecules assisting therefore the crystallization near the melting temperature. The combination of Celecoxib with PVP and TPGS resulted in interesting results, generally a decrease of molecular mobility similar to the effect of PVP alone, however the plasticizing effect of TPGS was able to increase the mobility, hence lower viscosity values near the glass transition compared to the effect of PVP. The viscosity curves showed mostly linear dependences with regard to temperature, however some deviation from linearity were observed most significantly for the samples containing PVP and PVP/TPGS. Fitting the viscosity curves with the Arrhenius and VTF equations evidenced these dependence. The hindering of local arrangements of the Celecoxib molecules (crystallization inhibition) by PVP was evidenced by the increase in activation energy of Celecoxib/PVP (115.96kJ/mol) compared to pure Celecoxib (92.28 kJ/mol) near the melt temperatures, whereas TPGS decreased the activation energy necessary for crystallization to occur (60.52kJ/mol), evidencing thus the plasticizing effect of TPGS. Interestingly, the combination of PVP and TPGS was not capable of reducing the activation energy (98.89kJ/mol), as significantly as PVP alone complementing the similar viscosity values near the melt temperature. Deviations from the Arrhenius equation were observed as the degree of undercooling progressed (10°C-80°C), evidenced by a decrease in the R^2 value of up to 12% for the Celecoxib/PVP/TPGS sample, compared with R^2 = 10% of Celecoxib/PVP sample. These deviations could be successfully corrected with the VTF model, increasing the R^2 value to 99.9% for both samples within the same temperature range. A marked VTF-to-Arrhenius transition near the glass temperature was not observed for these two samples within the temperature range studied, however Arrhenius equation resulted in higher R^2values compared to the VTF model (10% difference between both). VTF fitting demonstrated that all samples had a calculated strength parameter (D) of less than 10, confirming the kinetic fragility (crystallinity) of the liquids. However, no correlation could be observed between the impact in strength parameter, and the crystallization tendency and molecular mobility. The impact on the crystallization of Celecoxib by PVP (polymer) and TPGS (surfactant) was studied in terms of molecular mobility. Over all it was demonstrated that the polymer decreases the mobility of the drug, whereas the surfactant acts as a plasticizer. PVP hinders the mobility of Celecoxib molecules inhibiting to some degree the crystallization of the drug, evidenced by the significant increase in measured degree of undercooling, dynamic viscosity and calculated activation energy. The surfactant acted as a plasticizer aiding the mobility of the drug allowing therefore the crystallization of the drug to occur at earlier, and lowers the activation energy necessary for crystallization to occur. The dynamic viscosity of the Celecoxib with polymer and/or surfactant follows Arrhenius temperature dependence near the melting point (10°C of undercooling), however the VTF model results in higher agreement as the undercooling degree increases.

413132 Process Development for a Co-Crystal of Telaprevir (VX-950) and 4-Hydroxybenzoic Acid

Abstract:

When development work was initiated, the existing process for generation of a co-crystal of telaprevir (VX-950) and 4-hydroxybenzoic acid (4-HBA) was uncontrolled precipitation from a solvent mixture. Anecdotally, there was the existence of a "magic" co-crystal batch that had superior dissolution performance and had been generated from the uncontrolled process. For a time, technical work was driven by this knowledge and there was an ill-fated attempt to reproduce this batch. However, when it was determined that the precipitation led to co-crystal with variable properties and performance, focus shifted to obtaining a more fundamental understanding of the crystallization. The objective for developing the process was then to design a controlled crystallization that would lead to co-crystal material with reproducible physical properties and dissolution behavior. There is a significant difference in solubility of the API and the coformer which led to difficulty in choosing an appropriate solvent system. This presentation will review the candidate solvent systems and modes of crystallization that were attempted. After choosing dichloromethane and methyl *tert*-butyl ether as the solvent system, an equilibrium phase diagram was generated for the solvent composition at the seed point. This phase diagram indicated that the process was at risk to produce co-crystal with excess VX-950. By slightly changing the parameters in the crystallization, it was found that only co-crystal would be formed. Further work involved optimization of the solvent composition at the end-point to improve yield and minimize agglomeration. Development and understanding of the phase diagram was essential to reliably producing VX-950/4-HBA co-crystal.

416357 Primary Nucleation Under High Shear Conditions

Abstract:

The influence of external energy on the kinetics of primary nucleation has been investigated under high shear conditions. It is shown in literature, that shear-induced molecular alignment and in particular agitation-enhanced cluster aggregation are mechanisms that appear to deserve further attention [1]. To study this phenomenon further, the induction time has been measured at different supersaturation, and levels of energy input, at various scales using magnetic stir bar and rotor-stator based IKA wet mills. Since there is a distribution of induction time possible, a statistical view is proposed for the value of induction time at a given condition [2]. There is an overall tendency in the experiments that the primary nucleation is promoted by increased input of shear imparted due to mixing. In small vials agitated by magnetic stir bars and in a larger scale reactor with high shear wet mill, the induction time was found to decrease with increasing wet mill tip speed to the power 0.66, the induction time is inversely related to the shear rate. By fitting the parameters of the classical nucleation theory to experimental data, it is shown that the results can be explained as an influence on the pre-exponential factor. Interfacial tension is estimated between API and solvent. An empirical equation is proposed to quantify the energy input from wet milling, and to predict the induction time.

References:

1. Liu, J.; Rasmuson, A.C. Cryst. Growth Des., 2013, 13 (10), pp 4385–4394
2. Kulkarni, K.A.; Kadam, S.S.; Meekes, H.; Stankiewicz, A.I.; ter Horst, J.H. Cryst. Growth Des., 2013, 13 (6), pp 2435–2440

AbbVie contributed to the design, research, and interpretation of data, writing, reviewing, and approving the publication. Huayu Li, Moiz Diwan and Fredrik Nordstrom are AbbVie employees at the time that they worked on this project.

417728 Microfluidic Spherical Crystallization and Co-Formulation of Drugs and Excipients – Controlling Morphology and Polymorphism

Abstract:

Microfluidic Spherical Crystallization and Co-formulation of drugs and excipients – controlling Morphology and Polymorphism

Eunice W. Q. Yeap, Ambika Somasundar, Denise Z. L. Ng, Yiji Wang, and Saif A. Khan*

Department of Chemical and Biomolecular Engineering,

National University of Singapore, SINGAPORE

There is considerable interest in the creation of "designer" pharmaceutical microparticles with controlled morphology and polymorphism, as drug product intermediates for facile formulations. Emulsion-based crystallization is an attractive route for controlling crystal size in downstream processing of active pharmaceutical intermediates (API), yet crystals formed in batch processes show limited control over key attributes like size and shape.[1] We circumvent these limitations, by coupling capillary microfluidics based emulsion generation, with evaporative crystallization for the production of monodispersed spherical API agglomerates. This platform has allowed us to produce exquisitely designed spherical agglomerates (of drug or drug mixtures), formed in uniform size, composition, morphology and polymorphism.[2-3] We study how precise control over key attributes of drug or drug-excipient microparticles can be achieved by a combination of process and formulation parameters.

In this work, we demonstrate the co-formulation of model hydrophobic API drugs – ROY and carbamazepine – with excipient, ethyl cellulose (EC). We generate monodisperse emulsion drops containing the API-excipient mixture, and subject these droplets to evaporative crystallization in thin films of continuous phase. We find that the co-formulation of API with EC not only shortens the API crystallization process, but also allows for the direct creation of highly monodisperse spherical microparticles containing API crystals and macromolecular excipient, with nearly complete and tunable control over both morphology and polymorphic outcome. Specifically, we control polymorphic selection of both ROY and CBZ in the API-excipient microparticles, by implementing different film thicknesses for the evaporative crystallization; the film thickness mediates the rate of solvent evaporation from the droplets, controlling the kinetics of nucleation and crystallization. We obtain two completely specific and different polymorphic outcomes for each drug when crystallized in the presence of EC under different film thickness. In summary, our methods pave the way for the synthesis and formulation of novel drug product intermediates that bridge the considerable gap between primary and secondary pharmaceutical manufacturing.

Word Count: 308

CONTACT

* S. A. Khan, Tel: +65-6516-5133; saifkhan@nus.edu.sg

[1] Chadwick, K.; Davey, R. J.; Mughal, R.; Marziano, I., Org. Process Res. Dev. 2009, 13, 1284 – 1290.

[2] Toldy, A. I.; Badruddoza, A. Z. M.; Zheng, L.; Hatton, T. A.; Gunawan, R.; Rajagopalan, R.; Khan, S. A., Cryst. Growth Des. 2012, 12, 3977 – 3982.

[3] Leon, R. A., Badruddoza, A. Z. M., Zheng, L., Yeap, E. W., Toldy, A. I., Wong, K. Y., Hatton, T. A. Khan, S. A., Cryst. Growth Des., 2015, 15 (1), pp 212–218

423568 Morphology Evolution of Quadrangular Dendrites of Erythritol

Abstract:

The change of the morphology of erythritol was investigated in this paper. Offline tools such as scanning electron microscope (SEM), XRD, DSC and optical microscope and online tools such as PVM and FBRM were used to test the effects of the factors including temperature, additives, solvents, cooling rate and so on. It was found that when the erythritol solution is added the same anti-solvent, its habit changes from dendritic structure to octahedron with the rising of the initial temperature, and the transition temperature changes with the kind of the anti-solvent. A possible explanation of this phenomenon has been proposed on the basis of experiments. The dendritic erythritol is the early stage forms of the octahedron. The morphology of erythritol experiences a transformation process of dendritic—snow-like with four corners—octahedron. The influences of other factors on this phenomenon have also been discussed in detail.

414966 Implementation of a Reduced System Model to a Continuous Direct Compression Manufacturing Process

Abstract:

The continuous direct compression manufacturing process consists of several loss-in-weight (LIW) feeders, a horizontal mixer, a surge hopper, and tablet press equipped with a near infrared (NIR) probe at the die-filling station of the tablet feed frame. A system-wide deterministic dynamic model was used to fit dispersion coefficients within the various elements for a given formulation/throughput using the feed frame NIR to measure the system response. This partial differential non-linear algebraic system model consisting of ~8,000 equations is sufficiently complex that it cannot be solved real-time with the current technology. This necessitates a second, reduced system model based upon the original deterministic model and uses thousands of digital experiments to create a set of response surfaces. The result is a data-driven version of the model that can be executed real-time using a PAT controller (e.g. SynTQ) to assess the impact of feeder variability on the final product within the distributed control system (e.g. DeltaV). The real-time reduced system model provides prediction of the feed frame concentration, the time when a disturbance will reach the die-filling station, and the time when the disturbance will clear the system.

427926 Flowsheet Modeling for Oral Solid Drug Product Manufacturing

Abstract:

QbD is a systematic approach to pharmaceutical development based on scientific knowledge and risk management [1]. Continuous manufacturing is a promising alternative and important for the implementation of QbD because it reduces the need of scale-up studies [2]; it requires smaller equipment size and thus minimizes the plant footprint [3]; and it decreases human factor using automated operation and thus decreases labor cost [4]. To facilitate the development of continuous pharmaceutical manufacturing, each unit operation must be well understood in terms of the effect of different material properties and operational conditions on the final product quality attributes. This goal can be facilitated by the utilization of computer-aided process systems engineering (PSE) approaches. Flowsheet modeling is an efficient way to understand manufacturing process dynamics. An integrated flowsheet model can be used in the design, analysis, and optimization of a chemical process [3]. Simulation results can help identify the possible process integration requirements, study multiple designs and test control strategies. In pharmaceutical process modeling, a computationally efficient equation-oriented process simulator (gPROMS) has been used to develop models for unit operations and build flowsheet models [3] [4].

Among different modeling methods, residence time distribution (RTD) theory has been commonly used in developing pharmaceutical unit operation models. RTD is defined as the probability distribution of the time that the particles stay inside one or more unit operations in a continuous flow system [5]. With RTD, process changes (e.g., material changes, process disturbances) can be traced throughout the system. This helps determine disturbance dissipation along the process and thus help decide when to collect final products with desired quality. Consequently mathematical models based on RTD theory are very useful for modeling the continuous pharmaceutical manufacturing process.

We focus on building validated integrated flowsheet model with RTD-based unit operation models for direct compression and granulation lines. Major unit operations such as feeders, comil, blenders, transfer pipes, feed frames, tablet presses, and granulation equipment were modeled and properly integrated. These flowsheet models are used to simulate and predict the dynamics of entire continuous direct compression processes given input such as material properties and operational conditions. Moreover, the behavior of different processes in response to disturbances can also be simulated. Flowsheet models can also be easily customized. Users can drag and drop necessary unit models and connect them based on their process design or to test different designs and different processing strategies. Graphic user interfaces (GUIs) have been implemented in our flowsheet simulations to provide users with an icon-based interface (i.e., no exposure to model code) and set parameters for each unit. Multiple process configurations can be easily set with GUIs, which makes it convenient to run different simulations.

The objective of this presentation is to highlight the use of flowsheet modeling tools in the pharmaceutical tablet manufacturing process and to demonstrate their applications for process design. Because the process models can have thousands of differential and algebraic equations (DAEs) and integral expressions, they can be difficult to solve with deterministic solvers. In these cases, surrogate modeling methodologies are adopted to address the challenges. However with the high-dimension problems, the application of surrogate modeling is limited. For example, Kriging method is computationally expensive for high dimensional problems because it needs to numerically solve a maximum likelihood estimation problem to fit the model [7]. In contrast, Radial Basis Functions (RBF) is a promising alternative because it can provide simple and efficient model to approximate the original model [8]. We focus on solving high-dimensional design problems by using RBF-based surrogate modeling strategies. We introduce flexibility analysis for the design problem to quantify the ability of a process to maintain feasible operation in the presence of inherent variability or external disturbances [6]. The final goal is to achieve a more robust design for the continuous pharmaceutical manufacturing process.

References

[1] Lionberger, Robert A., Sau Lawrence Lee, LaiMing Lee, Andre Raw, and X. Yu Lawrence. "Quality by design: concepts for ANDAs." The AAPS journal 10, no. 2 (2008): 268-276.

[2] Leuenberger, Hans. "New trends in the production of pharmaceutical granules: batch versus continuous processing." European journal of pharmaceutics and biopharmaceutics 52, no. 3 (2001): 289-296.

[3] Boukouvala F, Niotis V, Ramachandran R, et al. An integrated approach for dynamic flowsheet modeling and sensitivity analysis of a continuous tablet manufacturing process[J]. Computers & Chemical Engineering, 2012, 42: 30-47.

[4] Rogers A J, Inamdar C, Ierapetritou M G. An integrated approach to simulation of pharmaceutical processes for solid drug manufacture[J]. Industrial & Engineering Chemistry Research, 2013, 53(13): 5128-5147.

[5] Gao Y, Muzzio F J, Ierapetritou M G. A review of the residence time distribution (RTD) applications in solid unit operations[J]. Powder Technology, 2012, 228: 416-423.

[6] Biegler L T, Grossmann I E, Westerberg A W. Systematic methods for chemical process design[J]. 1997.

[7] Regis R G. An initialization strategy for high-dimensional surrogate-based expensive black-box optimization[M]//Modeling and Optimization: Theory and Applications. Springer New York, 2013: 51-85.

[8] Shan S, Wang G G. Metamodeling for high dimensional simulation-based design problems[J]. Journal of Mechanical Design, 2010, 132(5): 051009.

427688 A Multi-Fidelity Mechanistic Modeling Framework for Wet Granulation Processes: Characterization and Technology Transfer

Abstract:

Although wet granulation processes are widely used in a range of industries, they are typically designed empirically and operated inefficiently. In order to improve these processes, predictive modeling tools are desired for process design, validation, and optimization. Recent advances have expanded the mechanistic understanding of these processes, enabling quantitative descriptions to be formed using population balance modeling techniques.

Although they share a common name, wet granulation processes in practice are operated using a range of configurations, resulting in different process outcomes. Further, various mechanisms drive wet granulation processes, including wetting, drying, particle flux, agglomeration, drop nucleation, breakage, consolidation, and layering. The dominant phenomena are often governed by equipment configuration and process conditions, and different types of granulators exhibit different behavior.

Consequently, there is no single population balance model for wet granulation processes. Instead, custom models must be configured to meet the needs of the application. Some considerations in model development include population balance dimensionality and grid resolution, representation of particle composition, use of compartmental models.

Oversimplified models are unable to capture realistic behavior and provide little additional insight into process outcomes, only representing idealized cases. In contrast, overly complex models are inefficient, and numerical challenges can mask the true results. These high-fidelity models can often be simplified by identifying and eliminating irrelevant details. Despite the broad range of wet granulation models developed in academia and industry, a unifying framework is needed to ensure consistency, establish best practices and facilitate technology and knowledge transfer.

In this study, various alternatives for modeling wet granulation are discussed. These configurations are unified in a flexible modeling framework that encompasses varying levels of fidelity. Through this framework, population balance dimensionality, compartmental configuration, and governing rate mechanisms can be customized to simulate a variety of wet granulation processes. Case studies of twin screw and fluid bed granulation processes are presented to demonstrate the implications of model alternatives and establish best practices for model-based design and operation of efficient and robust wet granulation processes.

425191 Prediction of Lubrication-Based Tensile Strength Reduction Using a Compartment Modeling Approach

Abstract:

In the production of pharmaceutical tablets and capsules, a lubricant such as magnesium stearate is frequently added to the powder blend or granulation to reduce friction between the powder and the tablet press or encapsulator components. While the presence of the lubricant is generally necessary to improve manufacturability of the dosage form, if the lubricated blend is exposed to excessive shear strain during processing and handling prior to tableting or encapsulation, adverse effects on quality attributes of the final dosage form may be observed. These effects can include an increase in wetting contact angle, a slowdown in disintegration and/or dissolution, and a reduction in tensile strength.

In this work, the extent of lubrication in a powder feeding process is studied experimentally and with a compartment model approach. Since a placebo blend is used to study this phenomenon, the tensile strength is the quality attribute used to quantify the extent of lubrication. Here, a compartment modeling approach is proposed to model the powder flow pattern and predict the lubrication-based tensile strength reduction that may occur during the production of a large batch. Parameter estimation is conducted through the use of a separate experiment utilizing an input step change in powder feed from undyed to dyed material. Model predictions are compared with experimental results as well as predictions from a separate discrete element method (DEM) modeling approach. Finally, additional predictions from the compartment model are made to assess the impact of various processing modifications.

15AIChE
Annual Meeting, Salt Lake City, UT

Finite Element Simulations on Strength and Impact of Pharmaceutical Tablets

Jie Ren, Jerry Klinzing, Jennifer Cislo, Adam Procopio

Merck & Co.

MERCK
Be well

Public

Tablet Chipping

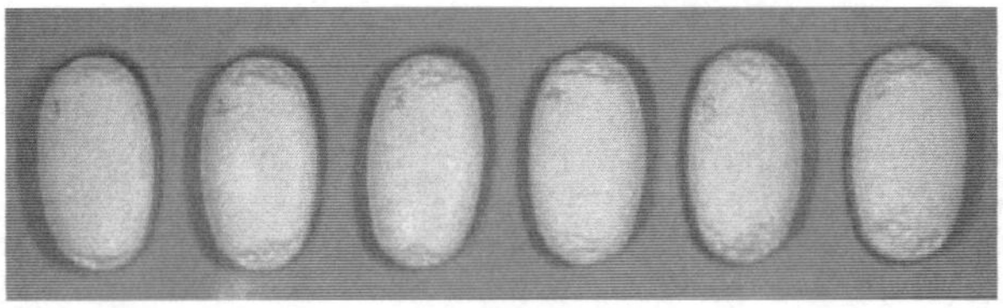

- Tablet chipping: among the most common modes of failure during solid oral drug development.
- Usually occurs during the post-processing steps of tablet manufacture – film-coating, packaging, transportation, patient handling
- In order to prevent/mitigate chipping, compressed tablets need to meet hardness and friability criteria.
- Hardness testers and friability testers

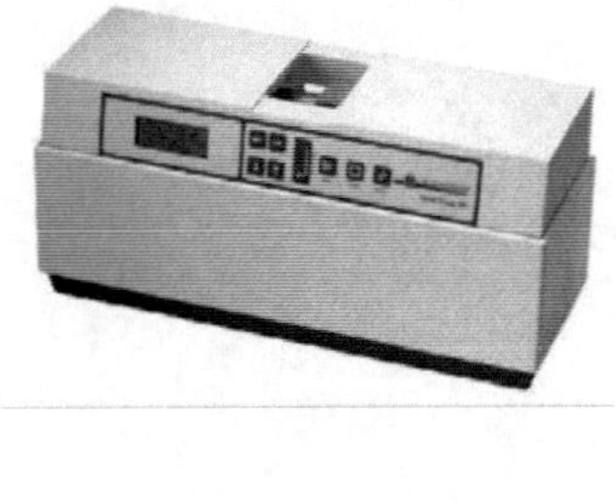

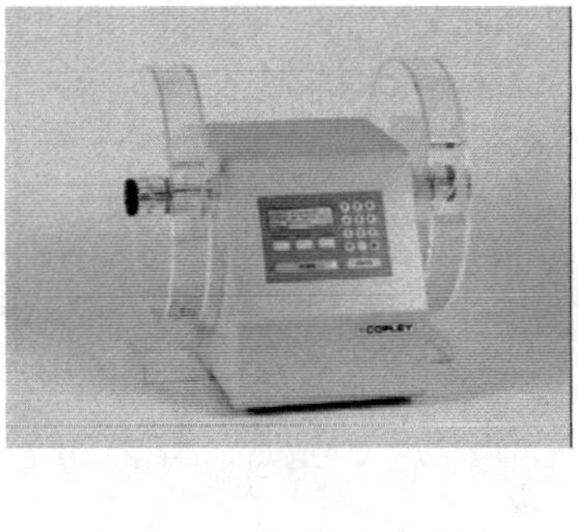

MERCK
Be well

Public

Limitation of hardness & friability testing

- Hardness and friability values can be hard to interpret
  - Hardness can be different for different shapes, even with the same material property.
  - Friability reveals weight loss after repetitive drops; not a direct reflection of elegance.
  - Neither hardness nor friability testing represent the real failure mode in a film-coating drum.

Hardness test

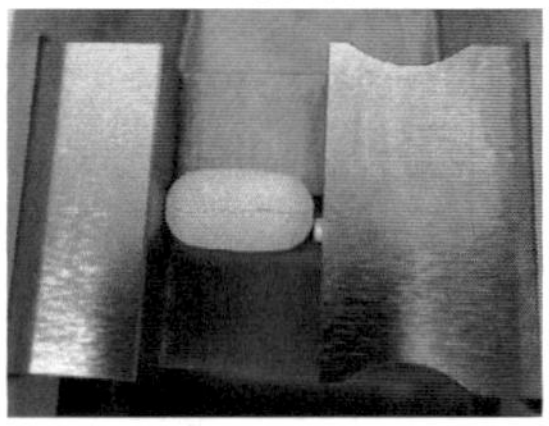

Tablet geometry

Break force (hardness)

Tablet material property

→ ? → Will the tablet break in friability test, and then, will it break during film coating?

Friability test

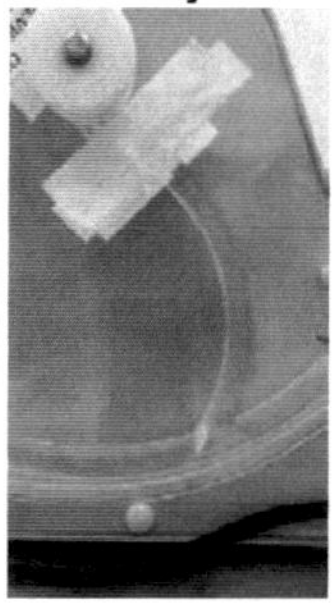

Ground-up understanding needed on the strength and impact response of tablets!

MERCK *Be well*

Public

Tablet Hardness vs. Tensile Strength?

Which shape is ideal?

Shape	Kp Value
	67
	44.8
	47.5
	41
	41

Compute Tensile Strength! →

Pitt's equation:

Round Flat Tablets

$$T.S. = \frac{2P}{\pi D t}$$

Round Concave Tablets

$$T.S. = \frac{10P}{\pi D^2}\left(2.84\frac{t}{D} - .126\frac{t}{W} + 3.15\frac{W}{D} + .01\right)^{-1}$$

Other shapes?!

No analytical solution!

FEM Simulation

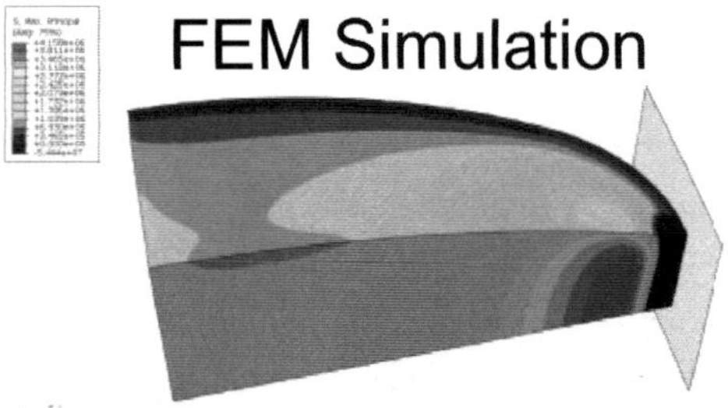

MERCK *Be well*

Public

Calculate tensile strength: diametrical test

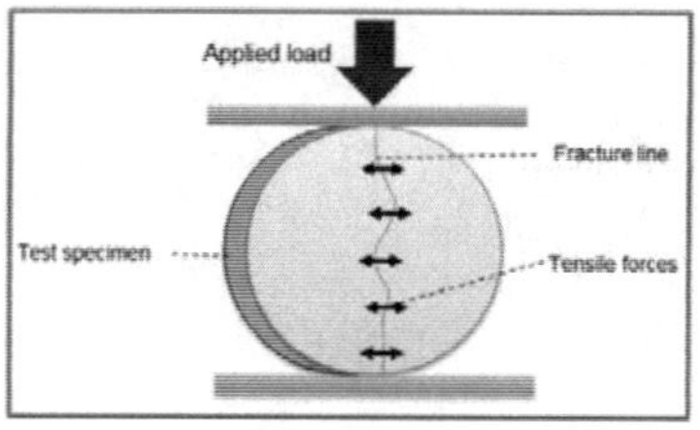

- Simulate 1/4 of a tablet based on symmetry
- Rigid plate moves towards the tablet at constant speed

Stress distribution along Z direction (transverse to compression direction)
This is the tensile stress that pulls the tablet apart

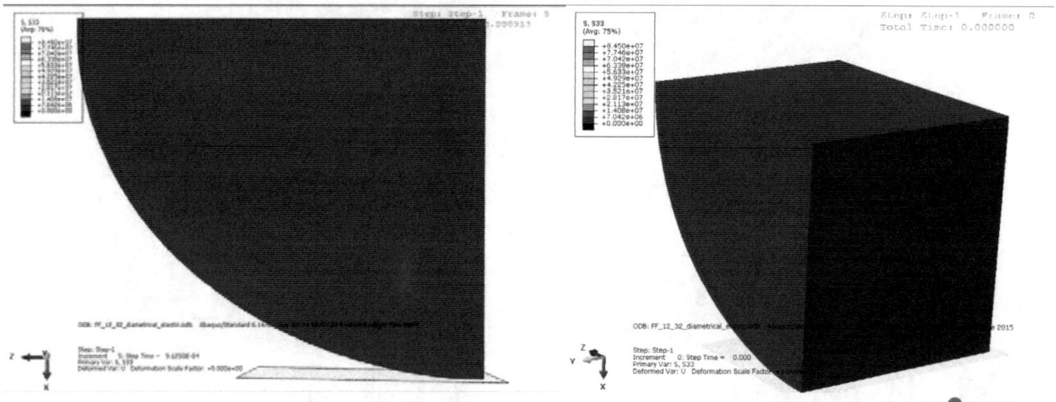

Public

MERCK
Be well

Calculate tensile strength: diametrical test

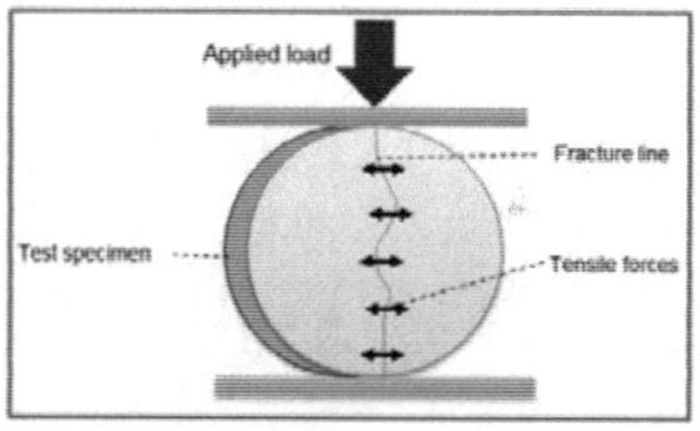

- Simulate 1/4 of a tablet based on symmetry
- Rigid plate moves towards the tablet at constant speed

Stress distribution along Z direction (transverse to compression direction)
This is the tensile stress that pulls the tablet apart

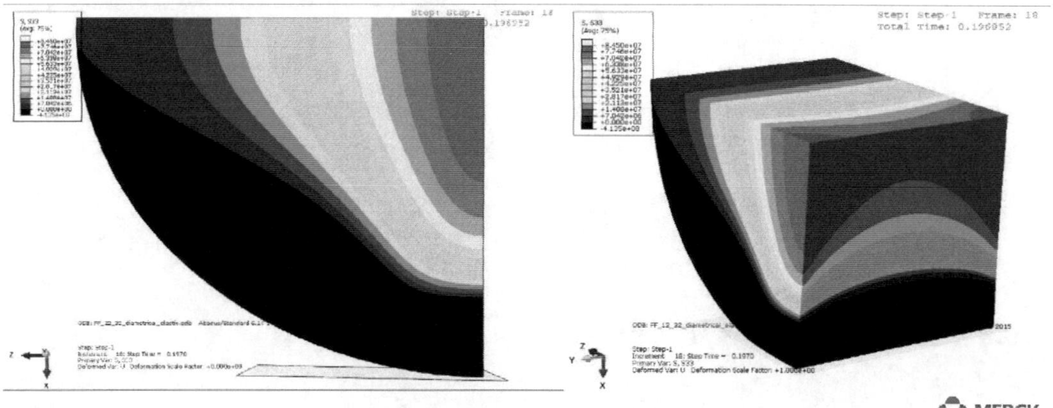

Public

MERCK
Be well

Diametrical test: Material Property Matters

- Compare Linear Elastic and Elastoplastic materials
  - Linear Elastic (typical for brittle materials: Lactose, Dicalcium Phosphate)
    - ρ = 1.2g/cc, E = 3.65GPa, v = 0.25; frictionless contact with the plate
  - Elastoplastic (typical for plastic materials: Microcrystalline Cellulose)
    - All the above properties, plus yield strength σ_y = 36.5MPa.

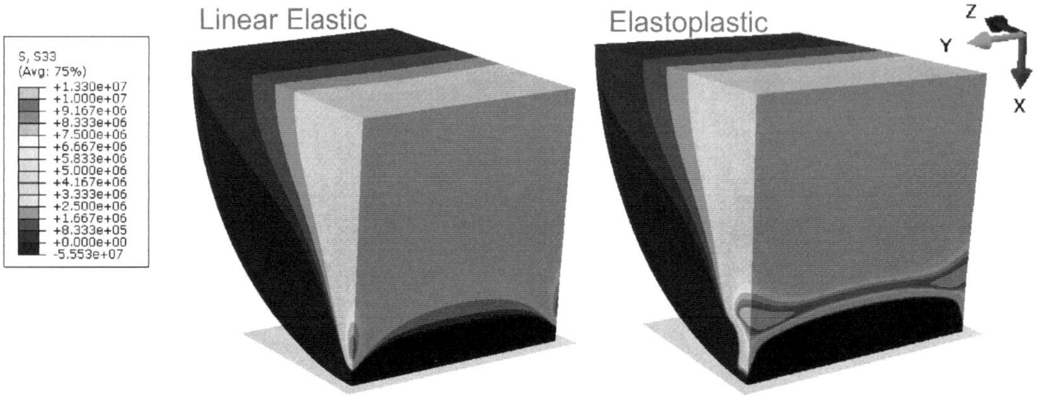

Principal stress along Z direction (tensile stress)

Public

MERCK Be well

Diametrical test: Material Property Matters

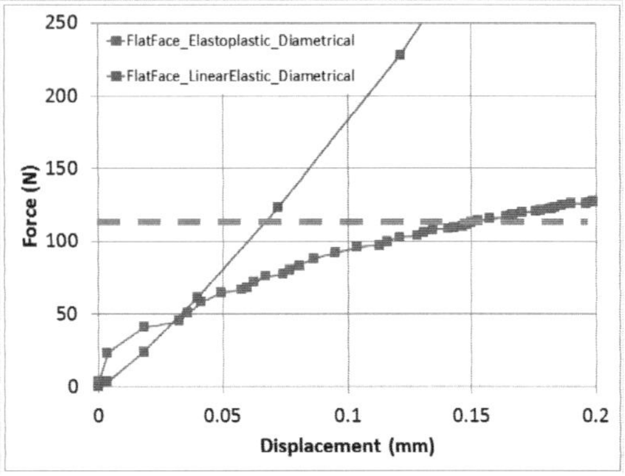

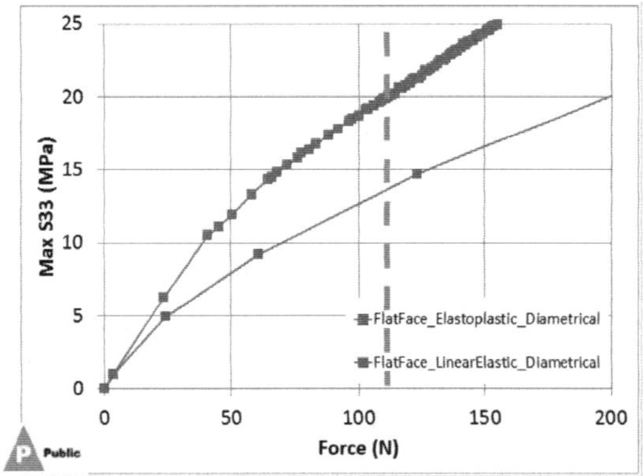

Big difference between Linear Elastic and Elastoplastic materials

- Linear Elastic – typical for brittle material: Lactose, Dicalcium Phosphate
- Elastoplastic – typical for ductile material: Micro-crystalline Cellulose

Force vs. Displacement:
It takes an elastoplastic tablet much larger deformation to reach the same diametrical force

Max Tensile Stress vs. Force:
Same diametrical force generates higher maximum tensile stress inside an elastoplastic tablet

Recall Pitt's equation: Round Flat Tablets

$$\text{T. S.} = \frac{2P}{\pi D t}$$

Green dashed line: experimental break force of a typical 330mg tablet

Public

MERCK Be well

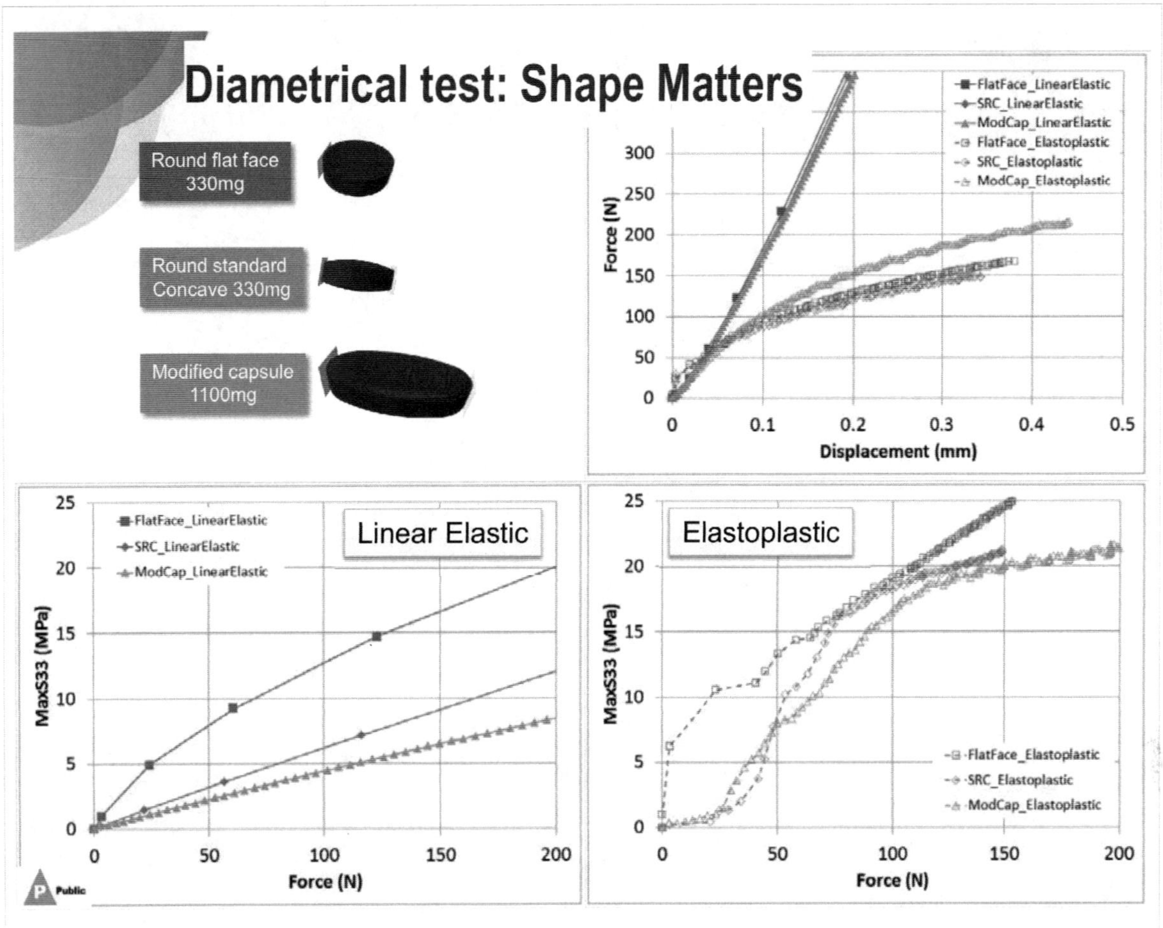
Diametrical test: Shape Matters
Round flat face 330mg
Round standard Concave 330mg
Modified capsule 1100mg
FlatFace_LinearElastic
SRC_LinearElastic
ModCap_LinearElastic
FlatFace_Elastoplastic
SRC_Elastoplastic
ModCap_Elastoplastic
Force (N)
Displacement (mm)
Linear Elastic
Elastoplastic
MaxS33 (MPa)
Force (N)
Public

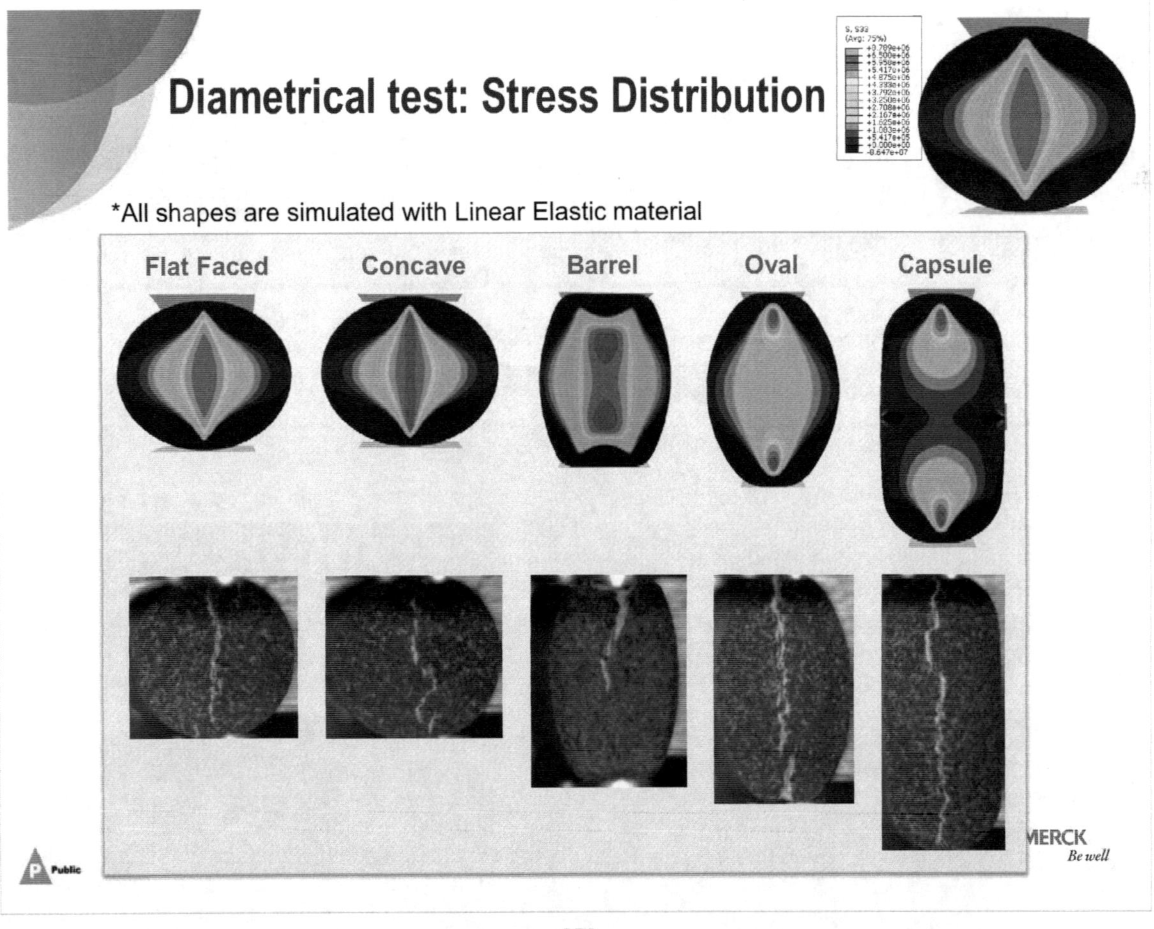
Diametrical test: Stress Distribution
*All shapes are simulated with Linear Elastic material
Flat Faced
Concave
Barrel
Oval
Capsule
MERCK
Be well
Public

Diametrical test: from Hardness to Tablet Strength

Tablet geometry

Material property (E, ν, σ_y)

→ FEM model → Break force (hardness) from experiment → Threshold tensile stress

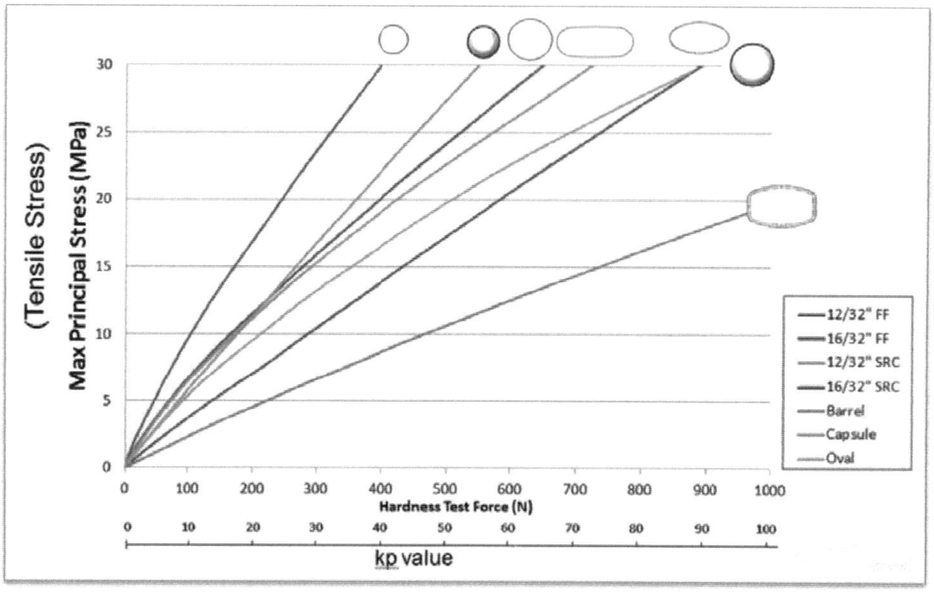

Public

Dynamic impact response?

Hardness / Tensile Strength → ? → Dynamic impact response

Hardness Testers or Instron machines: ~0.4mm/s ($\dot{\varepsilon}\sim10^{-2}$ s^{-1}) test speed

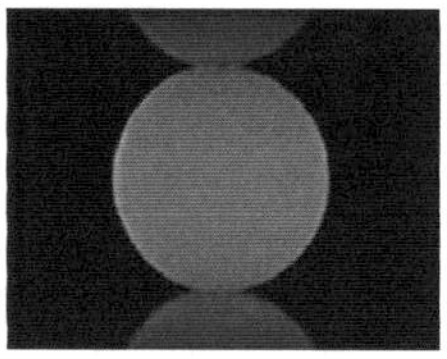

A tablet dropped in friability drum (156mm drop height): 1.75m/s ($\dot{\varepsilon}\sim10^{2}$ s^{-1})

A tablet dropped from 1m high: 4.4m/s

Drop test in FEM

We model a tablet impacting a hard surface

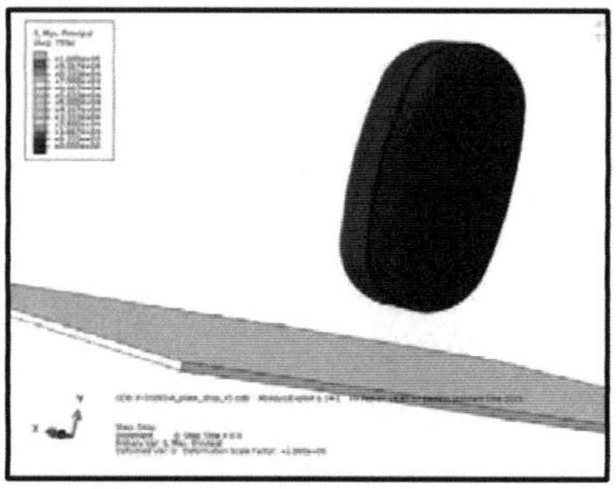

…with very fine spatial/temporal resolution:

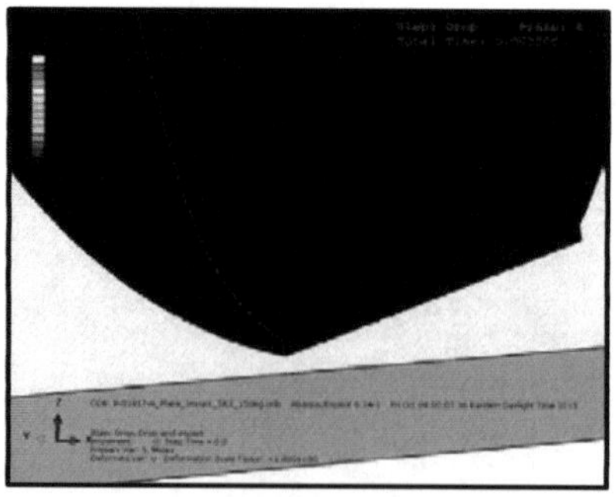

Public

Drop test: Critical Factors

Impact Angle

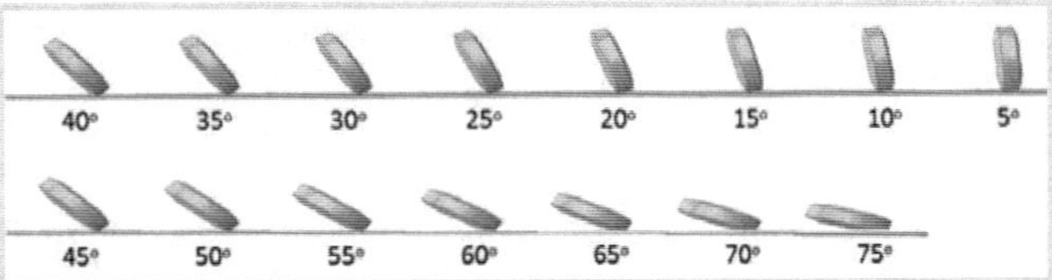

Material Property

- Linear Elastic
- Elastoplastic

Impact Speed

0.5m/s (0.013m drop height – friability drum) → 4.5m/s (4.5m drop height – film coating / patient handling)

Tablet Shape

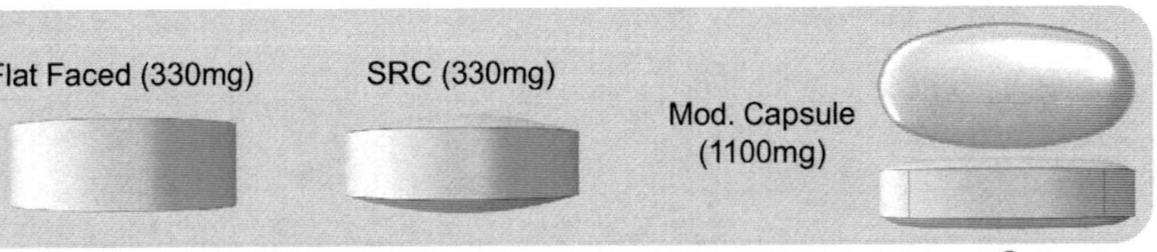

MERCK *Be well*

Public

Drop test: Impact Angle

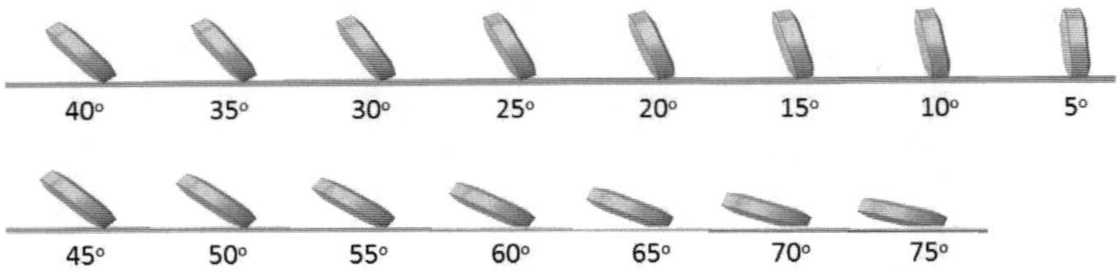

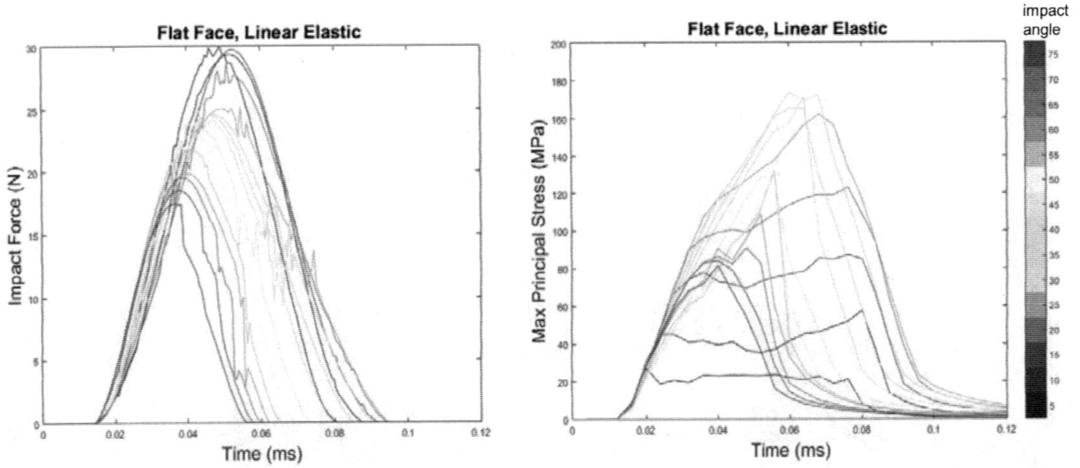

1.75m/s impact speed – typical for friabilily tester

MERCK *Be well*

Public

Drop test: Impact Angle

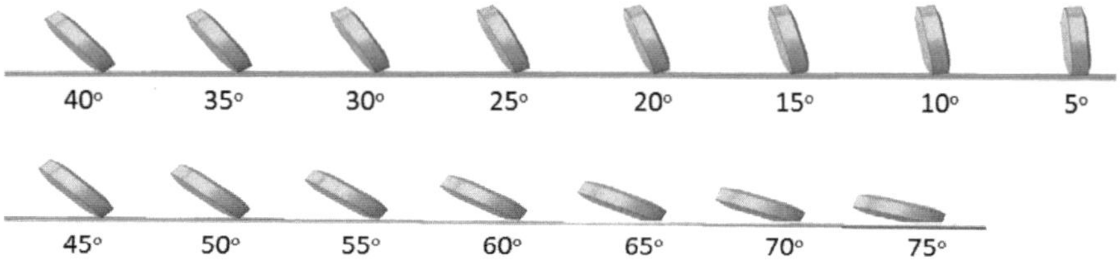

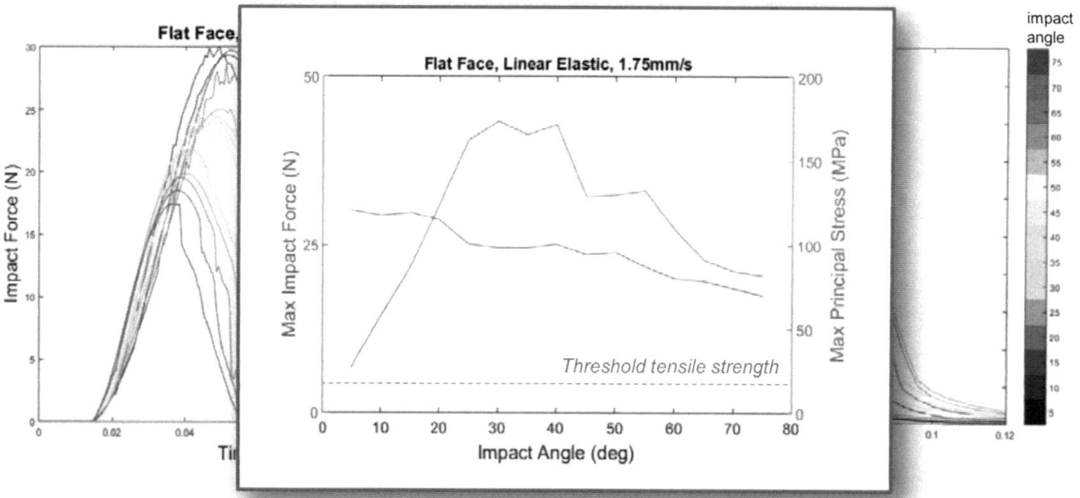

Public

MERCK Be well

Drop test: Material Property

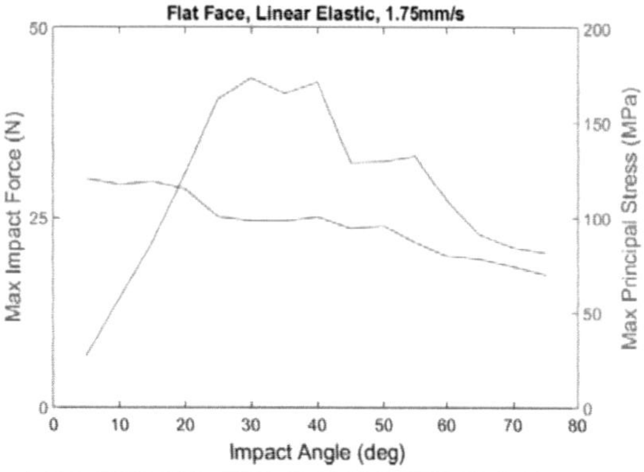

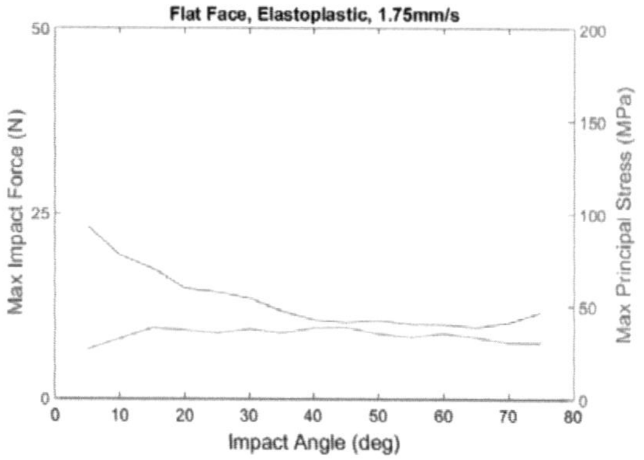

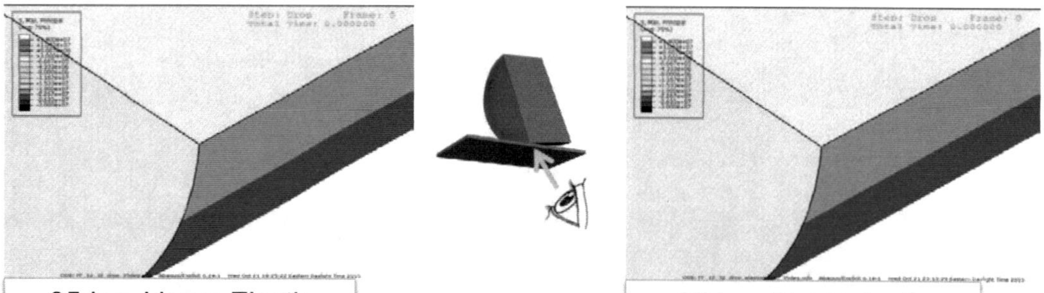

35deg, Linear Elastic

35deg, Elastoplastic

MERCK Be well

Public

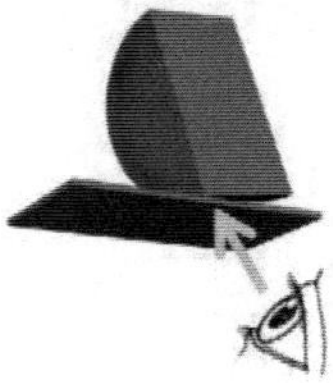

Drop test: Stress Distribution near Contact

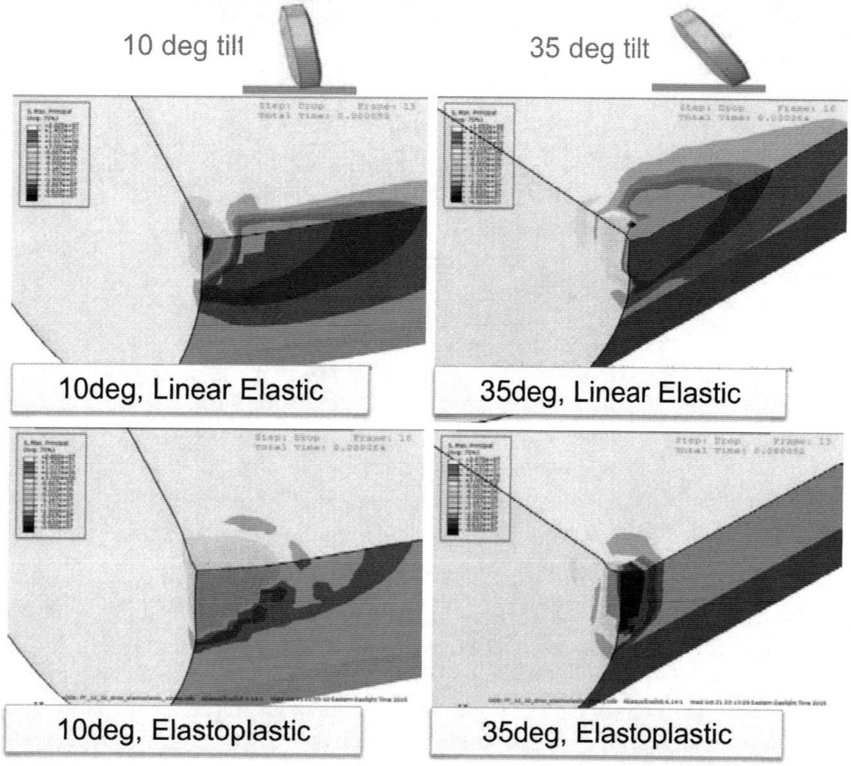

Linear Elsatic

Elastoplastic

Gray area: max. principal stress above 14MPa

Public

Drop test: Impact Speed

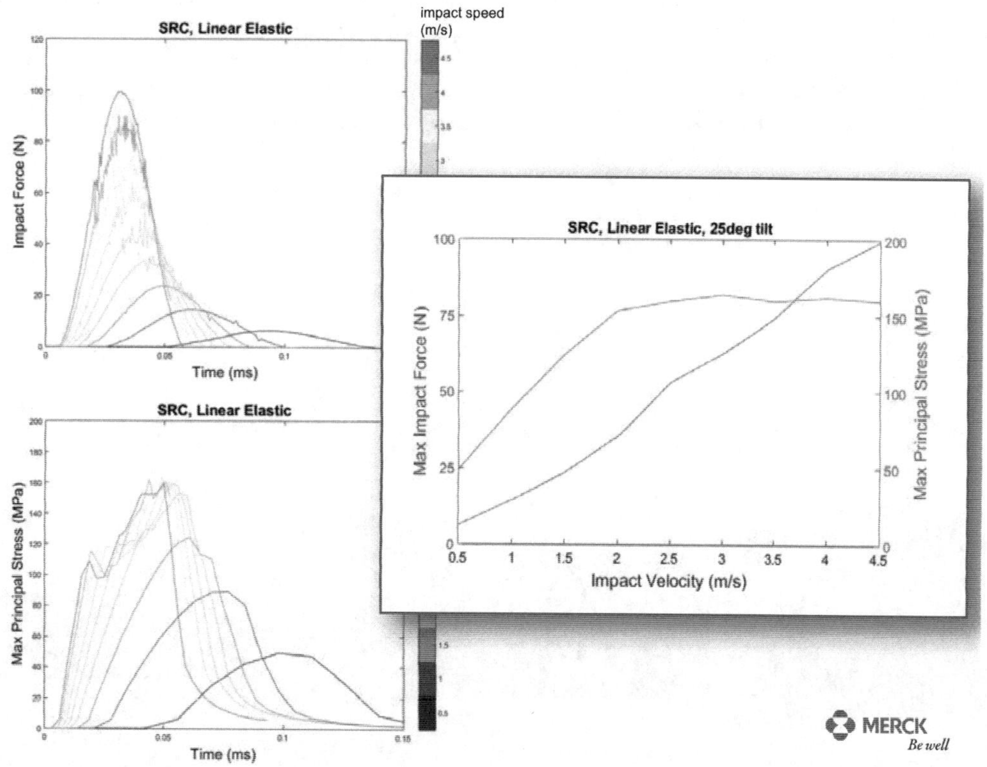

Public

MERCK
Be well

Drop test: Tablet Shape

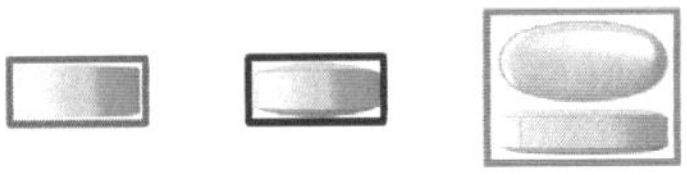

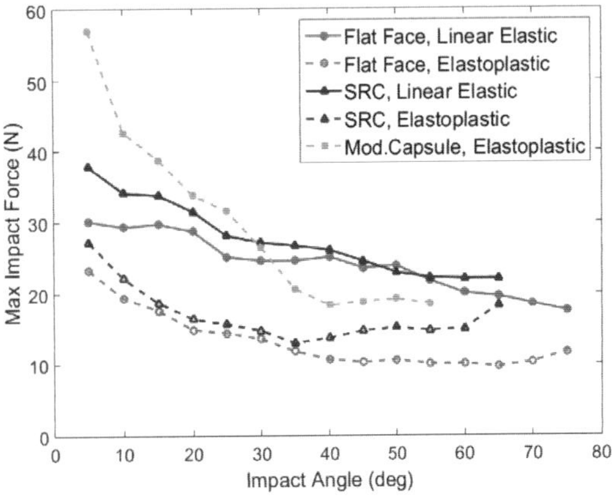

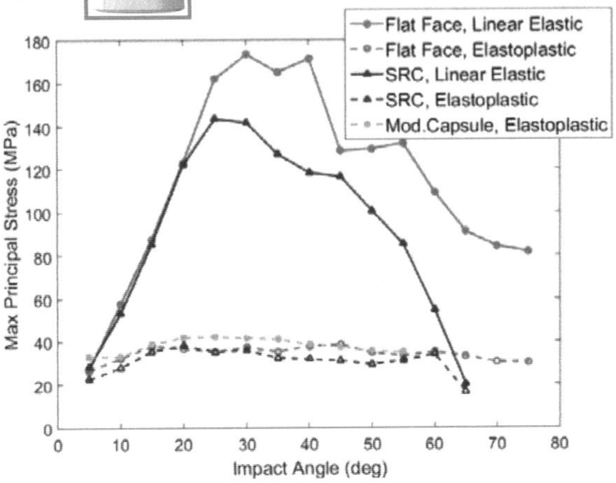

- Tablets with different shape => different angle dependencies of max impact force and max principal stress

Also:

- Linear Elastic material: higher max impact force; max Principal Stress peaks at 20-40deg
- Elastoplastic material: lower max impact force; max Principal Stress does not change with impact angle

MERCK Be well

Public

Conclusion

- We are now a step forward in predicting impact damage of pharmaceutical tablets
  - There are still liabilities: simplified material model, no fracture dynamics
- Next step: experimental confirmation of the impact modeling

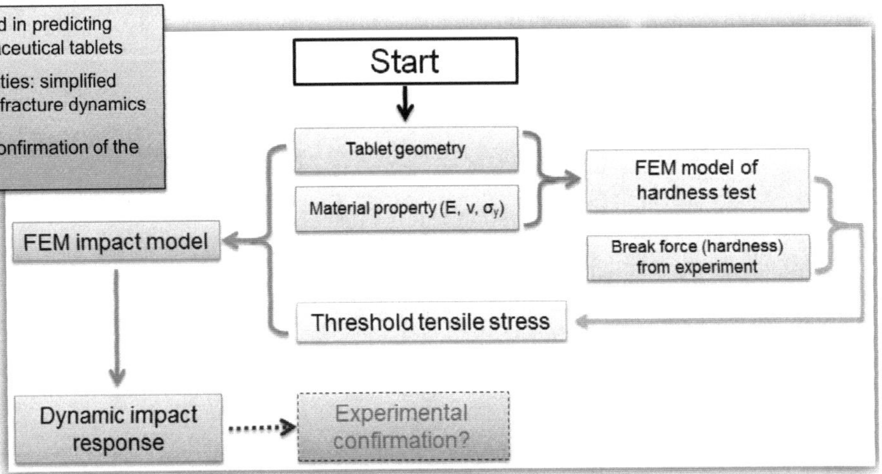

Hardness test

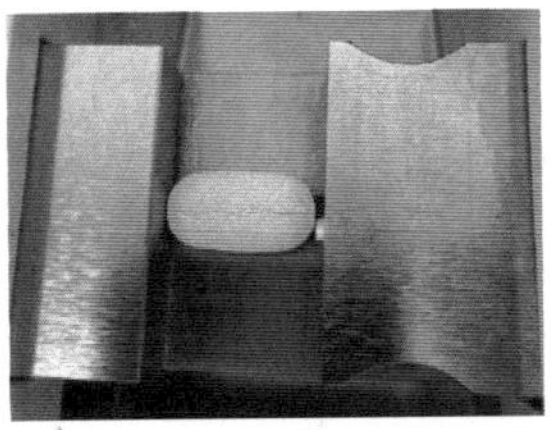

Impact test

Friability test

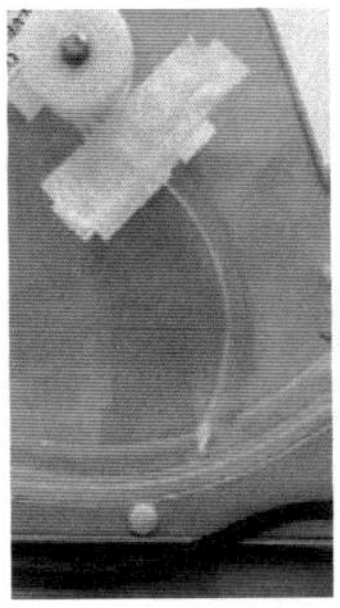

Chipped tablets

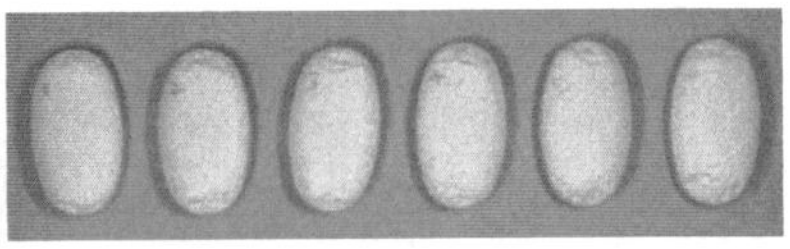

Public

MERCK Be well

Questions?

Public

MERCK Be well

Diametrical test: Shape Matters

Round flat face 330mg

Round standard Concave 330mg

Modified capsule 1100mg

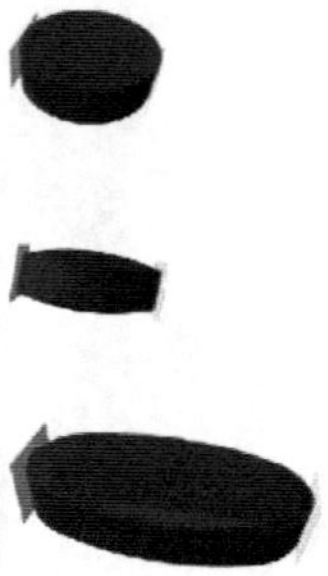

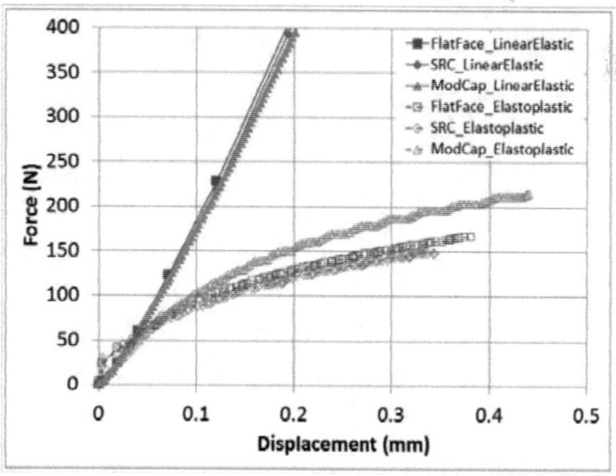

Public

MERCK Be well

Drop test: Impact Speed

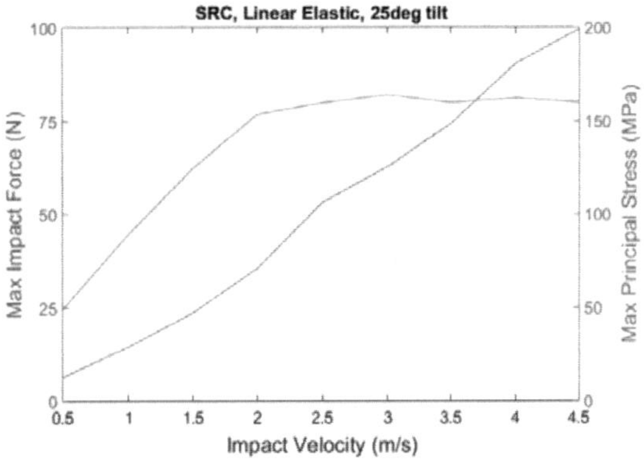

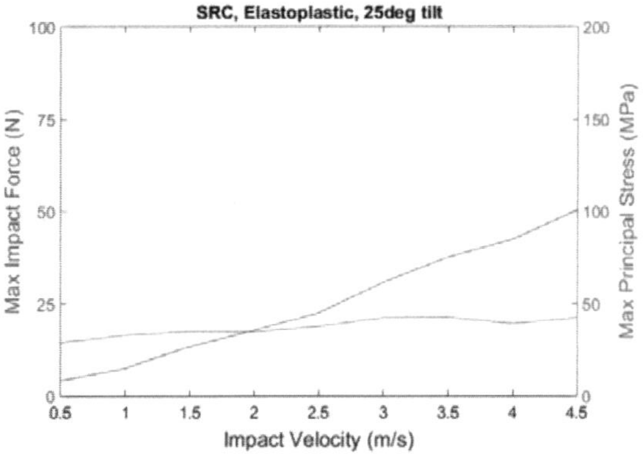

Public

MERCK
Be well

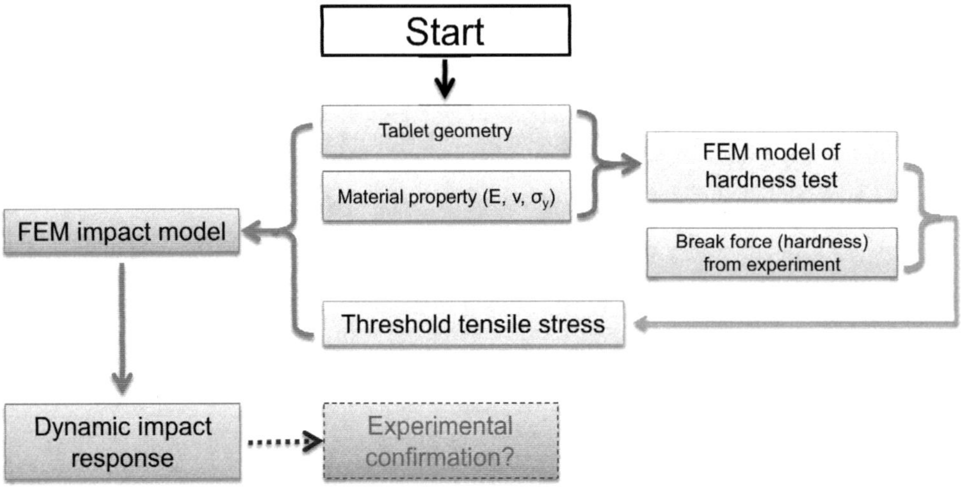

Public

MERCK
Be well

Impact Breakage

Tablet tensile strength or Hardness are currently measured by Hardness Testers or Instron machines

...with speed ~0.4mm/s ($\dot{\varepsilon}\sim10^{-2}$ s^{-1})

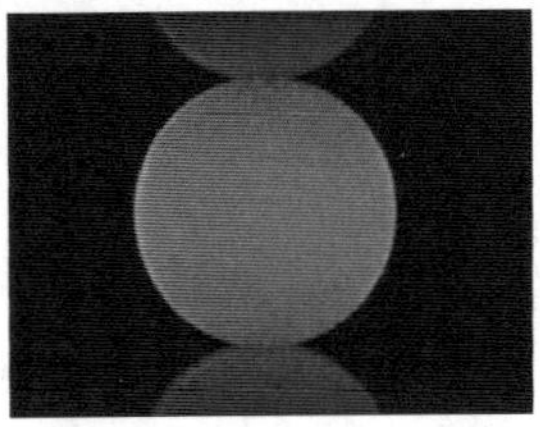

A tablet dropped in friability drum (150mm drop height): 1.7m/s ($\dot{\varepsilon}\sim10^{2}$ s^{-1})

A tablet dropped from 1m high: 4.4m/s

Tablet tensile strength or Hardness are currently measured by Hardness Testers or Instron machines

...with speed ~0.4mm/s ($\dot{\varepsilon}\sim10^{-2}$ s^{-1})

Drop test in FEM

We model a tablet impacting a hard surface

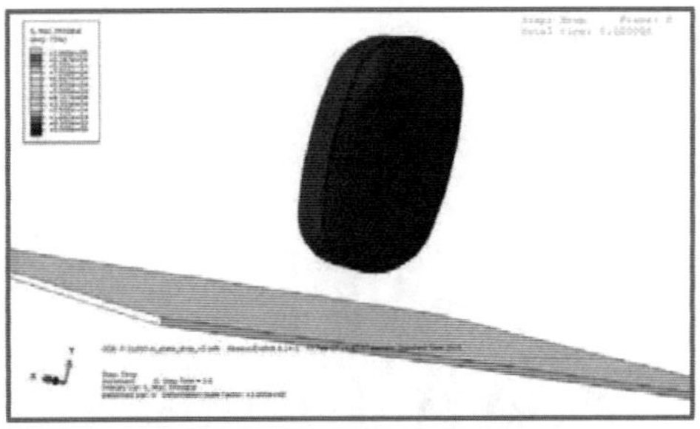

...with very fine spatial/temporal resolution:

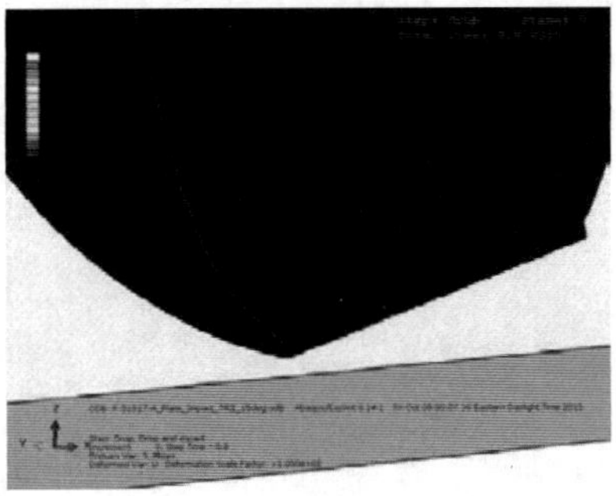

MERCK
Be well

Public

Drop test: Material Property

Linear Elastic

Elasto-plastic

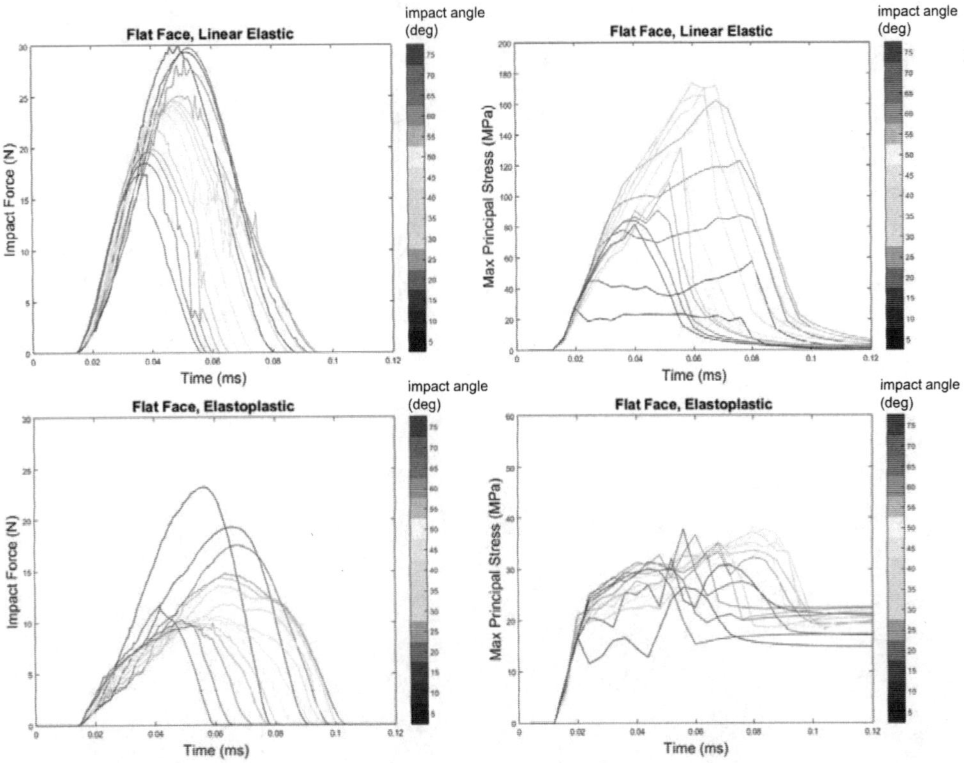

Public

Drop test: Impact Speed

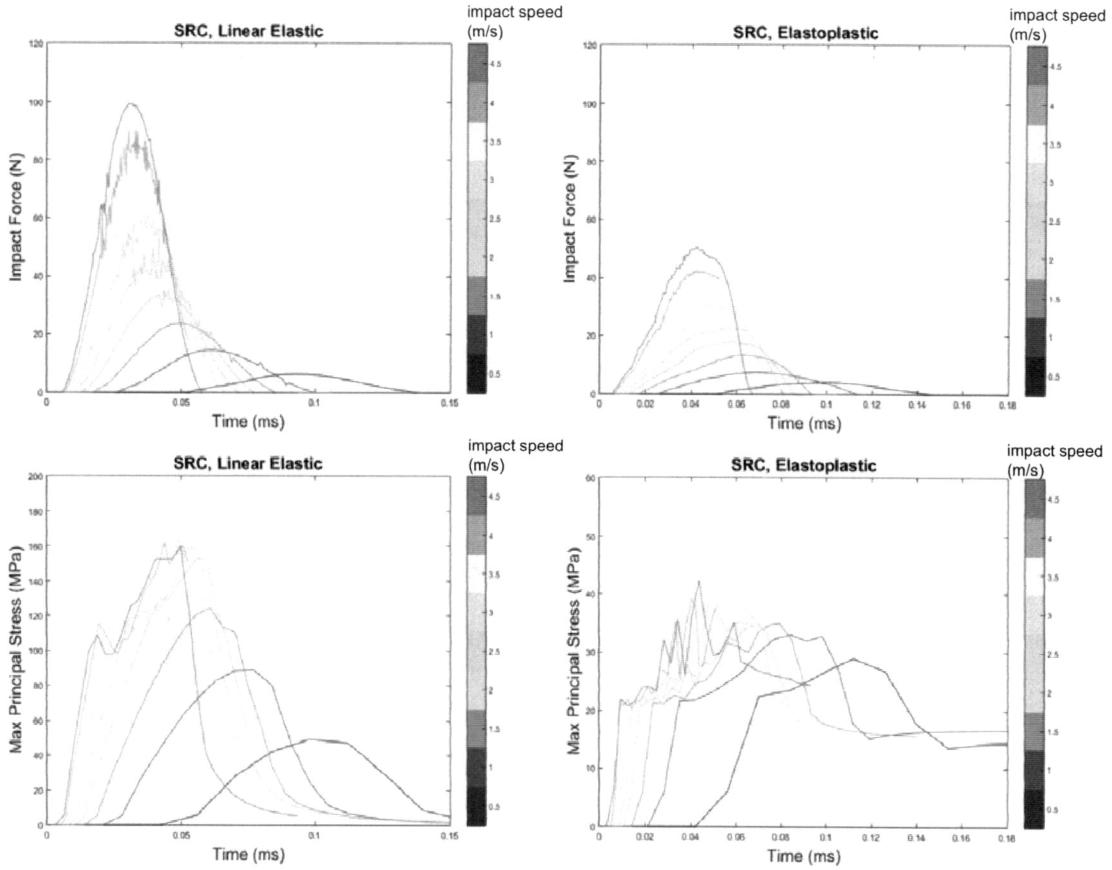

Public

425643 Dynamic Modeling and Advanced Control of Tablet Press

Abstract:

The tablet press is one of the most critical but least understood unit operation involved in both batch and continuous tablet manufacturing process. Various operations that take place simultaneously in a tablet press, including flow in a chute or a hopper, feeding of powder blend/granules into dies via feed frame, and tablet compression, are characterized by complex phenomena that are difficult to model. Moreover, the control system of existing tablet presses has not changed much in several decades and needs to be improved significantly. For example, with current state of the art techniques and tools, the tablet weight and the tablet hardness can not be controlled simultaneously in real time and raw materials and process variability can not be mitigated proactively. Powder raw material characteristics, feeder level, shear history, upstream milling and blending operation, and powder transport can all have significant impact on powder bulk density, flowability, and lubricity and thereby on critical tablet quality attributes. Therefore, dynamic modeling to understand tablet press performance, and an advanced control system is highly desirable. The Rutgers continuous pharmaceutical manufacturing platform provides an opportunity to improve the tablet press control system by allowing integration of meaningful real time measurements with feed-forward and feedback control systems.

In this work, dynamic models of tablet press, including a detailed feed frame model as well as a compression model, are developed and applied for the design of an advanced control system. The feed frame operation can have significant impact on powder mixing and die filling and thereby on critical quality attributes such as tablet weight, hardness, dissolution and API potency. One of the objectives of feed frame modeling was therefore to provide a predictive tool to understand the complex mixing and segregation phenomena occurring at this stage and to understand tablet die feeding behavior. The integrated tablet compression model will then allow studying the effects of feed frame speed and raw material properties on critical quality attributes. The tablet press model is integrated with the direct compaction line to study the effects of feeding, milling and blending effects on critical quality attributes. The developed mechanistic dynamic model has been employed to design an advanced tablet press control system and to evaluate its performance prior to implementation to the pilot-plant situated at Rutgers University. First, the feedback tablet weight and hardness control loops are decoupled so that these variables can be controlled independently and simultaneously. Then a feed-forward control loop is integrated with the feedback control loops to take the proactive mitigation actions based on powder bulk density variability.

The objective of this presentation is to demonstrate the dynamic model of tablet press, and to illustrate its applicability for design and evaluation of advanced tablet press control system.

419773 Data Driven Predictive Modeling of Formulation and Process Conditions of Pharmaceutical Spray Dried Intermediates

Abstract:

Data driven predictive modeling of formulation and process conditions of pharmaceutical spray dried intermediates

Elise Miller*, Gregory M. Troup, Justin Moser

Merck & Co., Inc., Formulation Sciences, West Point, PA

* E-mail: elise.miller@merck.com, gregory.troup@merck.com, justin.moser@merck.com

The pharmaceutical industry has experienced a shift in the physicochemical properties of pre-clinical drug candidates rendering them less soluble and bioavailable. Spray drying is used to enhance solubility and oral bioavailability by stabilizing the amorphous phase of a drug in a polymer matrix. However, the interaction of formulation and process variables is extremely complex. To fully simulate the process from atomization to final particle attributes based on first principles would require multiple computationally heavy models that still do not adequately capture formulation aspects.

In this work, a series of data driven models is developed and operationalized to relate formulation and process inputs to final spray dried particle properties. A user-friendly simulation tool will be showcased that enables *a priori* prediction of spray dried intermediate (SDI) particle properties as a function of formulation and process parameters. It combines a set of common formulation and process inputs, thermodynamic calculations of outlet process conditions, and SDI particle attributes in partial least squares projection to latent structure (PLS) models. The tool enables simultaneous evaluation of multiple hypothetical formulations and process conditions on the predicted physical attributes of particle size, bulk density, and tapped density. The multi-Y PLS models utilize data from 54 historical spray drying batches. The variables include SDI composition, solution composition, process inputs, process dependent variables, and SDI particle attributes. This exploratory analysis shows that the complexities of the spray drying process are readily captured and that the multivariate model results are consistent with process fundamentals. Then, separate optimized PLS models are developed for the key quality attributes: particle size, bulk density, and tapped density. The resulting PLS models have R2 values greater than 0.8, and Q2 values greater than 0.6, with root mean square estimation errors that are between 8% and 12% of the range. The optimized PLS models allow for prediction of these particle attributes from a limited set of formulation and process inputs, allowing proposed Design of Experiments to be simulated to ensure that a meaningful range of particle attributes can be achieved. Further, the user interface incorporates multivariate diagnostics to access the similarity between the proposed hypothetical batches and the model calibration data. The diagnostics also highlight input parameters that are outside of historical operating ranges. The tool has added utility as a training platform for scientists new to the technology and requires no modeling expertise.

427565 Development of a Reactive Slurry-Slurry Approach to Improve Crystallization Process Performance and Scalability

Abstract:

The crystallization of a potent compound, hereafter refered to as compound A for the sake of confidentiality and currently in development at GlaxoSmithKline proved to be challenging in terms of obtaining acceptable yield, consistent and desirable API powder properties, and a good process robustness and scalability. In order to address these issues, the salt formation crystallization was redesigned and a relatively novel approach was developed that allowed significant improvement in yield (from ≈70% to ≈95%), consistent API powder attributes within the desired range and an improved process manufacturability. This presentation describes the approaches and the process design of the crystallization recipe which utilizes a seed bed of the base as mitigating factor to prevent spikes in supersaturation with respect to the salt. The new process design resulted in a relatively novel crystallization path, which is a direct result of the solubility behaviors of both the base and the salt. The process was scaled-up successfully in the pilot plant to manufacture material for clinical supply.

427844 Effect of Seed Type on the Resistance of API Agglomerates to Agitated Drying

Abstract:

Control of the Active Pharmaceutical Ingredient's (API) powder properties is often necessary to ensure successful drug product development and manufacture. This is, in part, accomplished by designing a crystallization process that can deliver consistent batch-to-batch powder properties. When an API forms agglomerates during the crystallization, the resulting powder often exhibits enhanced flow properties and bulk density, which are of benefit to the formulation. However, development of a crystallization to consistently produce agglomerates that can withstand downstream processing, particularly agitated drying, represents a significant challenge. Weak agglomerates that are prone to break during agitated drying increase the risk of batch-to-batch powder property variability and scale dependence.

This presentation describes the development of a crystallization process that consistently yields API agglomerates which are robust to agitated drying. During development, the impact of several factors on agglomerate formation and strength was evaluated, and a mechanism of agglomeration was proposed. These studies showed that seed type has a strong impact on agglomerate properties and, subsequently, on the extent of agglomerate breakage during drying. Seeds with varying properties were generated via wet milling, jet milling, and hammer milling, and evaluated in the crystallization on laboratory and pilot scale. Filter, conical, and helical dryers were used on pilot and manufacturing scale to investigate the propensity of agglomerates made from the different seed types to withstand agitation. Results from the studies enabled definition of a robust manufacturing process and powder property control.

429782 Continuous Nanoparticle Production Using Focused Acoustics

Abstract:

Continuous nanoparticle production using focused acoustics

Many drugs currently in development are either water insoluble or lipophilic, thus creating formulation challenges to administer these drugs at a high dose. In some situations, nanoparticles are an appropriate formulation strategy as a result of their intrinsic properties such as small shape and size (e.g., radius) and large surface area to volume ratios. Nanoparticles can be generated by grounds up crystallization, which is typically processed in batch mode. One basic approach of crystallization is to vary the solubility of the solute of interest with different methods, such as vary the ratio to non-solvent vehicle (anti-solvent).

Low frequency, unfocused ultrasound has previously been demonstrated for improving crystallization kinetics. However, unfocused ultrasound has efficiency limitations due the inability to focus the energy, creating issues in thermal management such as non-uniform mixing, and uncontrolled thermal fields. Additionally, these unfocused solutions are not able to provide sufficiently tight, uniform nucleation control at the chemical solvent boundary interface that is required for nanoparticle production. The Covaris Adaptive Focused Acoustics™ (AFA) technology is a highly controlled mechanical based process that provides a practical tool to precisely control both the nucleation formation and the rate of crystal growth. The high energy efficiency of AFA avoids excessive heat generation, reducing potential thermal damages to the sample and allowing greater control over thermally dependent chemical reactions. Moreover, the precisely controlled energy delivery improves the reproducibility of the sample processing for higher quality control.

As a model of nanoparticle formation, a Solvent:Anti-solvent crystallization method with an active AFA energy field was tested in different batch volumes as well as in a continuous flow format. Cinnarizine was used as a model

th

compound. Nanoparticles were measured using a Malvern Mastersizer 3000, and particle sizes were reported for 10 percentile, 50th percentile, and 90th percentile populations (d10, d50, and d90 populations respectively).

Initial studies were performed in triplicate on batch scale processing of 2 mL samples. Parameters evaluated include:

Solvent	Dimethylacetamide (DMA)
Drug solvent concentrations	50 mM, 125 mM, 250 mM
Stabilizer	0.2% PVP30 and 0.25 mM SLS
Stabilizer percentage	0.2% 1.5%, 3.0%
Solvent:Anti-solvent ratio	1:29, 1:14, 1:6.6

For comparison, a control was run at each setting. The control was mixing 2 mL samples via a magnetic stir bar in a 20 mL scintillation vial for the same time duration of 5 minutes.

In all cases, the particle sizes of the AFA-processed samples were lower and more reproducible than the control. A typical comparison is shown in the following tables:

S220x	Covaris Focused-ultrasonicator
300 W	Peak Incident Power (PIP)
50%	Duty Factor (DF)
1,000	Cycles Per Burst (CPB)
12x24	2 mL process vessel
300	processing time (seconds)
6°C	water bath temperature

Table 1. Results from the stir bar method (Control).

125mM-Control				
Ratio	**d10-Control**	**d50-Control**	**d90-Control**	**Span**
1:6.6	9.0	29.9	1730.0	57.5
1:6.6	8.0	26.8	66.5	2.1

1:6.6	8.0	26.5	64.2	2.1
Average	8.3	27.7	620.2	20.6
SD	0.45	1.54	784.70	26.1

All values are in microns

Table 2. Results from the AFA method.

125mM-AFA				
Ratio	**d10-AFA**	**d50-AFA**	**d90-AFA**	**Span**
1:6.6	0.0406	0.122	0.325	2.335
1:6.6	0.0406	0.122	0.333	2.387
1:6.6	0.0407	0.123	0.334	2.395
Average	0.041	0.122	0.331	2.372
SD	0.000	0.000	0.004	0.027

All values are in microns

d50 particle size populations for the control sample averaged 27.7 microns while the same method processed with AFA yielded average d50 particle size populations of 122 nanometers.

A continuous manufacturing process was evaluated with the following settings:

Drug solvent concentration 125 mM (46 mg/mL)

Solvent flow rate 800 μL/min

Anti-solvent flow rate 10 mL/min

Stabilizer 0.2% PVP30 and 0.25 mM SLS

Stabilizer percentage 0.2%

Solvent:Antisolvent ratios 1:12.5 (10 mL/min anti-solvent, 800 μL/min solvent)

With 125 mM at a 800 μL/min solvent flow rate, the production rate was 36.5 mg/min.

Higher total acoustic power input (Peak Incident Power) yielded both smaller particle sizes and repeatable results. At PIP values of 500 Watt, particle size distributions are listed in the following table.

Table 3. Particle size distributions at PIP = 500 W.

	Sample 9	**Sample 10**	**Sample 11**	**Sample 12**
d10	0.085	0.091	0.098	0.090
d50	0.180	0.191	0.202	0.187
d90	0.415	0.409	0.423	0.399
Span	1.834	1.663	1.611	1.654

All values are in microns

Control of particle size was demonstrated by varying PIP. Particles sizes at 300 PIP are listed in the following table.

Table 4. Particle size distributions at PIP = 300 W.

	Sample 5	**Sample 6**	**Sample 7**	**Sample 8**
d10	2.15	1.79	1.89	1.78
d50	3.48	3.23	3.03	2.86
d90	5.55	5.01	4.95	4.73
Span	0.98	1.32	1.01	1.03

All values are in microns

The current study demonstrates the technical feasibility of using Adaptive Focused Acoustics (AFA) to control nucleation and crystal growth for continuous flow Solvent:Anti-solvent crystallization. Nanoparticles were successfully and reliably produced in a continuous flow process with d50 particle size populations below 200 nm. By using the AFA system to control the crystallization process, a choice of nano- or micron-sized crystals can be easily generated with a simple change of PIP process parameter. In addition, the isothermal, reproducible AFA process enables control of fluid shear effects on secondary nucleation to minimize crystal size, accelerate the overall rate of crystal formation, and improve yield. The effect of AFA on the morphology of the crystal is currently under investigation.

430222 Crystal Growth Inhibition of a Poorly Water-Soluble Drug Using Polymeric Additives

Abstract:

A large percentage of active pharmaceutical ingredients (APIs) in the drug development pipeline demonstrate poor aqueous solubility as crystalline solid forms. This poses a problem for oral therapeutics due to decreased bioavailability. Utilizing the amorphous form of the drug rather than the crystalline form is an effective method to alleviate solubility issues, due to its greater overall free energy. Much emphasis has been placed on methods to stabilize the inherently unstable amorphous form and prevent crystallization. One such method utilizes polymers to act as a mechanical barrier, occupying growth sites. The effectiveness of this method, however, varies for each polymer-drug combination.

In this work, we study inhibition of crystal growth of the drug felodipine in solution using a variety of polymers. Using atomic force microscopy (AFM), we can determine the fractional surface coverage as well as the conformation of adsorbed polymers on crystalline felodipine. These results reveal insights into the mechanisms by which polymers can effectively inhibit crystal growth, as well as elucidate important polymer properties that impact their effectiveness

430469 Dynamic Modeling and Control of Multi-Stage Slug-Flow Crystallization

Thursday, November 12, 2015: 4:40 PM
Ballroom D (Salt Palace Convention Center)

Michael L. Rasche[1], Mo Jiang[2] and Richard D. Braatz[2], (1)Department of Chemical and Biomolecular Engineering, University of Illinois at Urbana-Champaign, Urbana, IL, (2)Department of Chemical Engineering, Massachusetts Institute of Technology, Cambridge, MA

The design of robust pharmaceutical crystallization processes has been of high interest in academia and industry for the last decade (Nagy and Braatz, 2012). Continuous-flow tubular crystallizers have shown potential for high reproducibility and process efficiency at low capital and production cost (e.g., Alvarez & Myerson, 2010; Lawton et al., 2009; Vacassy et al., 2000). A recent advance that combines the advantages of continuous and batch crystallizers is the air/liquid slug-flow crystallizer (Eder et al., 2010; 2011; 2012; Jiang et al., 2014; 2015) which has unique advantages including narrow residence time distribution, no stirrer generating secondary nucleation, and easy post-crystallization separation. The potential application of slug-flow crystallization in the final stage of pharmaceutical manufacturing is the focus of this design study.

Similar to batch crystallization, the design of a robust slug-flow crystallization process comes from a good understanding of how the supersaturation profile is affected by the design variables, such as the method and rate of supersaturation generation, and the corresponding implementation. For example, supersaturation should be minimized so as to avoid impurity incorporation and secondary nucleation (e.g., Jiang et al., 2012; 2014), which is important for slug-flow crystallization with its possibly short residence time (on the order of minutes) (Eder et al., 2012; Jiang et al., 2014) and with fast heat transfer associated with the large surface area-to-volume of tubular crystallizers.

Unlike past studies that involve mathematical modeling or analysis of similar continuous crystallizers (e.g., Eder et al., 2010; Kubo et al., 1998), this study only analyzes an experimental demonstration (Jiang et al., 2014) but also investigates the effect of and optimizes numerous design variables (e.g., number of heat exchangers, length of tubing in each heat bath/exchanger, mass flow rate through the heat exchangers) while minimizing the total equipment/material use in the design. This presentation describes the first dynamic simulation of slug-flow crystallization for a series of counterflow, concentric shell-and-tube heat exchangers.

The method of moments is combined with the numerical method of lines with a finite difference approximation to reduce the model described by four partial differential equations (balances on population, solute, and energy in both the shell and tube) to a system of ordinary differential equations. The system is to be designed so that, under normal operating conditions, the temperature drop in each heat exchanger produces an environment to minimize the maximum supersaturation, to promote molecular purity, while applying a constraint that ensures a nearly maximum yield.

Operating so that the exit concentration is near the minimum concentration provides the desired yield. The optimal crystal size is specified by the upstream nucleation rate. Excessive length of the tubes should be avoided, as that would introduce a time delay in the process dynamics that is undesirable for closed-loop control. The control objective is to minimize the maximum supersaturation in the system by manipulating the flowrate and inlet temperature of cooling water in each heat exchanger to suppress the effects of process disturbances.

Optimal steady-state spatial profiles for the four-stage slug-flow crystallizer are shown in Figure 1. The supersaturation reaches the minimized maximum value in less than 5 meters, and returns to that maximum once in each heat exchanger. The drop in supersaturation near the outlet also occurs in less than 5 meters.

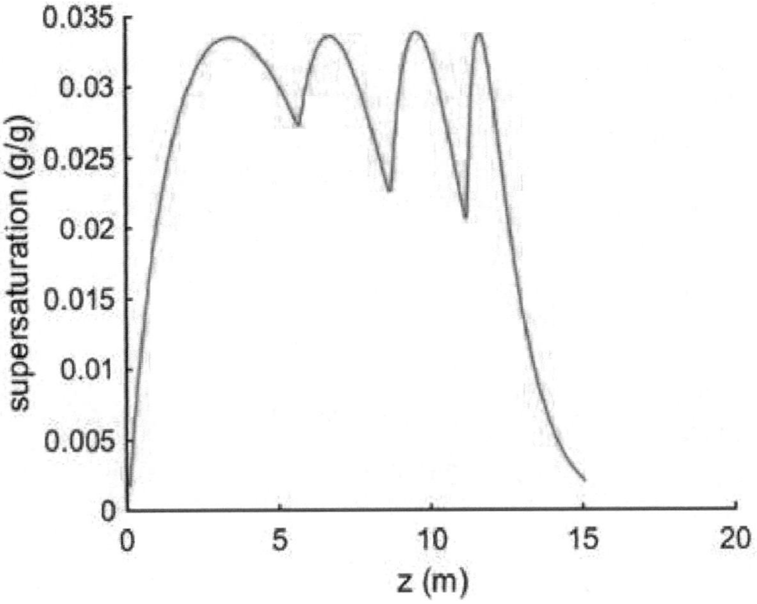

Figure 1: Supersaturation profile under steady, optimal conditions in the slug-flow crystallizer.

Investigating the dynamic response of the measured and product quality variables to disturbances, manipulations, and uncertainties provides a great deal of information and insight not available from steady-state models. The solute concentrations in the slug as it leaves each heat exchanger ("exit concentrations") are observed to be highly sensitive to cooling water temperatures and variation in the growth kinetics. The cooling water flowrates have relatively low effectiveness as manipulated variables for rejection of some disturbances. Increasing the cooling water flowrate or decreasing the cooling water temperature decreases the exit concentrations. The model shows that the effect of most input disturbances on concentration, yield, and average particle size are dampened by the process design automatically.

The response of maximum and average supersaturation is of vital importance as the driving force for growth as well as nucleation. While this system is designed to minimize secondary nucleation, supersaturation levels must be kept under control to avoid the incorporation of impurities into the crystalline matrix. Cooling water temperature and flowrate are seen to have a large effect locally. Inlet conditions for the slug stream have a decreasing effect with each subsequent heat exchanger. Persistent sources of relatively large deviation are again variation in the growth kinetics.

Outlet temperatures are also compared and evaluated in terms of value as measurements for feedback control, due to the relative ease and accuracy at which temperature can be measured. Not surprisingly, the greatest effect by far on the outlet slug stream temperature is the temperature of the cooling water.

Manipulating only the cooling water flowrates is shown to be insufficient to reject process disturbances that result from the deviation of growth kinetic parameters from nominal values. Additionally, without a tight control of the inlet temperature of cooling water, manipulation of the cooling water flowrates will be insufficient to suppress that resulting deviation as well. Changes in the inlet temperature and concentration for the slug stream are less problematic; the system design is inherently robust to disturbances which occur far upstream. The robustness allows the on-line measurement of particle size, a relatively expensive and inaccurate endeavor, to be eliminated from the design of a control algorithm; particle size is not observed to be affected greatly by upstream disturbances.

In an effort to increase the output response to changes in manipulated variables, the cooling water inlet temperature should be a manipulated variable rather than allowing it to be a disturbance. That is, a lower level regulatory control system should be implemented so that the cooling water inlet temperature is controlled, with the setpoint to the cooling water inlet temperature being a manipulated variable of the higher level control system. The extra degrees of freedom from controlling not only the cooling water flowrates, but also the cooling water temperatures, greatly increase the controllability of the slug-flow crystallization process. From a design standpoint, sensitivity analysis indicates the tubing length should be increased beyond nominal design conditions to serve as a buffer to disturbances.

Several control strategies motivated by the dynamic analysis are investigated for the multivariable control of the slug-flow crystallizer (e.g., Figure 2). Neglecting particle size measurements for the reasons mentioned above, there are two measurements of interest that can be taken at the outlet of each heat exchanger: temperature and solute concentration. There are also two manipulated variables for each heat exchanger: cooling water flowrate and cooling water temperature. A variety of factors are considered in selecting the best control strategy in terms of the number of measurements, number of controllers, and which controllers receive which measurements (and control which manipulated variables). Of the eight possible measurements, the four temperature measurements are inexpensive and accurate while four measurements of solute concentration would be relatively expensive and difficult. The design of this process does not provide evidence that intermediate solute concentration measurements would be useful process control elements, and it is unlikely that the final implemented control systems design would contain those measurements. With respect to the eight manipulated variables, the inlet temperature of cooling water could be altered by applying energy to a thermal bath or, perhaps more likely, combining multiple feeds from multiple baths at Y-mixers. The cooling water temperature displays a greater effect on process dynamics than the cooling water flowrate, but incompressible fluid flowrates have the advantage of being easily and quickly controlled. After comparing the pros and cons of different control strategies, an optimal control system design will be described.

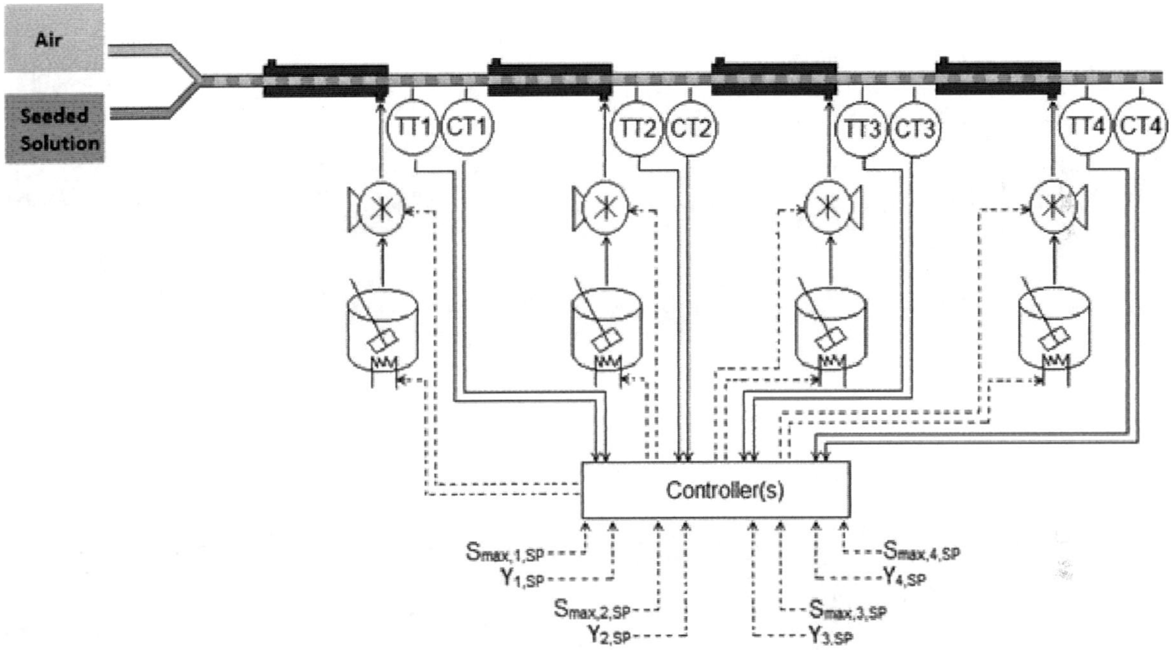

Figure 2: Example control strategy for the slug-flow crystallizer, which uses the complete set of eight measured and eight manipulated variables.

References:

Alvarez, A.J., Myerson, A.S., 2010. Crystal Growth & Design, 56, 349-369.

Eder, R.J.P., Radl, S., Schmitt, E., Innerhofer, S., Maier, M., Gruber-Woelfler, H., Khinast, J.G., 2010. Crystal Growth & Design, 10(5), 2247-2257.

Eder, R.J.P., Schmitt, E.K., Grill, J., Radl, S., Gruber-Woelfler, H., Khinast, J.G., 2011. Crystal Research and Technology, 46(3), 227-237.

Eder, R.J.P., Schrank, S., Besenhard, O., Roblegg, E., Gruber-Woelfler, H., Khinast, J.G., 2012. Crystal Growth & Design, 12(10), 4733-4738.

Jiang, M., Zhu, Z., Jimenez, E., Papageorgiou, C.D., Waetzig, J., Hardy, A., Langston, M., Braatz, R.D., 2014. Crystal Growth & Design, 14(2), 851-860.

Jiang, M., Papageorgiou, C.D., Waetzig, J., Hardy, A., Langston, M., and R. D. Braatz, R.D., 2015. Crystal Growth & Design, 15(5), 2486-2492.

Kubo, M., Kawakatsu, T., Yonemoto, T., 1998. Chemical Engineering Research & Design, 76(6), 669-676.

Lawton, S., Steele, G., Shering, P., Zhao, L., Laird, I., Ni, X.-Y., 2009. Organic Process Research & Development, 13(6), 1357-1363.

Nagy, Z.K., Braatz, R.D., 2012. Advances and new directions in crystallization control, Annual Review of Chemical and Biomolecular Engineering, 3, 55-75.

Vacassy, R., Lemaitre, J., Hofmann, H., Gerlings, J.H., 2000. AIChE Journal, 46(6), 1241-1252.

(730f) Polymorphism in the Development and Scale-up of Reactive-Crystallization of an Active Pharmaceutical Ingredient

Zhang, H., Sunovion Pharmaceuticals Inc.
Saranteas, K., Sunovion Pharmaceuticals Inc.
Originally presented on: 11/12/2015 17:00:00 - 17:20:00

Polymorphism in the Development and Scale-up of Reactive-crystallization of an Active Pharmaceutical Ingredient

Haitao Zhang, Kostas Saranteas

Chemical Process Research & Development, Sunovion Pharmaceuticals Inc., 84 Waterford Drive, Marlborough, Massachusetts 01752, U.S.A.

Abstract

Polymorphism is widely observed in pharmaceutical compounds and continues to be an important issue in drug development because of its impact on the physicochemical properties of drugs. It is well recognized that polymorphism and solvate formation affect the various pharmaceutically important physicochemical properties, such as stability, solubility, dissolution rate, crystal habit (shape), tableting behavior. Crystallization is a crucial step in the pharmaceutical industry in terms of the understanding and further control over the polymorphism during the manufacturing of active pharmaceutical ingredients (APIs).

Herein is described the observed polymorphic behavior of the API SEP-363856-01 in the reactive-crystallization process. This API is crystallized out from isopropanol (IPA) as the HCl salt through the addition of HCl/IPA to an IPA solution of the SEP-363856 freebase. As the API crystallizes it displays two distinct morphologies, the first a block like crystal (Form A) and the second a needle like crystal (Form B). Single crystal x-ray diffraction has shown Form A has a monoclinic crystal system while Form B has an orthorhombic crystal system. The experimental results indicate that the supersaturation control during the addition of HCl/IPA is critical in terms of the polymorphism control. Polymorphism transformation of kinetically preferred Form B to thermodynamically stable Form A is observed during a solvent-mediated phase transition in isopropanol. It has been shown that the transformation of Form B needles to Form A blocks is influenced by the quantity of Form B present and is accelerated with increasing temperature. A controlled HCl/IPA addition strategy has been proposed and successfully implemented in the lab development and scale-up manufacturing campaign. With this, the generation of unstable form B crystal can be avoided in the first place.

Solid phase polymorphism transformation is observed as well. The isolated samples of Form B needles held in the laboratory for a period of time yield a more rod like morphology, however XRPD analysis shows this morphology is consistent with the Form A blocks. Single crystal x-ray diffraction has shown Form A has a monoclinic crystal system while Form B has an orthorhombic crystal system.

429113 Hybrid Cell Culture Models for Efficient Process Development

Abstract:

Cell culture processes are dynamic systems governed by a large number of influential and highly interacting process variables. Despite the stringent control of several process variables, considerable variation in the process outcome is often observed. Moreover, generalized deterministic models for dynamic process prediction based on the impact of the process variables on the product quantity and quality attributes do not exist. According to the Process Analytical Technology (PAT) initiative, multivariate data acquisition and analysis should be implemented to achieve advanced process knowledge and to gain effective control on risk in manufacturing. While several studies have shown that such multivariate statistical models provide promising possibilities for process monitoring and prediction based on the process history, they are very limited regarding their capability for the support of experimental design.

This work demonstrates how the flexibility of statistical models combined with the formalized bioengineering knowledge of simplified deterministic models can improve the efficiency in cell culture process development. Based on a rigorous case study, the results obtained with the hybrid approach will be compared to the ones from the purely statistical and deterministic models to visualize its promising capability of providing significant process information using the least number of experiments. Different kinds of hybrid cell culture models will be discussed and the possibility to incorporate such models into a sequential experimental design methodology to define the process design space will be presented.

420830 The Impact of Glycation on Protein Color

Abstract:

Health Authorities expect that color of liquid formulations of monoclonal antibodies be monitored and controlled. Subcutaneous drug formulations often require high drug substance concentrations which may result in more intense product color. A potential mechanism of coloration is through the Maillard reaction, where reducing sugars in the cell culture media react with the secreted protein product in a series of non-enzymatic reactions that result in formation of glycated protein and advanced glycation end products (AGEs). We varied the glucose feeding strategy and cell culture media across multiple cell lines to study and delineate the effects. Presence of AGEs was also measured using mass spectrometry. We observed good correlation between the glucose fed to the culture and the level of the glycated protein. However, the glucose feeding strategy tested did not significantly impact color, and we observed no strong correlation between glycated protein and color. Changes in the levels of the AGEs for different culture conditions will also be discussed.

433606 Increase Cell Culture Development Value Using the Modular Automated Sampling Technology (MASTTM) Platform in an Integrated Bioprocess Lab Environment

Abstract:

Biopharmaceutical companies are continually striving to fundamentally understand and optimize cellular performance within their bioreactors. While on-line process analytical technology (PAT) tools like dielectric and Raman spectroscopy are helping to provide insight into cell processes that impact titer and antibody quality, the integration of these data with off-line measurements such as cell density, viability, and titer remains an elusive goal.

Bend Research Inc., in collaboration with Pfizer Inc. and other major biopharmaceutical companies, is advancing the Modular Automated Sampling Technology (MAST^{TM}) platform and integrating this system with automated analytical and data retrieval systems. This approach allows application of novel experimental methods providing greater insight into the bioreactor environment. In this process, aseptically collected bioreactor samples are delivered to multiple analytical devices with resulting data being automatically retrieved and curated. This data is presented at a graphical user interface (GUI) for real-time data analysis and predictive model development. Early studies with the MAST sample delivery system coupled with our data-processing software increased system observability and real-time process understanding by the end user.

This presentation describes how Bend Research has used the MAST platform as a cornerstone of an integrated lab environment. We will describe how the MAST system works and the features that set it apart from other autosampling solutions. Applications will be presented where MAST facilitated the use novel development approaches yielding enhance process understanding

From this new level of insight into the bioreactor, scientists can obtain enhanced data-driven guidance for key activities like optimizing process operation. With this new technology, the bioprocess industry can make major advances toward advanced real-time testing, predictive control, and overall enhanced bioprocess design and operation.

428887 Application of Raman Spectroscopy in Cell Culture Process Development

Abstract:

In recent years, the European Medicines Agency (EMA) and the United States Food and Drug Administration (FDA) have stressed the need of greater process understanding and control in the area of pharmaceutical development, manufacturing and quality assurance, all which is summarized in the Process Analytical Technology (PAT) initiative. Therefore, in bioprocesses, and particularly in cell cultures, multivariate data acquisition and analysis are likely to provide versatile and significant real-time information. Lately, Raman spectroscopy has been characterized as a promising tool to detect multiple process variables simultaneously.

This work presents several cases studies, where Raman spectroscopy was successfully applied for the quantification of significant variables in monoclonal antibody (mAb) cell culture processes. These studies include online monitoring of fed-batch cultures, offline analysis of (partially spiked) cell culture samples as well as the utilization in perfusion cell culture processes. Besides the evaluation of those applications of Raman spectroscopy, the central role of multivariate spectra processing techniques including pretreatment, spectral range selection and data reduction will be discussed with regards to robust calibration and prediction of the components of interest. Finally, the role of those spectral data with regards to overall process prediction, monitoring and control will be highlighted within the scope of all acquired online and offline data in bioprocesses.

411442 Density Surrogate for Fine Tuning Dissolution of an Extended Release Tablet to Increase Manufacturing Process Robustness

Abstract:

Tablet dissolution can be affected by several factors including raw materials, formulation and processing conditions. For a tablet with a fixed formulation, such as that of a commercial drug product, there remain few levers by which to improve tablet critical quality attributes (CQA). Tablet dissolution is one such known critical quality attribute to be affected by tablet hardness, which can be controlled by varying the compression force, typically in a directly proportional relationship. Beyond a certain threshold of compression force value, however, hardness reaches a maximum value and plateaus. Continuing to increase compression force after this threshold value could lead to loss in tablet cohesion due to high internal stresses or eventually tooling and equipment damage. Near the plateau threshold, there is a space over which tablet hardness may not adequately capture the tablet properties controlling its dissolution. This work explores this space for a polymer matrix extended release tablet where improved dissolution release robustness is desired. A parameter of weight/thickness ratio is suggested as a density surrogate to fine tune dissolution performance near and within the hardness plateau region.

431023 Can R&D be Standardized? Implementation of a Recipe Framework Supporting Lab to Patient Execution and Analytics

Abstract:

The benefits of a recipe framework have been recognized in manufacturing organizations for several years, but implementation in a development environment brings different challenges – and measurable benefits to drive appropriate scientific and engineering knowledge. To accomplish this task in a development environment a significant amount of structure needs to be implemented in the underlying laboratory systems, ensuring data can be captured with appropriate context. In this presentation, we will outline how we are implementing a flexible S88 recipe model across the organization to enable process definition, drive execution with context, and provide visualization of experiments/batches to compare this contextualized data across sites and scales. Detailed examples of the approach, resulting analysis tools, and challenges encountered, will be provided. This presentation will also outline how this framework can help a research and development organization easily analyze scale up and manufacturing processes, and enable transfer of process knowledge through development and commercialization.

414967 Enabling Rapid Elemental Analysis of Highly-Automated, Parallel Screening Studies for Pharmaceutical Process Development

Abstract:

In the pharmaceutical industry, utilization of lab automation for parallel, statistically designed experiments to optimize reagent and processing parameters allows for accelerated development of chemical processes. The resulting generation of comprehensive, high-fidelity data sets provides in-depth knowledge which ultimately informs a robust chemical manufacturing process. For a given chemical step that may require downstream removal of a metal catalyst, the use of X-ray fluorescence (XRF) enables rapid elemental analysis of samples for pharmaceutical process development. Together, automated parallel experimentation and XRF technology have proven to be a valuable partnership for the high-throughput analysis of large sample arrays for obtaining process knowledge and advancing pharmaceutical process development. Herein, we exhibit a case study where the combination of highly-automated design of experiments (DoE) studies and XRF allowed for rapid metals analysis and, subsequently, significant improvements in processing a late stage asset to remove trace metals.

423647 Pharmaceutical Process Development Using Computer Aided-Methods and Tools

Abstract:

A majority of pharmaceutical production systems involve batch processes, which have functioned well for pharmaceutical industries as well as the regulatory bodies. Batch vessels provide the flexibility of handling different unit operations in a multipurpose plant. However, batch processes are not very efficient for product quality assurance and have a number of drawbacks. Over the last decade the regulatory bodies required the pharmaceutical companies to demonstrate more process understanding. Moreover, financial drivers require profit expectations to be met due to increasing R&D cost and high competition [2]. Consequently, the pharmaceutical industry is expected to look for opportunities to rapidly evaluate the different alternatives for process improvement. To achieve this objective, Process Systems Engineering (PSE) methods and tools can play an important role, for example, (a) apply computer aided-methods and tools that are mature for other industries (such as chemical and petrochemical) also in solving problems in the pharmaceutical industries, (b) in the evaluation and implementation of alternative solutions and/or designs and (c) to evaluate opportunities for continuous manufacturing [1-2].

An integrated computer-aided framework for the development of pharmaceutical processes has been developed to tackle key research areas for pharmaceutical processes (such as continuous processing, separation and reaction processes, solvent selection/recycle/optimization and process intensification/integration [3]). The objective of the framework is to improve pharmaceutical process operation/design and to evaluate opportunities for continuous manufacturing through better process understanding of pharmaceutical processes by using systematic model-based methods. Computer aided-methods and tools are important for such a framework as they can assist in solving various problems such as property prediction, mixture separation (ternary and binary azeotropic separations, separation sequence and type), solvent selection and process modeling and in providing data through knowledge libraries. The developed framework consists of four main parts: reaction mechanism identification, reaction analysis, separation synthesis and design and finally process simulation, evaluation, optimization and operation.

To extend the applicability of the framework a solvent swap method has been developed and implemented in a computer-aided tool. The solvent swap method is based on the method proposed by Gani et al. 2006 [4] and is treated as a special solvent selection problem with additional criteria consideration. The objective of the method is the quick and reliable identification of swap solvents which are suitable for the following process step and it can increase the process efficiency (minimize solvent and product loss and reduce energy demands).

The application of the proposed integrated framework with the added features will be highlighted through the BHC (Boots and Hoechst-Celanese) patented synthesis pathway for ibuprofen and the production of region-someric products through Heck reactions [5]. The improved BHC synthesis of ibuprofen consists of three reaction steps [6] and it is used to verify the proposed integrated framework. Finally, the application of the solvent swap method using the computer-aided tool for both examples will be illustrated.

References:

[1] K. Plumb, 2005. Chem. Eng. Res. Des., 83, 730.

[2] K.V. Gernaey, A. E. Cervera- Pardell, J. M. Woodley, 2012. Comput. Chem. Eng., 42, 30

[3] C. Jiménez-González et al., 2011. Org. Process Res. Dev. 15(4), 900.

[4] R. Gani, M. Jones, l. Powell, J. H. Atherton, J. L. Cordiner, 2006. Chem. Eng. 30.

[5] R. L. Hartman, J. R.Naber, S. L. Buchwald, & K. F. Jensen, 2010. Angew. Chem. Int. Ed. Engl., 49(5), 899.

[6] Elango, V. Murphy, M., Smith, B.L., Davenport, K.G., Mott, G.N., Zey, E.G., Moss, G.L. (Hoechst Celanese Corp.), 1990. European Patent 4.981.995

Accuracy of Micro Powder Dosing and Feeding via a Vibratory Sieve-Chute System

M.O. Besenhard[1,2], E. Faulhammer[1,2], S. Fathollahi[1,2], G. Reif[2], V. Calzolari[3], S. Biserni[3], A. Ferrari[3], S.M. Lawrence[4], M. Llusa[1], A. Paudel, J.G. Khinast[1,2*]

1: Research Center Pharmaceutical Engineering (RCPE) GmbH, 8010 Graz, Austria
2: Graz University of Technology, Institute of Process and Particle Engineering, 8010 Graz, Austria
3: MG2, Via del Savena 18, I-40065, Pian di Macina di Pianoro, Bologna, Italy

4: GlaxoSmithKline, New Frontiers Science Park, Third Avenue, Harlow, Essex CM19 5AW, UK

* Corresponding author:

khinast@tugraz.at

Abstract:

This presentation describes our current studies on powder dosing [1] and feeding via a vibratory sieve-chute system, i.e., the MG2's Microdose®. Figure 1 shows the operating principle and parts of the set-up. The chute is tilted at a fixed angle of 5° and sieves with 5-10 holes of e.g. 0.7mm in diameter are fixed on its top. The powder was discharged from the sieve into the chute and the capsule body using gravity. Vertical vibration occurred with a broad range of frequencies and amplitudes. The fill weight (*fw* fw) during dosing events was recorded via a capacitance sensor, which had two parallel electrode plates encompassing the capsule body. The electrical field and the capacitance varied depending on of the powder quantity in the capsule. Due to this scale, fill weights could be recorded dynamically facilitating the analysis of filling characteristics, i.e., the fill rates and their robustness.

First, the range of frequencies and amplitudes was screened for settings that facilitated reasonable (no blocking of spilling in the sieve) fill rates for three lactose powders, serving as model substances. The filling characteristics were studied in detail within this operating space.

The filling rate was determined by averaging the slopes (*Δfw(t) Δt [mg s]* $\Delta fw(t)/\Delta t$ $[mg/s]$) of *30* 30 filling profiles. The higher the probabilities of an unsteady filling behavior (as shown in Figure 5 b) or fluctuations in the filling rate are, the higher is the standard deviation ([mg/s]) of the evaluated slopes. Therefore, the latter can be used to characterize the robustness of the fill rate, which is of prime importance for selecting the settings for continuous (micro) feeding using such a system.

The results reveal similar operating spaces for all powders and the fill rate robustness varied distinctly in the operating space. Figure 2 depicts the results for one of the studied powders. Utilizing the gained knowledge on filling characteristics micro dosing studies have been performed, dosing capsules with 2.5 mg. These studies compared accuracy of dosing via a constant frequency and amplitude or a constant frequency but the amplitude manipulated by a PID controller. The results reveal that dosing was the most accurate when using constant settings, since the usage of PID control caused inhomogeneous powder beds on the chute, as shown in figure 3.

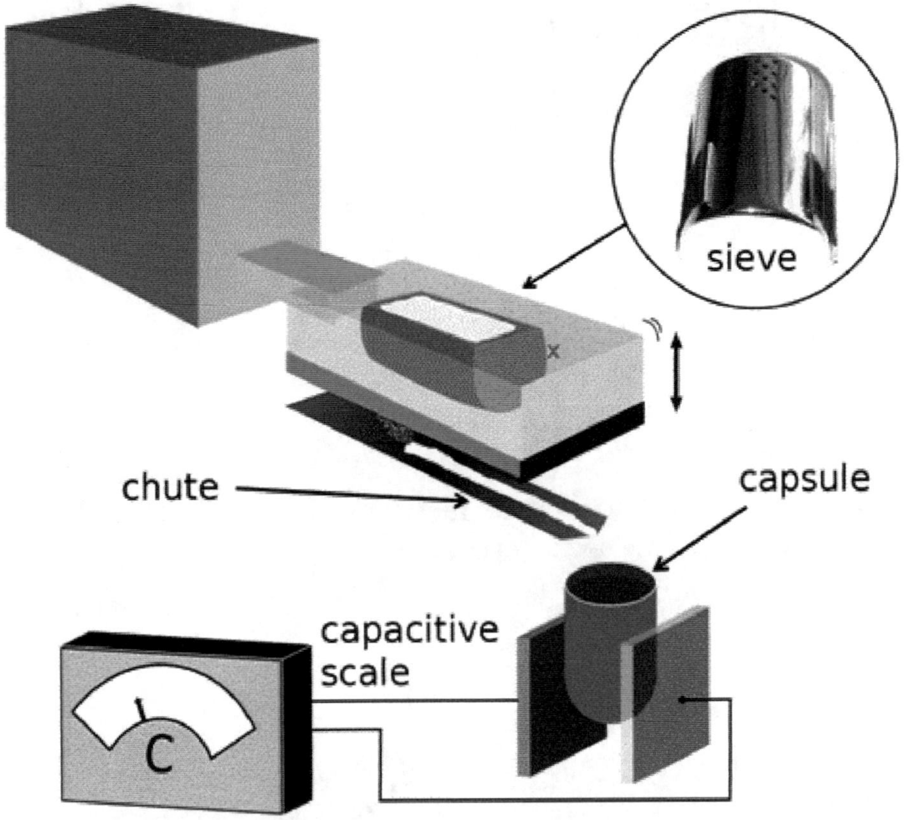

Figure SEQ figure * ARABIC 1: Principle of powder dosing via a vibratory sieve-chute system. The inlet shows a photo of the sieve used, and the red x depicts the position of an acceleration probe used to study the oscillation patterns.

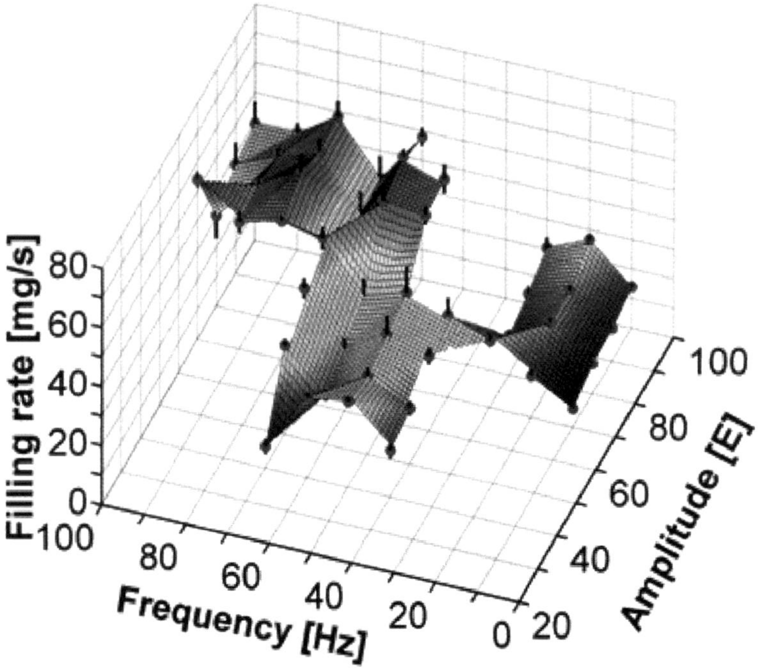

Figure 2: Filling rate and its standard deviations (vertical bars) for InhaLac 230 in the amplitude and frequency settings of the operating space.

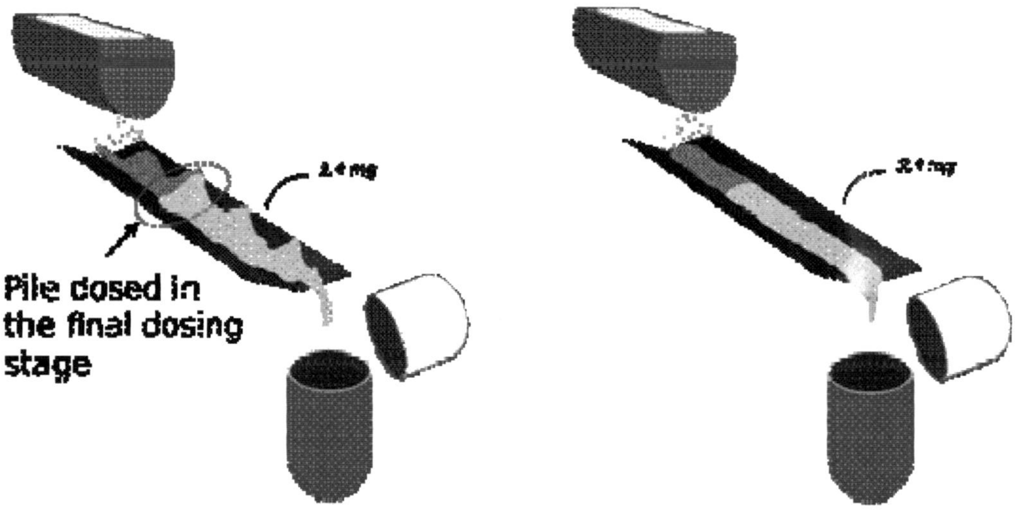

Figure 3: Powder bed in the chute during a PID controlled dosing (left) and dosing using a constant amplitude (right).

[1] Besenhard, M.O., Faulhammer, E., Fathollahi, S., Reif, G., Calzolari, V., Biserni, S., Ferrari, A., Lawrence, S.M., Llusa, M., Khinast, J.G. (2015). Accuracy of Micro Powder Dosing via a Vibratory Sieve-Chute System. *European Journal of Pharmaceutics and Biopharmaceutics,* accepted.

431162 Chromatographic Tools to Predict the Stability of MAbs for Faster Identification of Therapeutic Candidates

Abstract:

Chromatographic Tools to Predict the Stability of mAbs for Faster Identification of Therapeutic Candidates

S. Hedberg*, J.Y.Y Heng*, D.R. Williams* and J. Liddell**

*Surface and Particle Engineering Laboratory

Department of Chemical Engineering

Imperial College London, SW7 2BY,

United Kingdom

**Research and Development

Fujifilm Diosynth Biotechnologies

Billingham, TS23 1LH, United Kingdom

ABSTRACT

Protein-protein molecular interactions are known to be involved in protein solution aggregation behaviour and are a common issue for the manufacturing of therapeutic proteins such as mAbs. Much effort has been employed to gain a better understanding of aggregation, however the mechanisms leading to protein aggregation are still not fully understood. The osmotic second virial coefficient (B_{22}) is a fundamental physiochemical property that describes protein-protein interactions solution, which can be a useful tool to predict aggregation propensity of proteins.

One way of predicting aggregation propensity is self-interaction chromatography (SIC), which recently have shown to be a promising tool for better understanding of phase behaviour of proteins. Another technique, cross-interaction chromatography (CIC), has shown to be an even more high-throughput technique than its predecessor with possibly the same capabilities. This work consists of two experimental studies with therapeutic mAbs to improve SIC and CIC as a tool to predict protein aggregation.

The first part includes a 10 times scale-down study of therapeutic mAbs from laboratory scale macro-columns to micro-scale columns, which will enable the determination of B_{22} for the individual protein as well as the cross-virial coefficient, B_{23}, between two proteins. Micro SIC and CIC uses only a few milligrams of mAb in order to obtain a complete formulation study. The results from the first part of the study proved to give good comparable results between the micro and macro scales enabling the use of micro SIC for B_{22} determinations.

The second part of this work presents an extensive formulation study of mAbs, varying pH and salt, as well as the presence of different stabilisers as well as different external factors known to induce aggregation. The B_{22} and B_{23} values determined from the formulation study are then correlated with aggregation data obtained from size-exclusion chromatography. It was shown that over all test conditions, good correlations could especially be found between B_{22} and aggregation rate.

In summary this work hopes to lead to a better insight into the mechanisms of aggregation by looking at the extent of tailing of the peaks in the chromatograms and comparison with conformational studies.

REFERENCES

1. Dumetz, A.C., Snellinger-O'Brien, A.M., Kaler, E.W., and Lenhoff, A.M. (2007) *Patterns of Protein–Protein Interactions in Salt Solutions and Implications for Protein Crystallization.* Protein science. **16**(9): p. 1867-1877.

2. Quigley, A. (2012) *Protein Aggregation Behaviour and the Second Virial Coefficient,* Ph.D. Thesis, Imperial College London.

3. Tessier, P.M., Sandler, S.I., and Lenhoff, A.M. (2004) *Direct Measurement of Protein Osmotic Second Virial Cross Coefficients by Cross-Interaction Chromatography.* Protein science. **13**(5): p. 1379-1390.

4. Jacobs, S.A., Wu, S.-J., Feng, Y., Bethea, D., and O'Neil, K.T. (2010) *Cross-Interaction Chromatography: A Rapid Method to Identify Highly Soluble Monoclonal Antibody Candidates.* Pharmaceutical Research. **27**(1): p. 65-71.

GlaxoSmithKline

Advancements of Accelerated Stability Studies and Modeling through Use of Nonlinear Kinetics and Bootstrapping

Don Clancy, Neil Hodnett, Rachel Orr, Martin Owen, and John Peterson

Primary corresponding author email: Donald.J.Clancy@gsk.com

What is Accelerated Stability? – 4 Steps

- Conduct a short term designed experiment at elevated temperature with varying humidity.

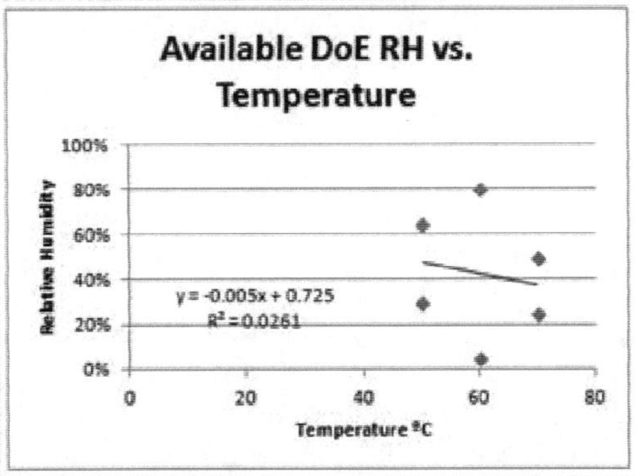

- Fit kinetic models for impurities and physical changes using short term data under accelerated conditions.
- Develop a long term model to predict container humidity and moisture content for a typical medicine bottle or blister pack.
- Extrapolate kinetic models from short term studies and integrate them with the long term humidity model to predict long term stability.

Background and Current Available Approach

- Currently, the most widely published approach for modeling drug product degradation based on accelerated stability data is that described by Waterman (Waterman, 2012), (Adami & Waterman, 2004). This is based on the modified Arrhenius equation shown in Equation (1):

$$ln(k) = ln(A) - \frac{Ea}{R \cdot T} + B \cdot (RH) \quad (1)$$

Where "A", "Ea", and "B" are fitted terms, "R" is the gas constant, "T" is temperature in Kelvin, and "k" is defined as the change in impurity with respect to time.

Rearranging into a typical kinetic rate expression:

$$\frac{dy}{dt} = A \cdot exp\left(\frac{-Ea}{R \cdot T}\right) \cdot exp(B \cdot RH) \quad (2)$$

Makes the implicit assumption the rates of reaction is an exponential function of RH, which implies moisture affects molecular mobility similar to temperature.

Background Continued - Isoconversion

- Equation 1 is further reduced to predict time to failure (t.alpha), or "isoconversion."

$$ln\left(t_{\alpha}\right) = \frac{Ea}{R \cdot T} - B \cdot RH - C \quad (3)$$

- This approach often requires significant extrapolation or interpolation of data to a failure limit to estimate the time till failure for each condition.

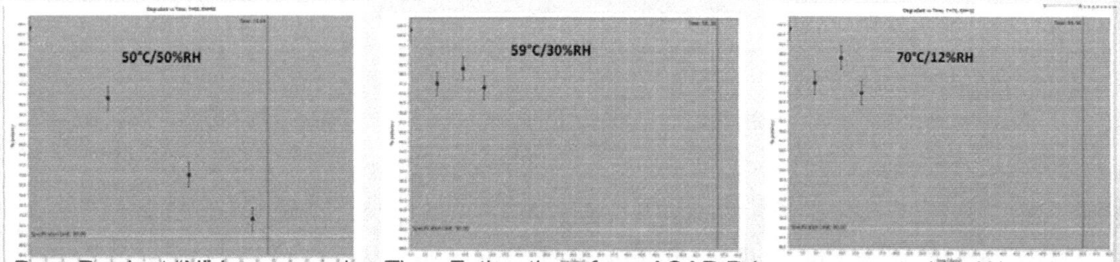

Drug Product "N" Isoconversion Time Estimations from ASAP Prime as reported by Waterman.

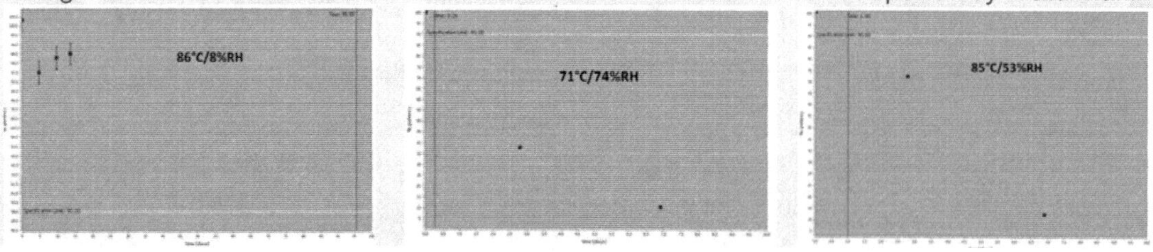

Kinetic Model Development With a Different Starting Assumption

Change in reaction conversion (called alpha in this analysis) as a solid API reacts with water is typically related to the "activity" of the reactants, raised to a power, as:

$$\frac{d\alpha}{dt} = k \cdot \left[H2O_{activity}\right]^{N1} \tag{4}$$

The "k" term in these equations is an Arrhenius factor relating an exponential relationship between temperature "T", the activation energy of the reaction "Ea" and a gas constant "R". Expanding on the Arrhenius term for equation (5) yields (7):

$$\frac{d\alpha}{dt} = K1 \cdot exp\left(\frac{-Ea}{R \cdot T}\right) \cdot RH^{N1} \tag{7}$$

Where, K1 is a pre-exponential term called the "frequency factor." For simplicity of nomenclature and units in the models described in this report, the activation energy divided by the gas constant (Ea/R) will be simplified to a fitted parameter "K2".

$$\frac{d\alpha}{dt} = K1 \cdot exp\left(\frac{-K2}{T}\right) \cdot RH^{N1} \tag{8}$$

Kinetic Model Development - Continued

Fitting Equation 8 can be challenging as K1, and K2 are correlated in an exponential relationship. A way of reducing this exponential correlation between K1 and K2 is to scale the exponent to a value of 1.0 at the average temperature in the study (Tref) by modifying equation 8 to create equation 9.

$$\frac{d\alpha}{dt} = K1 \cdot exp\left(\frac{-K2}{T} + \frac{K2}{T_{ref}}\right) \cdot RH^{N1} \tag{9}$$

What is actually fitted to the experimental data is the integrated form of the kinetic models, as the final impurity value at time "t" is what is observed. Integrating (9) with respect to time yields (10):

$$\alpha = K1 \cdot exp\left(\frac{-K2}{T+273.15} + \frac{K2}{Tref}\right) \cdot (H)^{N1} \cdot \Delta t + C \tag{10}$$

Where "C" is the value of alpha at time zero, and "H" is the humidity expressed as either RH or Vapor pressure. The type of humidity dependence will need to be tested within the experimental analysis as two alternate models.

Accounting for the Potential Curvature With Respect to Time

Model Type	Description	Typical Application	Nonlinear Model (Integrated Form) Used for Fitting:	Limitations	Terms Needed to Fit
A	Reaction slows exponentially with time	Reaction rate proportional to a limiting reagent that is close to being consumed	$\alpha = C + K3 \cdot \left(1 - exp\left(-K1 \cdot exp\left(\frac{-K2}{T+273.15} + \frac{K2}{Tref}\right) \cdot H^{N1} \cdot \Delta t\right)\right)$		4
B	Reaction slows down with time	Diffusion	$\alpha = K1 \cdot exp\left(\frac{-K2}{T+273.15} + \frac{K2}{Tref}\right) \cdot (H)^{N1} \cdot \Delta t^{N2} + C$	N2<1	4
C	Reaction is linear with time	Zero order reactions	$\alpha = K1 \cdot exp\left(\frac{-K2}{T+273.15} + \frac{K2}{Tref}\right) \cdot (H)^{N1} \cdot \Delta t + C$		3
D	Reaction speeds up with time	Power Law Induction Reactions	$\alpha = K1 \cdot exp\left(\frac{-K2}{T+273.15} + \frac{K2}{Tref}\right) \cdot (H)^{N1} \cdot \Delta t^{N2} + C$	N2>1	4
E	Reaction speeds up exponentially with time	Nucleation, or reaction rate dependent on product concentration	$\alpha = exp(C) \cdot exp\left(K1 \cdot exp\left(\frac{-K2}{T+273.15} + \frac{K2}{Tref}\right) \cdot (H)^{N1} \cdot \Delta t\right)$		3

Table 1 Kinetic Equations Available in the Accelerated Stability Dashboard

- Each of these 5 possible curvature types with respect to time is tested with either relative humidity or vapor pressure in the "H" term.
- All models are fitted and then compared.
- Data with large extents of conversion (>15%) may need an Avrami transform if exponential.

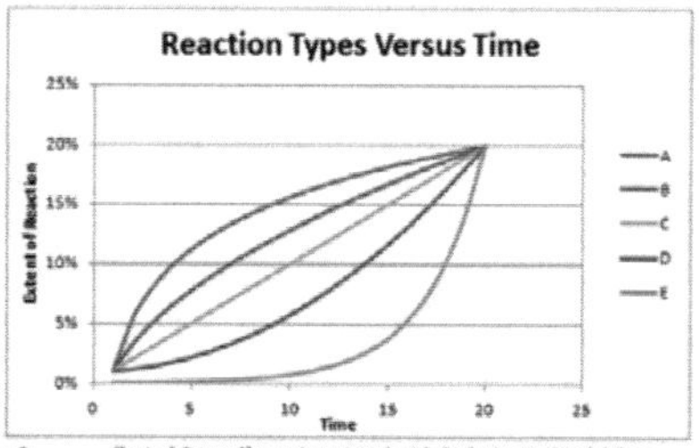

Figure 1 Typical Curve Shape Associated with Each Kinetic Model Type

Scientific Rational for Initial Model Selection

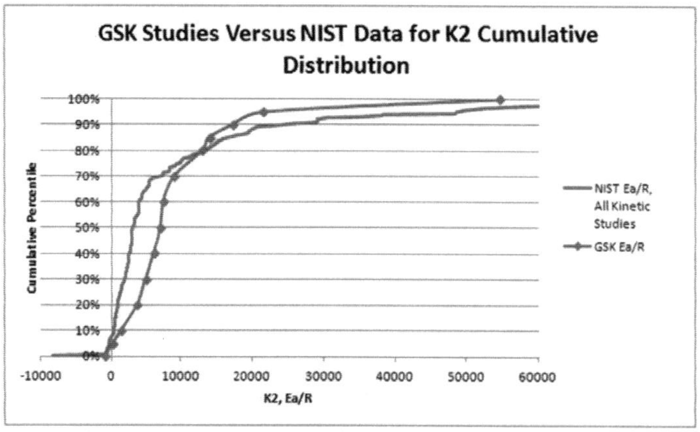

Figure 2 Cumulative Distribution of Ea/R for GSK studies Compared to All Organic Reactions in NIST Data Set

- To help decide if a potential kinetic model might be real, the activation energy K2 value is compared to the known potential distribution of activation energies to assess how unusual it is.
- Models with extremely unusual activation energy values are eliminated from further consideration.

Typical Power on Humidity Term (N1)

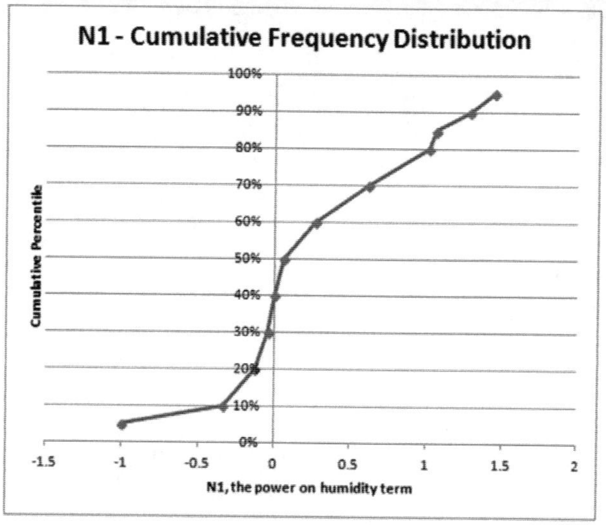

Figure 5 Cumulative Percentile Plot of N1 for GSK Accelerated Stability Studies

- From 120+ impurity analysis, the cumulative frequency distribution for the humidity power (N1) is found to typically fall between -1 and 1, which is the opposite of an exponential curve with respect to RH.
- There is a long tail to the right with N1 values as high as 6 being noted for some unusual hydrolysis reactions in the top 5% of the distribution.

Example Kinetic Model Selection in GSK Software within JMP

Potential Kinetic Models and Parameters

Model Type	K1	K2	K3	N1	N2	nParams	RMSE	Score	AICc	BIC
RH Linear Time Kinetic Model	0.69	8374	-	3.34	-	4	0.1115	-	-22.964	-21.742
AH Linear Time Kinetic Model	118	-8801	-	3.34	-	4	0.1115	-	-22.966	-21.744
RH Accelerating Kinetic Model	4.4	-17419	-	11.19	-	4	0.3514	-	25.237	26.459
AH Accelerating Kinetic Model	1	1000	-	1	-	4	1.3584	-	82.024	83.247
RH Decelerating Kinetic Model	0.33	12601	3.7	3.35	-	5	0.0221	-	-88.206	-87.939
AH Decelerating Kinetic Model	57.33	-4588	3.7	3.34	-	5	0.0221	-	-88.3	-88.032
RH Power Model	1.9	7500	-	3	0.5	5	0	-	-848.163	-847.896
AH Power Model	193.48	-7937	-	3	0.5	5	0.0001	-	-312.928	-312.66

Table 2 Fit of All Possible models to Data Set Created by an RH Model, with K2 Value of 7500, and N2 Value of 3.0, as applied to a typical Accelerated Stability study

- Potential kinetic models fit with highly unusual K2 and N1 parameters are first eliminated, as shown in red.
- Of the remaining models, the model with the best (lowest) Bayesian Information Criterion (BIC) is then chosen.

What Can Go Wrong in Accelerated Stability? Nitrazepam Stability (1977)

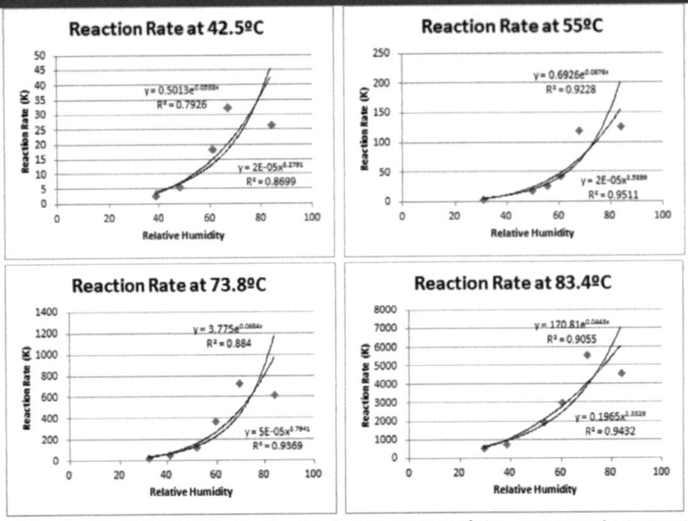

Figure 6 Data Visualization for Reaction Rate for Disappearance of nitrazepam at varying temperatures

The analysis of nitrazepam stability data from Kesselring's and Genton's 1977 paper (Kesselring & Genton, 1977) is of particular interest as it's an unusual case due to high level of humidity sensitivity and is a good example of what can go wrong in an Accelerated Stability study.

This study also has historical significance as it is referenced in Waterman's papers to support the hypothesis that the kinetics of degradation for drug products can best be modeled as an exponential function of relative humidity.

Analysis of Nitrazepam Stability

The data in the paper is expressed as reaction rate "K" which is the change in nitrazepam purity with respect to time. Our standard model given as equation 11 is simplified to equation (11) to predict K.

$$\frac{dy}{dt} = K = K1 \cdot exp\left(\frac{-K2}{T+273.15} + \frac{K2}{335.83}\right) \cdot (RH)^{N1} \quad (11)$$

Waterman's equation from equation 2, the "RH Exponential Model" is expressed as equation (12) using the standard nomenclature developed in this paper.

$$\frac{dy}{dt} = K = K1 \cdot exp\left(\frac{-K2}{T+273.15} + \frac{K2}{335.83}\right) \cdot exp(B \cdot RH) \quad (12)$$

- Since the Nitrazepam data set was the primary reference in prior work to the hypothesis that reaction rates are proportional to the exponential of RH, both equation forms will be tested.

Comparison of Model Fits

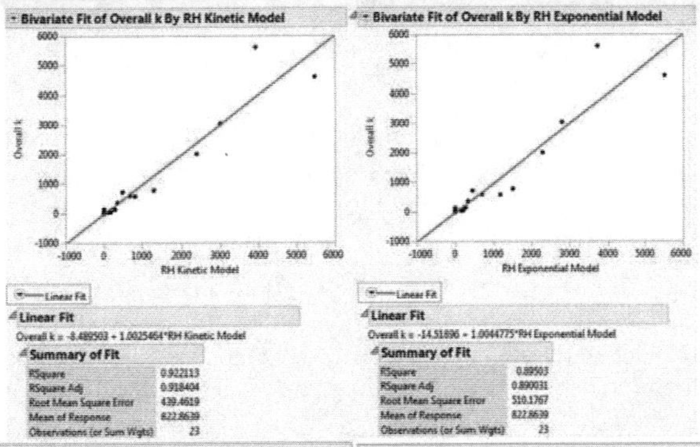

Solution (Left)

SSE	DFE	MSE	RMSE
4056908.012	20	202845.4	450.38361

Parameter	Estimate	ApproxStdErr
K1	0.0167416035	0.02574611
K2	26757.744805	5939.75581
N1	1.8220013208	0.27078848
C	0	0

Solution (Right)

SSE	DFE	MSE	RMSE
5469488.3438	20	273474.42	522.94781

Parameter	Estimate	ApproxStdErr
K1	4.8687759394	6.08294001
K2	26807.829412	6997.3303
C	1e-100	0
B	0.0285954924	0.00467036

Figure 7 Fits for Linear Kinetic Model as Derived versus Waterman's Exponential RH Model (Right)

- This data does not support the hypothesis that this reaction has an exponential rate with respect to humidity. A typical kinetic equation fits the data slightly better.

Dropping Highest Temperature Data Due to Change in Activation Energy

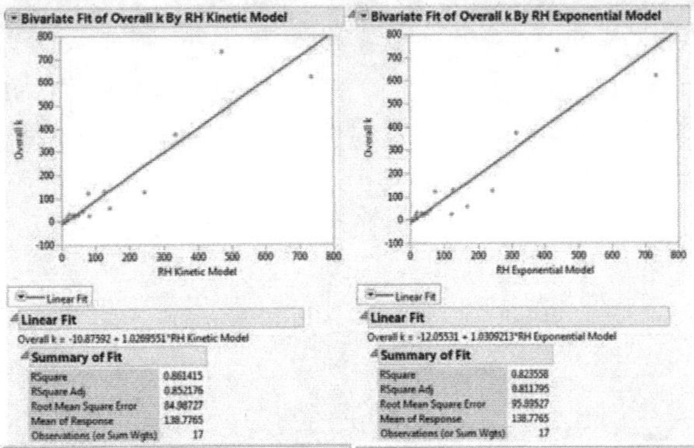

Solution (Left)

SSE	DFE	MSE	RMSE
109627.14446	14	7830.5103	88.490171

Parameter	Estimate	ApproxStdErr
K1	0.0106708776	0.02166786
K2	10458.983857	2823.98606
N1	2.2911430108	0.4702406
C	0	0

Solution (Right)

SSE	DFE	MSE	RMSE
139482.01185	14	9963.0008	99.814833

Parameter	Estimate	ApproxStdErr
K1	14.846054948	9.50787154
K2	10364.598576	3210.1629
C	1e-100	0
B	0.0348019883	0.00768587

Figure 8 Fits for Normal Kinetic Model by Equation 11 (Left) versus Waterman's Exponential Model (Right) for Overall Disappearance of nitrazapem, excluding data at highest temperature in study

- If the activation energy markedly changes as the highest temperature data is dropped, then likely there was a phase change, form change, or some other physical change impacting the results. Here the estimate changed from K2 value of 26,800 to 10,500.

- **RECOMMENDATION:** Get DSC and/or TGA data up to 90 degrees prior to the study.

Final Model Fit – Dropping Highest Humidity Data

Solution			
SSE	DFE	MSE	RMSE
3620.9308018	11	329.17553	18.143195

Parameter	Estimate	ApproxStdErr
K1	7.6732252e-8	8.45784e-8
K2	10836.303303	770.225387
N1	5.1846287097	0.26325205
C	0	0

Solution			
SSE	DFE	MSE	RMSE
4957.4557942	11	450.6778	21.229173

Parameter	Estimate	ApproxStdErr
K1	0.6966261868	0.22797375
K2	10723.079571	899.058984
C	1e-100	0
B	0.086179662	0.00482877

Figure 9 Parameter Fits for Normal Kinetic Model by Equation 11 (Left) versus Waterman's Exponential Model (Right) for Overall Disappearance of Nitrazapem, excluding data at highest temperature and highest humidity in study

- Dropping highest humidity in this study also results in significant changes in the estimate of the humidity sensitivity. Likely some change in physical behavior above 75% RH.
- Leads to conclusion that experimental designs need to have the flexibility to test for changes in humidity sensitivity at extremes in humidity.
- Finally, while the impact of RH may have looked exponential in this case, it fits standard kinetics consistently better. High values of N1 as high as 6 have been observed for some unusual hydrolysis reactions.

Initial Experimental Designs - Hexagon

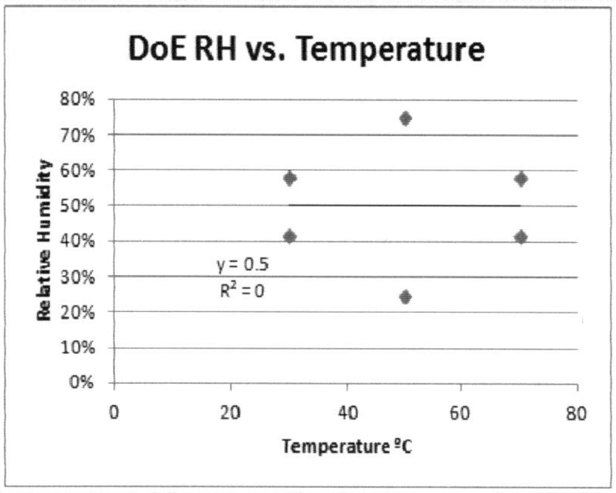

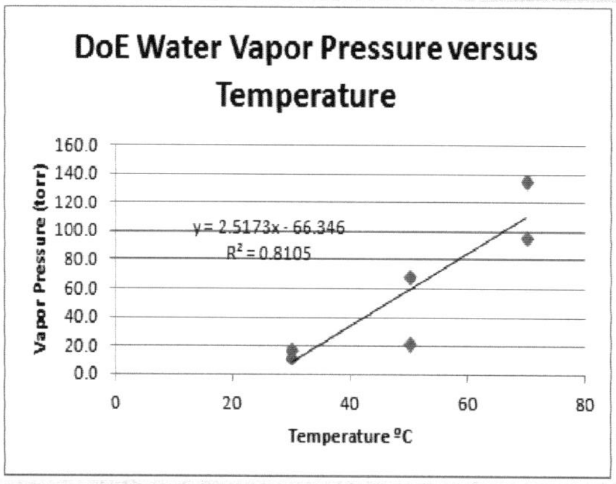

- A mathematically "optimized design" for fitting kinetic models looks like a square, and only tests a low and high humidity value in the design. This is a risky design due to the observed potential changes in kinetics at low or high humidity.
- A Hexagon is an attractive design, as it allows for 3 levels of temperature, and 3 levels of humidity within the design.
- Problem with this design is the potential correlation between temperature and vapor pressure makes fitting vapor pressure based models prone to error in parameter estimation.

Typical Initial Experimental Design - Offset Hexagon

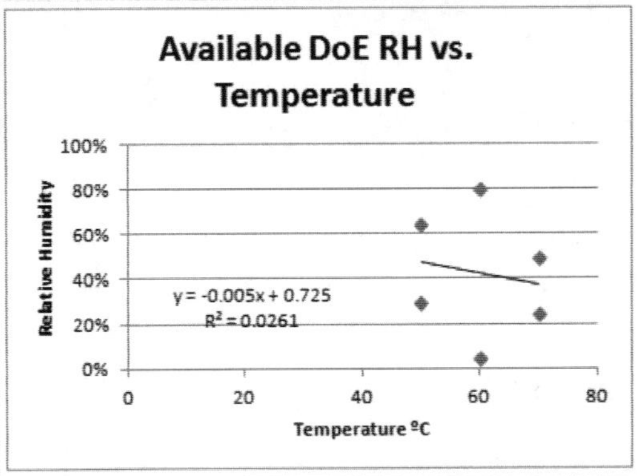

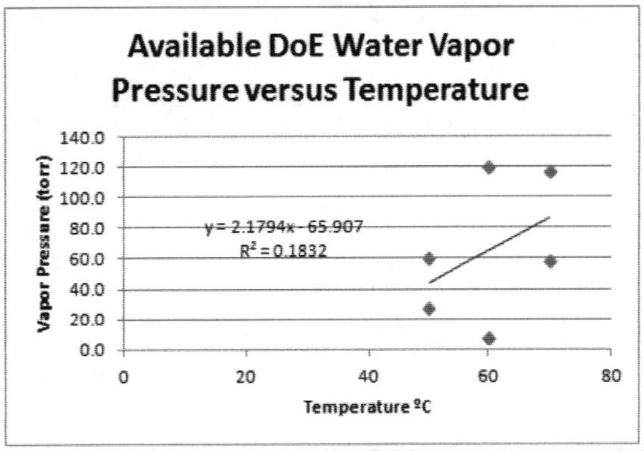

- Skewing the run points slightly reduces the resulting correlation between vapor pressure and temperature in the study without compromising the RH based design.
- Allows either RH or vapor pressure to be fit as the humidity term within the kinetic equations.

Estimation of Relative Humidity in Long Term Storage

- Moistures permeate into plastic storage bottles, following a standard permeation model, same process as polymer membranes:

$$Flux = P \cdot \left(\frac{SA}{thickness}\right) \cdot \left(VP_{exterior} - VP_{interior}\right) \cdot \Delta t \tag{3}$$

Where

"Flux" is the water vapor transmission rate,
"P" is the permeability coefficient,
"SA" is the surface area,
"thickness" is the wall thickness,
"$VP_{exterior}$" is the water vapor pressure on the exterior of the bottle or capsule film,
"$VP_{interior}$" is the water vapor pressure in the interior of the bottle or capsule film and,
"Δt" is the change in time.

Permeability coefficients follow an Arrhenius relationship with temperature, and can be calculated at different temperatures using equation 4 (Abdel-Bary, 2003).

$$P = P0 \cdot exp\left(\left(\frac{Ea}{R}\right) \cdot \left(\frac{1}{273.15 + T_{ref}} - \frac{1}{273.15 + T}\right)\right) \tag{4}$$

Solve for Equilibrium RH on Each Day

- Need to experimentally measure and define models for:
  - Moisture permeation of bottles at varying temperatures.
  - Moisture versus RH relationship of desiccant
  - Moisture versus RH relationship of tablet

- With known initial water content, and known permeation of water for a given day, four equations and four unknowns can be solved to find the equilibrium RH on each day.
  - Water Fraction in air
  - Water fraction in tablet
  - Water fraction in desiccant
  - Overall mass balance

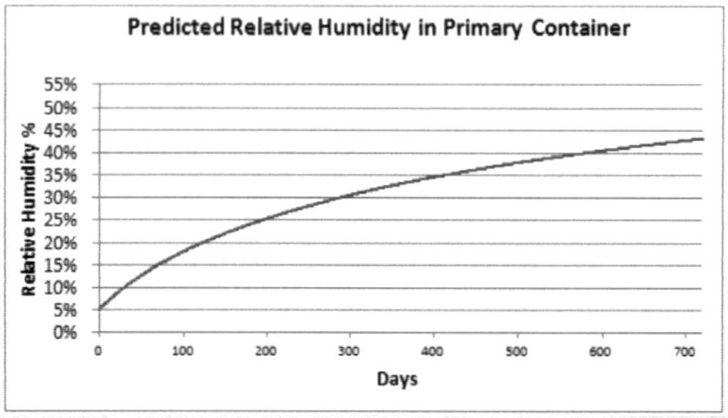

Estimation of Uncertainty in Prediction – Bootstrapping Approach

- The kinetic models can then be integrated over time with the estimated humidity curve to create a predicted line.

- To estimate the error of the prediction, the error in the kinetic model fit is assumed equivalent to the error in the experimental data.

- Each raw data point is perturbed with random noise equivalent to the error of the model. A new model is generated and plotted for the long term prediction.

- Process is repeated several thousand times to generate a distribution in the prediction.

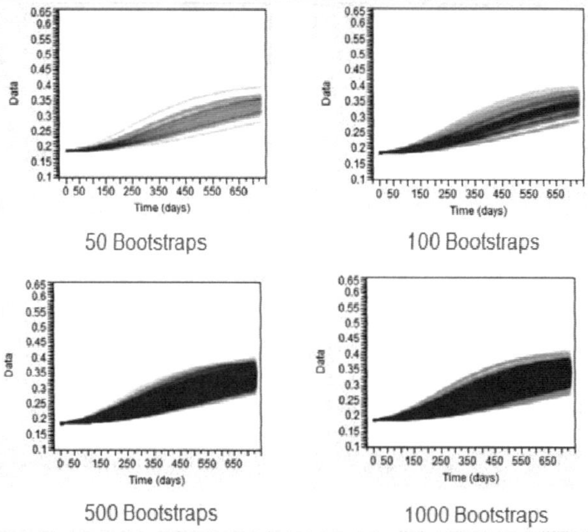

Example 1 from Drug Product 81:

	Specification Limit		35	35	35	34	60	80
	Assay/Imp		Impurity	Impurity	Impurity	Impurity	Impurity	Impurity
Temperature	Relative Humidity	Time	Imp101	Imp966	Imp545	Imp799	Imp283	Imp102
°C	%	days	parts per 10000	parts per 10000	parts per 10000	parts per 10000	parts per 10000	parts per 10000
71.3	77.6	4	6.7	14.8	9.3	7.7	38.2	13.2
79.3	27.7	4	8.3	13.5	9.5	6.9	36.8	14.0
61.8	39.9	7	7.4	6.0	9.4	6.1	34.1	13.1
71.5	51.3	7	7.5	12.9	9.7	7.2	35.6	13.7
71.3	77.6	7	7.2	13.2	8.8	8.8	37.8	14.1
79.3	27.7	7	9.8	16.3	9.9	7.5	37.7	14.0
61.9	6.6	10	9.2	5.7	9.4	6.0	34.9	13.8
61.8	39.9	10	8.1	6.9	9.4	6.4	34.6	14.1
71.5	51.3	10	8.2	14.6	9.7	7.8	36.1	13.4
50.4	74.8	14	6.8	6.0	9.5	6.6	34.4	13.7
61.9	6.6	14	10.1	5.5	9.6	6.5	34.6	14.5
50.4	74.8	10	6.9	5.4	9.4	6.6	34.3	14.3
61.8	39.9	0	5.2	2.0	9.5	5.8	29.8	13.1

Figure 13 Drug Product 81 Accelerated Stability Experimental Data

- Just eyeballing this accelerated data, surely Impurity 966 is the worst one?
- Eyeballing again, that impurity 102, can't possibly be a risk, it's nowhere near the specification limit.
- Lesson #1: Don't believe that eyeballing data is a good approach. Don't stop doing an analysis due to little response in the data.

Example #1: Comparisons of Model Predictions to Long Term Data for Drug Product 81

- Impurity 102 had a low activation energy, did not accelerate substantially at higher temperatures, but ended up the most critical impurity at lower temperatures.
- Impurity 966 was low risk with high activation energy, and showed little to no increase at low temperatures.
- Points out a problem with the isoconversion approach – low activation energy reactions.

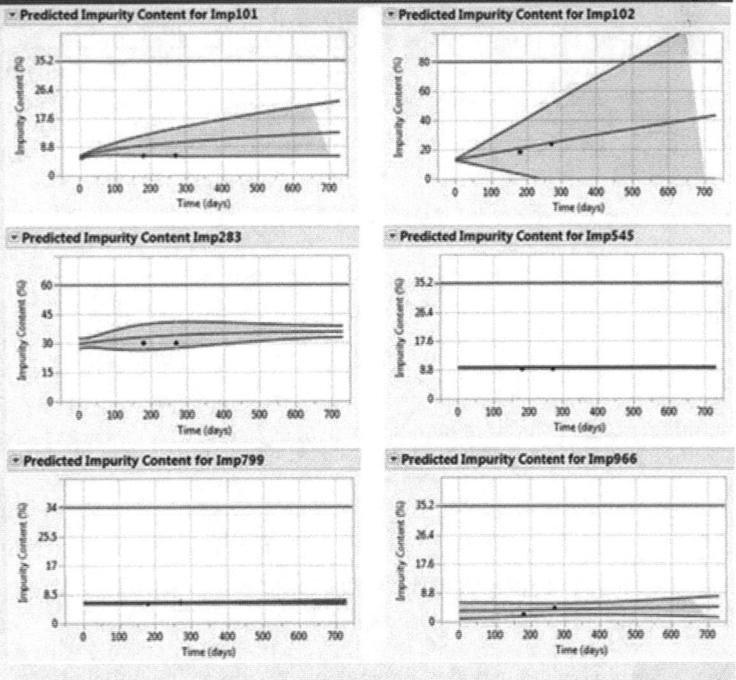

Drug Product 81 Impurity models overlaid with long term data at 30 deg C, 65% RH in HDPE Bottle with Desiccant

Example #2 – Drug Product "P"

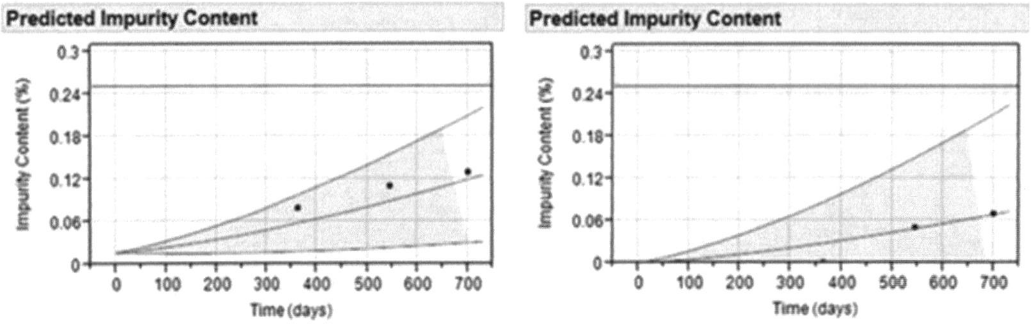

Figure 15 Drug Product P Impurities 0.94 (left) and 0.11 (right) models overlaid with long term data

•Predicted passing for both impurities of concern, and this proved true.

Example 3: Drug Product S

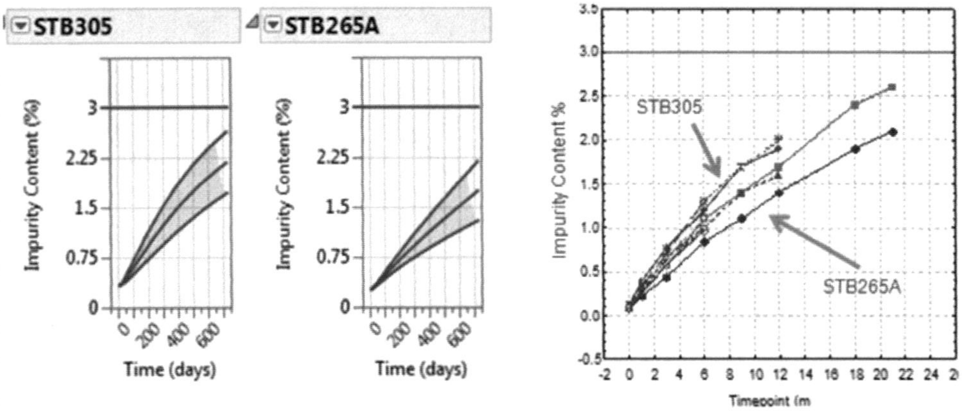

Figure 16 Drug Product S Batches 305 and 265A in Accelerated Stability Prediction (left and center) Compared to Long Term Observation (Right)

- Multiple formulations were compared in accelerated stability prediction (left), then run in long term stability (right).
- Accelerated stability predicted which formulation was going to have the most stable results.

Example 4: Desolvation of DMSO From an API

- Accelerated stability modeling can be used to predict some physical phenomenon, such as desolvation.
- Cannot predict other physical mechanisms, such as loss of free DMSO that became unbound from the API prior to the experiment start.
- The accelerated experiment must be run with similar DMSO vapor pressure as in final package in long term stability.

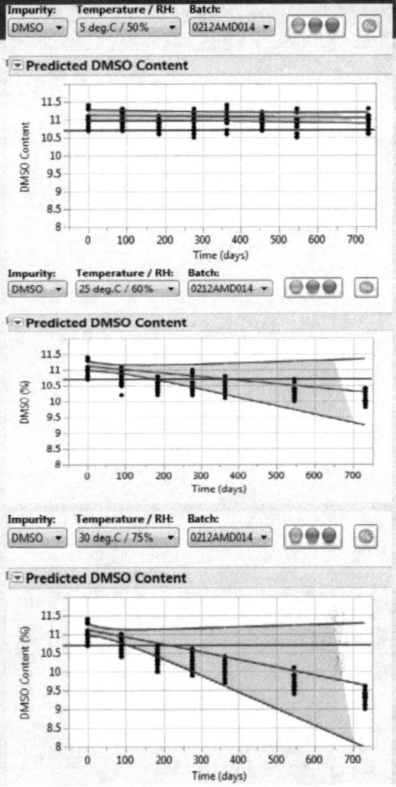

Example 5: Salt Disproportionation to Free Acid

- Can also predict salt disproportionation, and the subsequent nucleation and growth of free acid crystals.
- Note that this phenomenon was exponential in time, and followed an Avarami curve. What's predicted here is the Avarami transform of the data: Y*=ln(Y/(100-Y))

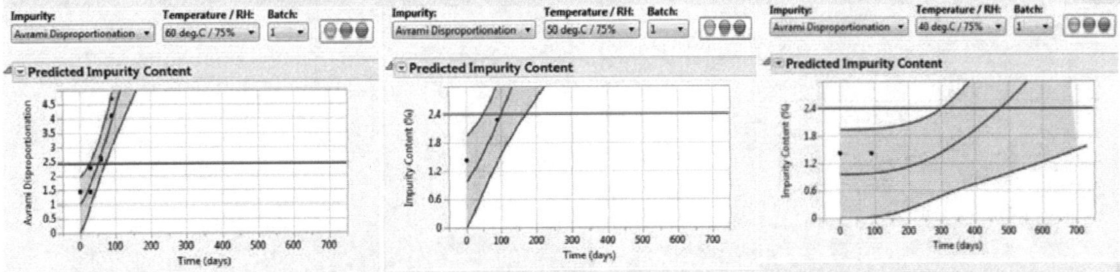

- The important prediction that the 30 degree 65% RH condition with desiccant will remain flat and undetectable, has proved true for the first year so far.

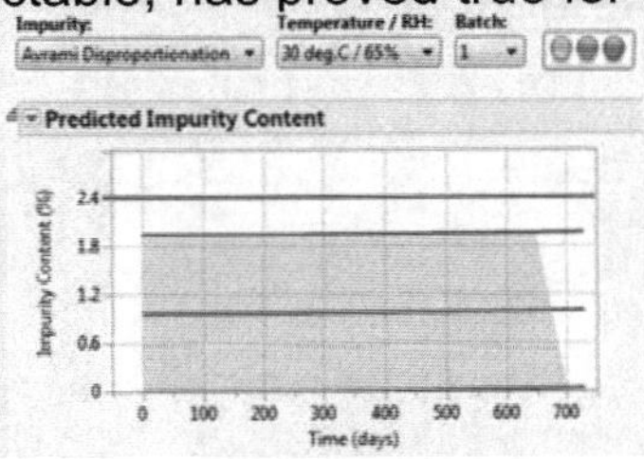

Example 6: Biological Product Main Isotherm Impurity

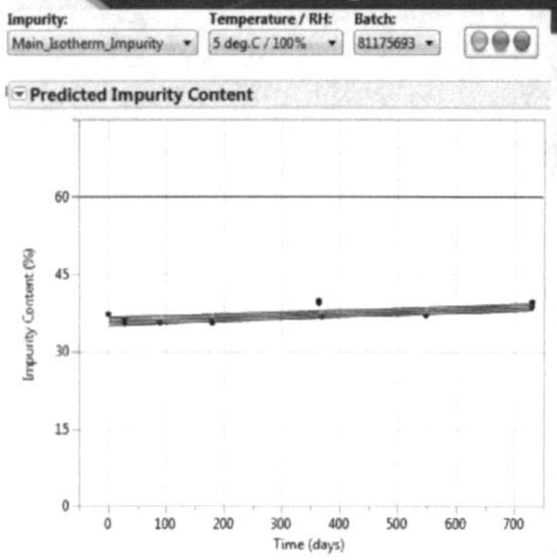

5 degrees

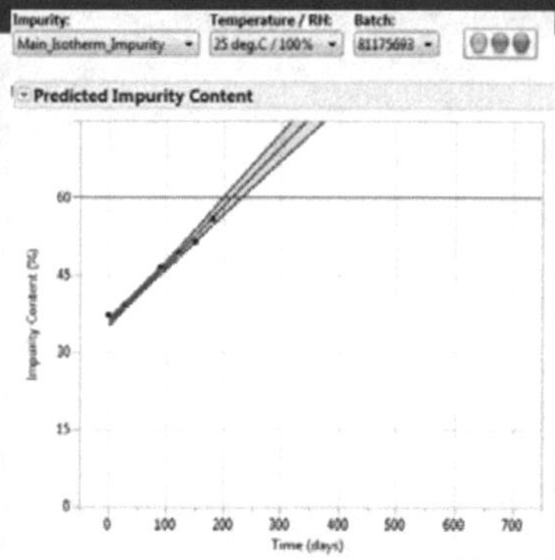

25 degrees

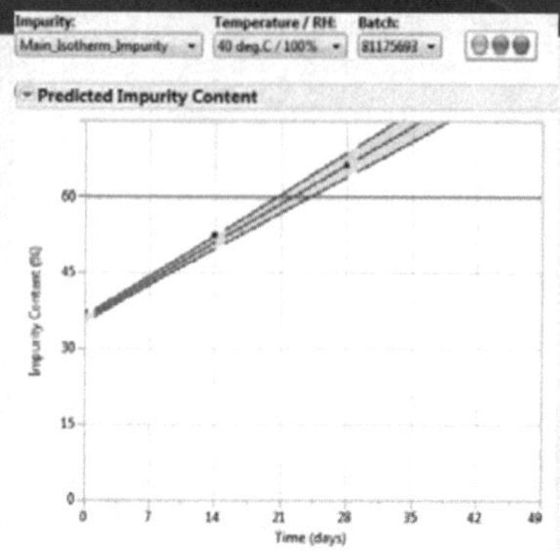

40 degrees

- Main isotherm impurities, and caliper reduced impurities are often well modeled as an Arrhenius model, linear with time, with no humidity impact since it's in liquid.
- Need lower temperatures in the study. Multiple years of shelf life at 5 degrees, reduces to 21 days at 40 degrees.
- Cannot predict below the freezing points generally around -15 degrees.

Example 7: Drug Product "N" Done in Both ASAP Prime and GSK Methodology

		Specification Limit	91.5	10	2.406846103
		RRT	impurity	Impurity	Impurity
Temperature	Relative Humidity	Time	Assay	Conversion	Avrami_Assay _Loss
°C	%	days	% a/a	% a/a	% a/a
50	50	0	100.3	0.7	-0.349720122
50	50	4.8	96.8	4.2	1.477558048
50	50	9.8	93	8	2.161962565
50	50	13.7	90.8	10.2	2.428848952
59	30	4.8	97.5	3.5	1.288031107
59	30	9.8	98.3	2.7	1.020348263
59	30	13.7	97.3	3.7	1.345654347
70	12	4.8	97.5	3.5	1.288031107
70	12	9.8	98.8	2.2	0.810480272
70	12	13.7	97	4	1.426703899
71	74	2.8	37.8	63.2	5.129115715
71	74	6.9	10.5	90.5	6.76909511
86	8	4.8	97.5	3.5	1.288031107
86	8	9.8	98.3	2.7	1.020348263
86	8	13.7	98.5	2.5	0.941354701
85	53	2.8	72.5	28.5	3.681438042
85	53	6.9	12.5	88.5	6.572394425

- Drug product "N" has an assay that declines with time.
- Since the extent of conversion is high (>15%) and exponential in time, an Avrami transform is used to linearize the data first.

Conversion = 101-Assay

Avrami Assay Loss = ln(100*(Conversion)/(101-Conversion))

Example 7: ASAP Prime Analysis

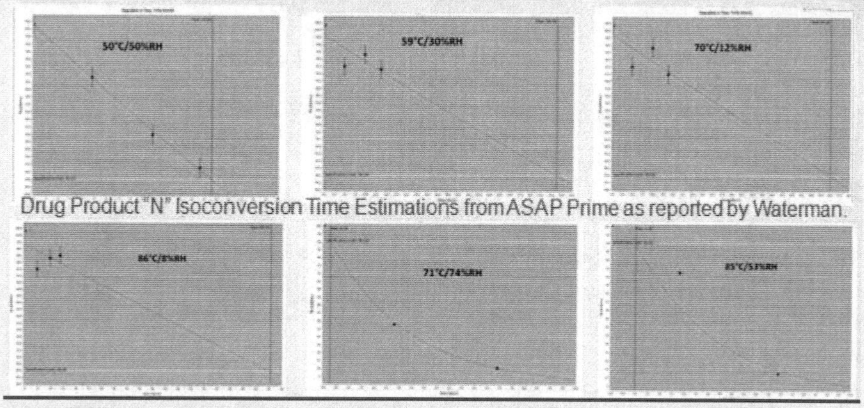

Drug Product "N" Isoconversion Time Estimations from ASAP Prime as reported by Waterman.

Temperature (°C)	%RH	Isoconversion Time (Days)	SD (Days)
50	50	14.6	1.2
59	30	56.4	18.4
70	12	55.0	17.5
71	74	0.3	0.1
85	53	1.0	0.0
86	8	95.6	44.8

Table 2 Calculated Isoconversion Time for Experimental Design Conditions

Example 7: ASAP Prime Predictions

ln A	18.7±3.1
E_a	15.6±2.0 kcal/mol (65.3±8.4 kJ/mol)
B	0.103±0.008
R^2	1.00

Table 3 Parameters for Moisture Modified Arrhenius Equation

Predictions were then made for shelf life in PVC blisters.

Condition	Measured Shelf-life (mos.)	ASAP*prime*™ Shelf-life (mos.) with PVdC
25°C/60%RH	17.9 (13.4-32.5)	11.2 (9.2-13.6)
30°C/65%RH	8.2 (5.7-11.3)	6.2 (5.3-7.3)
30°C/75%RH	4.8 (4.4-7.6)	5.2 (4.5-6.1)
40°C/75%RH	1.1 (0.8-1.3)	2.0 (1.8-2.2)

Table 4 Comparison between the measured long-term storage shelf-life and ASAP*prime*™ predicted shelf-life (to 91.5% potency)

- Due to the exponential assumption, an exponent cannot approach zero reaction rate at near zero humidity. The limit of an exponent of a small number is 1.
- As a result it tends to substantially overestimate reaction rates of humidity sensitive compounds extrapolated to low environmental humidity. These PVC blisters started with an RH of 6%.
- In this case the predicted failure time of the product in PVC blisters was 11 months, where the actual was approximately 18 months at 25 degrees C.

Example 7: GSK Methodology Prediction of RH in Blisters

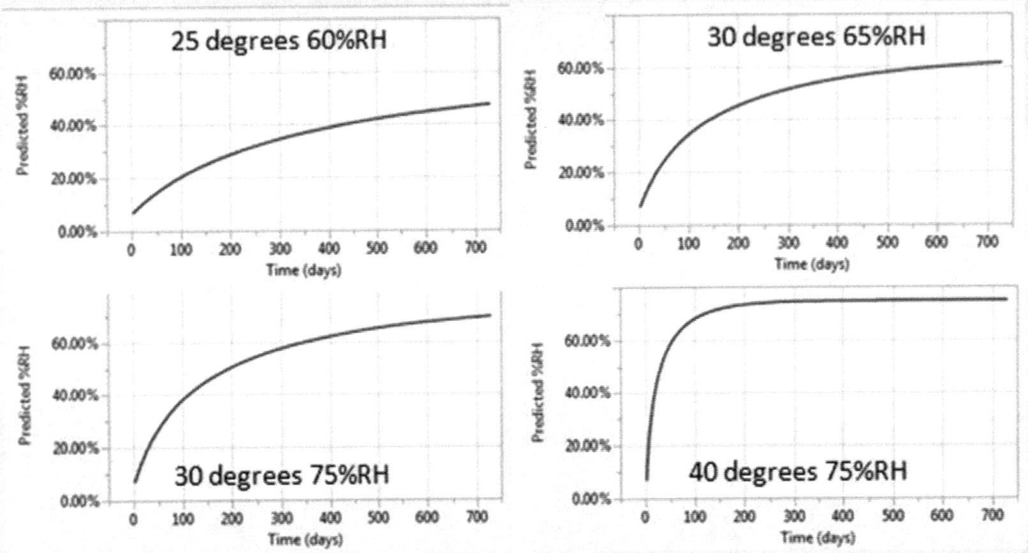

Example 7: Model Selection and Long Term Predictions versus Actual Long Term Data

Avrami_Assay_Loss Model Selection

Model Type	K1	K2	K3	N1	N2	nParams	RMSE	Score	AICc	BIC
RH Linear Time Kinetic Model	1.74	6562	-	2.61	-	4	0.6414	-	44.037	42.749
AH Linear Time Kinetic Model	45.74	-6632	-	2.59	-	4	0.6424	-	44.089	42.801
RH Accelerating Kinetic Model	-188386.9	15388	-	-3	-	4	3.0955	-	97.556	96.268
AH Accelerating Kinetic Model	-1	-7500	-	-3	-	4	3.197	-	98.654	97.365
RH Decelerating Kinetic Model	0.82	7344	6.82	3.67	-	5	0.4607	-	36.375	32.974
AH Decelerating Kinetic Model	80.08	-11170	6.82	3.63	-	5	0.4608	-	36.378	32.977
RH Power Model	3.77	4843	-	2	0.51	5	0.4789	-	37.688	34.287
AH Power Model	45.39	-5216	-	1.97	0.51	5	0.4786	-	37.665	34.264

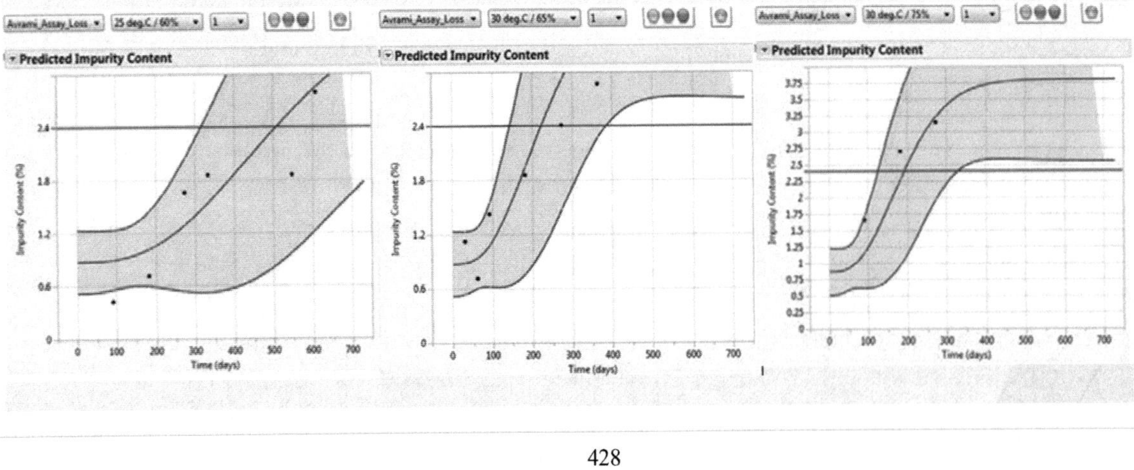

Example 7: Summary of Comparison of Results Between ASAP Prime and GSK's Methodology for Drug Product "N"

Storage Condition	Observed Time to Failure (months)	ASAP Prime Isoconversion method Prediction	GSK's Prediction
25 degrees 60% RH	17.9	11.2	16.8
30 degrees 65% RH	8.9	6.2	7.4
30 degrees 75% RH	5.3	5.2	6.3
40 degrees 75% RH	1.3	2	1.5
25 degrees 60% RH	percent error	37.4%	6.3%
30 degrees 65% RH	percent error	30.2%	16.7%
30 degrees 75% RH	percent error	1.2%	18.8%
40 degrees 75% RH	percent error	52.0%	12.5%
	Average percent error	30.2%	13.5%
	Slope actual vs. Predicted	1.84	1.10

- ASAP Prime's average error in prediction was 30% for this example.
- Methods described in this presentation resulted in an average of 13% error compared to actual long term data for this example.

Conclusions

- At GSK, we've observed that stability does not appear to be exponentially affected by increasing RH. Typically curvature is closer to cubed root or square root rather than exponential with power terms typically between -1 and 1, and the majority of data falling within -2 and 2 as would be typically expected for the order of reaction in reaction kinetics.

- We've chosen to pursue an approach using kinetics rather than focus on the isoconversion approach as it allows us to improve the logistics of doing a large number of accelerated studies.

- The aspiration of this model development would be to predict long term data ranging up to two years from a short 2 to 4 week accelerated stability study. The approach we've developed has thus far shown good fit with long term trends in available data. The number of studies is currently greater than 80, and impurities or physical phenomenon modeled greater than 300.

425091 Hydrodynamics and Mass Transfer Effects in Mini Vessel Dissolution Testing Systems

Abstract:

In the pharmaceutical industry, dissolution testing is routinely conducted to evaluate the *in vitro*performance of solid dosage forms in pharmaceutical development, determine the dissolution behavior of formulations and provide an optimization of drug release dosage forms, and as a quality control tool to insure that solid dosage forms have consistent dissolution properties. The USP Dissolution Testing Apparatus 2 is the device most commonly used for this purpose. In a typical test, a tablet is added to the vessel and samples are manually withdrawn from the vessels over time and analyzed for their drug content.

The USP 2 drug dissolution testing method is not always appropriate, especially for dissolution testing involving very small tablets or when not enough drug substance may be available for proper dissolution testing in the USP 2 Apparatus, especially during the initial stage of drug development. Small-volume dissolution testing apparatuses (mini vessel systems) can be advantageously used instead of the standard USP 2 system to overcome these limitations. The most common of these mini vessel systems is similar in shape and operation to the USP 2 apparatus, but it only requires a small volume vessel (200 mL) and a mini paddle impeller.

Before mini vessel systems can gain wide acceptance in industry, their hydrodynamics needs to be studied, and their operating conditions must be defined in order to compare the velocity distribution and mass transfer coefficient in a mini vessel system with those in the standard USP2 system.

Very few studies on mini vessel systems are available. Therefore, the objective of this work was to quantify the hydrodynamics and drug mass transfer rates in mini vessel systems, and determine the operating conditions under which mini vessel systems can be reliably used instead of USP 2 system for dissolution testing purposes.
The velocity distribution in a minivessel was obtained here using Computational Fluid Dynamics (CFD) at different agitation speeds. The predictions were compared to the velocity vector maps obtained using Particle Image Velocimetry (PIV). Four horizontal iso-surfaces were selected to be studied in detail. The CFC predicted and PIV measured flow patterns were found to be in good agreement. As expected, the 3D flow velocity magnitude increased as the agitation speed increased. However, the velocity in the inner core region below the impeller always remained very low.

The tablet-medium mass transfer process was additionally modeled, and the mass transfer coefficients at four different agitation speeds were obtained and compared to the standard system. The mini vessel system has a flow pattern similar to that of the standard USP 2 system. It was found that similar velocities at the tablet surface (on the iso-surface at y=7mm, i.e., at the vessel bottom) in the mini vessel system and in the USP 2 system were obtained when the mini vessel was operated a 100 mL, 125 rpm and the USP 2 system was operated at 900rpm and 100rpm, indicating that these two systems should result in similar dissolution profiles when operated under these conditions.

427772 Stability By Design (SbD): Utilizing Material Science Principles to Predict Physical and Chemical Stability of Pharmaceutical Solids

Friday, November 13, 2015: 8:30 AM
Ballroom D (Salt Palace Convention Center)

Michael Brunsteiner[1], Corinna Gressl[1], Johannes G. Khinast[2], Amrit Paudel[3], Adrian L. Davis[4], Margaret Landis[5], Klimentina Pencheva[4], Garry Scrivens[4] and Geoff Wood[6], (1)RCPE, Graz, Austria, (2)Research Center Pharmaceutical Engineering (RCPE), Graz, Austria, (3)Advanced Products and Delivery, RCPE, Graz, Austria, (4)Pfizer Worldwide R&D, Sandwich, Kent UK, Kent, England, (5)Pfizer Worldwide R&D, Groton, CT, Groton, CT, (6)Pfizer

Stability by Design (SbD): Utilizing material science principles to predict physical and chemical stability of pharmaceutical solids

M. Brunsteiner [1], C. Gressl[1], J.G. Khinast[1], A. Paudel[1]

Adrian L. Davis [2], Meg Landis [3], Klimentina Pencheva [2], Garry Scrivens [2], Geoffrey P.F. Wood [3]

[1]Research Center Pharmaceutical Engineering, Graz, Austria

[2] Pfizer Worldwide R&D, Sandwich, Kent UK

[3]Pfizer Worldwide R&D, Groton, CT, USA

Ensuring, both, physical and chemical long-term stability of the drug product is one of the key requirements in successful drug formulation development. While significant knowledge exists on solution-state stability of pharmaceuticals, the predictive knowledge of solid-state physical and chemical stability in solid oral dosage forms is limited. Here we present two examples of our research towards achieving stability by design (SbD) for solid oral dosage forms through the use of *in-silico* methods to accelerate and streamline formulation development.

As a model for chemical drug degradation we consider Carvedilol (CAR), a compound used in the treatment of congestive heart failure. The compounds secondary alcohol has been shown to undergo esterification when formulated with citric acid (CA)as an excipient.(1) The free base form of CAR can crystallize in at least two different polymorphic forms. Based on structural and energetic results from Molecular Dynamics (MD) simulations of the crystal surfaces of two different CAR polymorphs in contact with vacuum or citric acid, we calculated a range of descriptors, including surface energies and exposed reactive surface areas. A thorough analysis of these descriptors and their relative magnitudes for different crystal surfaces of the CAR polymorphs allows us to investigate the relative importance of degradation of molecules on crystal surfaces versus degradation of molecules that are highly mobile due to surface defects and partial amorphization. Accordingly, two different crystalline forms (included in MD simulation) and an amorphous form of CAR were generated of two different size ranges. Following the comprehensive solid-state and particulate characterization, CAR-CA blends prepared using different solid-states and sizes of CAR were subjected to accelerated condition of 50°C. The comparison of the experimental data on kinetics and extent of CAR-CA interaction products formation with the descriptors derived from MD simulations will provide the predictability of chemical stability by the computed descriptors.

As models for physical drug stability, we consider a series of API-polymer combinations that have been used to produce amorphous solid dispersions (ASD), (2) a commonly used strategy in the formulation of poorly soluble drugs. We performed extensive MD simulations of ASDs at various drug loads and calculated descriptors such as relative hydrogen boding propensities, free volumes, API mobilities, and API-polymer intermolecular interactions. A comparison of the obtained results with existing literature data for the relative stabilities of the different ASDs shows that both kinetic and energetic factors can play a role in determining the stability of ASDs. We suggest criteria that can be determined *in-silico* and used to estimate the relevance of these factors for a given API-polymer combination.

References

1. Larsen, J., Cornett, C., Jaroszewski, J. W., and Hansen, S. H. (2009) Reaction Between Drug Substances and Pharmaceutical Excipients: Formation of Citric Acid Esters and Amides of Carvedilol in the Solid State., Journal of Pharmaceutical and Biomedical Analysis 49, 11–17.

2. Eerdenbrugh, B. V., and Taylor, L. (2010) Small Scale Screening to Determine the Ability of Different Polymers to Inhibit Drug Crystallization upon Rapid Solvent Evaporation, Molecular Pharmaceutics 7, 1328–1337.

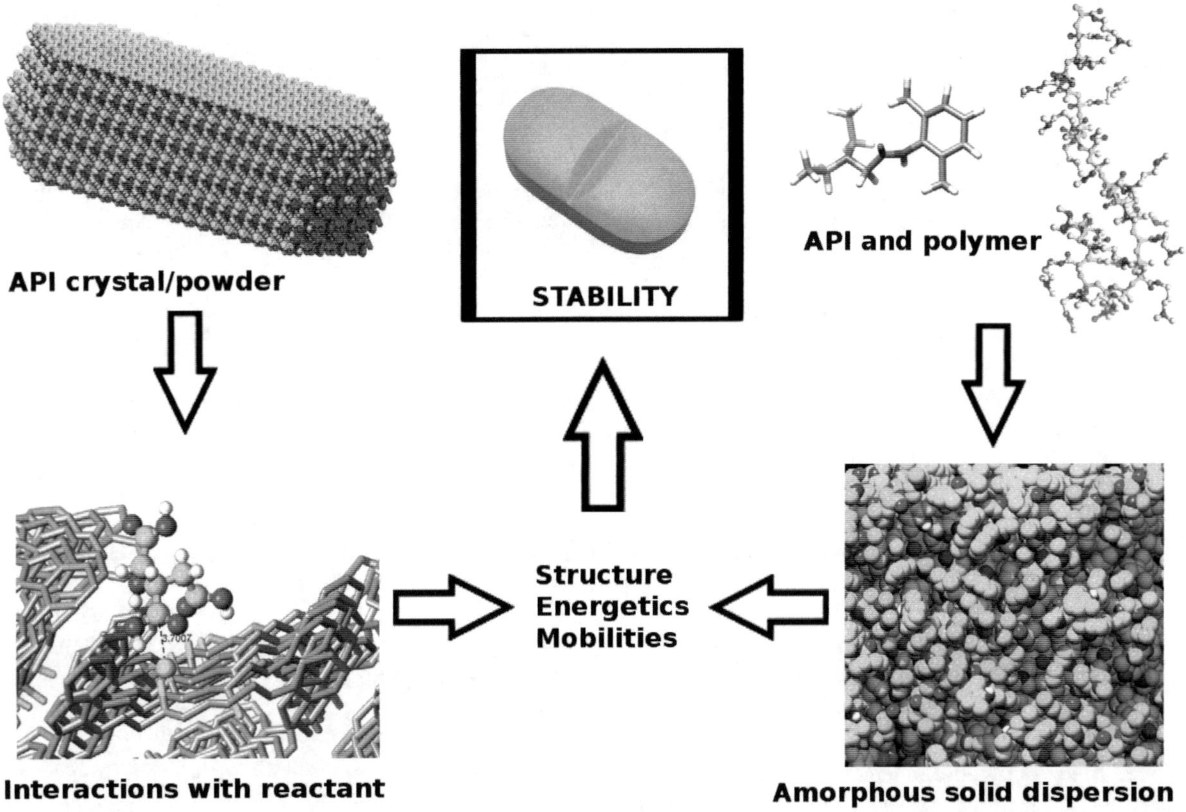
API crystal/powder
STABILITY
API and polymer
Structure
Energetics
Mobilities
Interactions with reactant
Amorphous solid dispersion

425319 Exploring Polymorph Free Energy Landscapes with Hamiltonian Reweighted Molecular Dynamics

Abstract:

The presence of multiple stable crystal structures in solid organic materials can limit their commercial viability when two or more observable polymorphs exhibit markedly different physical properties. Crystallography researchers estimate that 50% of all drug molecules exhibit polymorphism, but only 1% of the materials with a well-characterized crystal structure in national crystal structure databases contain a similar level of data for an alternative polymorph structure.

Current computational methods for predicting polymorphic behavior generate many possible crystal structures and rapidly evaluate the lattice energies of these candidate structures to determine the most likely to be observed experimentally. However, lattice energies alone do not fully capture the stability of systems that are at room temperature and sample a range of configurations due to thermal motion. In addition, *ab initio* quantum mechanical models or at minimum polarizable force fields are likely necessary to represent the physical system, but the extensive sampling required in molecular dynamics can currently only be performed with classical point-charge force fields.

In this work, we present an atomistic simulation approach to evaluate the free energy difference between crystal polymorphs in both computationally cheap and expensive Hamiltonians. Molecular dynamics simulations are conducted in a classical point-charge all-atom potential designed to overlap with the expensive Hamiltonian. Multistate Hamiltonian reweighting is then employed on these configurations post simulation to estimate the free energy difference in more expensive force fields without any additional sampling. We therefore access computationally demanding free energy differences by avoiding direct simulation within the polarizable potentials.

This simulation technique is applied to solid phase benzene, which is known to have multiple distinct polymorphs. We reweight to polarizable potentials and the simulation results indicate that the exact polymorph stability is strongly sensitive to the choice of Hamiltonian. Furthermore, the successful Hamiltonian reweighting technique demonstrates that the relevant phase space of polarizable Hamiltonians in this system can be approximated using only point-charge models. These preliminary results represent a significant step toward the ultimate goal of tractable thermodynamic calculations using more expensive many-body energies.

431785 Molecular Modeling to Design and Control (M2DC): A First Principles Approach to Polymorph Prediction and Crystallization Unit Design

Abstract:

Crystallization remains one of the most important industrial separation processes to date. Despite its wide spread use, a fully fundamental understanding of the nucleation process is lacking. The classical nucleation theory (CNT) rate expression, formulated by Gibbs, Weber, and Volmer, at the beginning of the 20th century, remains the main framework against which experimental evidence is compared, despite being overly simplistic, and failing to capture data well in a range of systems. Skepticism towards the degree to which thermodynamic bulk properties used in the CNT rate expression can be even applied to nano scale clusters is warranted. Rather, with the advent of modern supercomputers, the possibility to relax the suit of CNT assumptions in favor of fully atomistic calculations provided via molecular dynamics is becoming a more and more appealing alternative. From a solid-state physics perspective, polymorphism, the ability of a molecule to exist in multiple crystalline phases, remains a thorn in the side of both industrial crystallization experts, and molecular modelers alike. Landscape energy minimization techniques, although elegant in their own right, fail to include standard design variables such as solvent, temperature, or pressure. Molecular dynamics, which encapsulates the aforementioned design variables naturally, has suffered from time scale challenges associated with activated rate processes, such as nucleation. However, with the advent of modern supercomputers, and HPC hardware such as xeon phi coprocessors, these types of simulations are becoming more and more feasible. The crystallization community would more than like molecular simulations to become primetime, because in the absence of in silico polymorph specific nucleation rates, expensive solvent screening experiments must be performed. Furthermore, the question of "have we truly found all possible forms" is never truly answered by such experiments.

In this work, we investigate the potential of atomistic information provided by hybrid monte carlo molecular dynamics (MCMD) to be used to not only predict polymorphism as a function of operation conditions, but also the degree to which polymorph specific solubility, and metastable zone curves can be calculated. These calculations will serve as inputs into control models, allowing a fully atomistic design of a crystallization unit. As microfluidic devices provide a means to measure homogeneous nucleation rates using minimal reagents and in minimal time, they will serve as a benchmark for determining the degree of accuracy of our homogeneous nucleation rate calculations prior to further scale-up for industrial applications.

412913 PAT for Form Monitoring during Drying

Abstract:

PAT for Form Monitoring During Drying

M. Dimitri, S. Allwein, P. Mudipalli, and M. Christie

Teva Branded Pharmaceuticals, Inc.

Chemical Synthesis Center

383 Phoenixville Pike

Malvern, PA 19355

A.I.Ch.E. Annual Meeting 2015

Session: "PAT for Process Understanding and Elucidation of Fundamental Phenomena in Drug Product/Substance Development"

Abstract: Controlling and monitoring polymorphic form during drying of active pharmaceutical ingredients (APIs) is increasingly an area of interest particularly for specialized drying, such as humidified drying. The use of PAT instruments is often powerful in understanding the drying kinetics of an API, identification of the final polymorphic form in situ in the dryer, and determination of drying end point. This study examines the development and use of Raman Phat Probe technology to monitor a humidified drying process of an API and examines some of the key scale-up parameters involved. For this API, three different known polymorphs are possible during drying, with one polymorph being irreversible once formed. In addition, online analysis was determined to be critical as the API is extremely sensitive to relative humidity and form conversion occurs rapidly, making offline analysis challenging. This study demonstrates the use of raw spectra and chemometric analysis in monitoring and defining the drying kinetics of the two main forms of interest. The example of the use of this technology that is being presented shows its applicability both for drying process development and for confirmation of form at scale (ie, in situ in the dryer).

Keywords: Drying, Phat Raman, Form Monitoring, PAT

443949 Molecular Modeling of Additive Effects on Calcium Pyrophosphate Crystals

Abstract:

Calcium pyrophosphate (CaPP) deposition disease is a rheumatologic condition caused by the formation of calcium pyrophosphate crystals in the joints. CaPP crystals form in joint cartilage as the results of an unknown metabolic disorder. Eventually, CaPP crystals are ejected from the cartilage into the surrounding synovial fluid where they continue to grow. CaPP deposition disease becomes increasingly common with age, affecting approximately a quarter of people over 85 years old.[1] Despite its prevalence, there are no treatments to inhibit the growth or formation of CaPP crystals. Mechanistic understanding of how additives interact with crystal faces is important for the development of additives to inhibit CaPP crystal growth.

We use atomistic simulations to study the effects of additives on calcium pyrophosphate crystal faces. Accurate modeling of additive effects on CaPP crystal faces requires a molecular force field that can reproduce the experimental crystal structures of CaPP. We develop a new force field for pyrophosphate using the functional form of the CHARMM force field. Starting with parameters in the CHARMM General Force Field, we refit bond, angle, and dihedral parameters to match ab initio simulation data. Finally, the non-bonded parameters were optimized to minimize the root mean squared deviation between the experimental and force field derived crystal structure of the thermodynamically stable $Ca_2P_2O_7$Ÿ$2H_2O$. The resulting force field accurately reproduces the crystal structure of all solved hydrated CaPP forms.

Simple salts have been shown to affect polymorph selection and crystallization of CaPP. We study the effects of Ca^{2+}, Mg^{2+}, and Na^+ on CaPP, by calculating their binding free energies to CaPP surfaces using adaptive biasing force and free energy perturbation simulations. Ion binding energies are calculated for fast growing CaPP crystal faces for crystals grown homogeneously and heterogeneously on model biological surfaces. Positive ions that bind more strongly to fast growing crystal faces than Ca^{2+} may inhibit crystal growth. We find that Mg^{2+} binds more strongly than Ca^{2+} to the fast growing faces of β-$Ca_2P_2O_7$Ÿ$4H_2O$, while Ca^{2+} binds more strongly than Mg^{2+} to the fast growing faces of the thermodynamically stable $Ca_2P_2O_7$Ÿ$2H_2O$.

[1] Guerne, P. A., and R. Terkeltaub, “Calcium pyrophosphate dihydrate crystal deposition: epidemiology, clinical features, diagnosis, and treatment,” Chapter 21 in Terkeltaub, R., ed., “Gout and Other Crystal Deposition Arthropathies”, Elsevier Heath Sciences, Philadelphia, PA (2012).

AUTHOR INDEX

AUTHOR INDEX

AUTHOR INDEX

AUTHOR INDEX

AUTHOR INDEX

AUTHOR INDEX